Day and Thomson's Series.

PRACTICAL ARITHMETIC,

UNITING THE

INDUCTIVE WITH THE SYNTHETIC MODE OF INSTRUCTION.

ALSO, ILLUSTRATING THE

PRINCIPLES OF CANCELATION.

FOR SCHOOLS AND ACADEMIES.

By JAMES B. THOMSON, A.M.,

AUTHOR OF MENTAL ARITHMETIC; EXERCISES IN ARITHMETICAL ANALYSIS; HIGHER ARITHMETIC; EDITOR OF DAY'S SCHOOL ALGEBRA; LEGENDRE'S GEOMETRY, ETC.

FIFTY EIGHTH EDITION, REVISED AND ENLARGED.

NEW YORK:
PUBLISHED BY MARK H. NEWMAN & CO.,
199 BROADWAY.

1850.

DAY AND THOMSON'S MATHEMATICAL SERIES.

FOR SCHOOLS AND ACADEMIES.

I. MENTAL ARITHMETIC; *or, First Lessons in Numbers;*—For Beginners. This work commences with the *simplest* combinations of numbers, and *gradually* advances to more difficult combinations, as the mind of the learner expands and is prepared to comprehend them.

II. PRACTICAL ARITHMETIC;—Uniting the *Inductive* with the *Synthetic* mode of Instruction; also illustrating the principles of CANCELATION. The design of this work is to make the pupil thoroughly acquainted with the *reason* of every operation which he is required to perform. It abounds in *examples*, and is *eminently practical.*

III. KEY TO PRACTICAL ARITHMETIC;—Containing the answers, with numerous suggestions, &c.

IV. HIGHER ARITHMETIC; *or, the Science and Application of Numbers;*—For advanced classes. This work is complete in itself, commencing with the fundamental rules, and extending to the highest department of the science.

V. KEY TO HIGHER ARITHMETIC;—Containing the answers to all the examples, with many suggestions, &c.

VI. ELEMENTS OF ALGEBRA;—Being a School edition of Day's large Algebra. This work is designed to be a *lucid* and *easy* transition from the study of Arithmetic to the higher branches of Mathematics. The number of examples is much increased; and the work is every way adapted to Schools and Academies.

VII. KEY TO ELEMENTS OF ALGEBRA;—Containing the answers; the solution of the more difficult problems, &c.

VIII. ELEMENTS OF GEOMETRY;—Being our abridgment of Legendre's Geometry; with practical notes and illustrations.

IX. ELEMENTS OF TRIGONOMETRY, MENSURATION, AND LOGARITHMS.

X. ELEMENTS OF SURVEYING;—Adapted both to the wants of the learner and the practical Surveyor. (Published soon.)

PREFACE.

It has been well said, that "whoever shortens the road to knowledge, lengthens life." The value of a knowledge of *Arithmetic* is too generally appreciated to require comment. When properly studied, two important ends are attained, viz: *discipline* of mind, and *facility* in the application of numbers to business calculations. Neither of these results can be secured, unless the pupil *thoroughly understands the principle* of every operation he performs. There is no *uncertainty* in the conclusions of mathematics; there should be no *guess-work* in its operations. What then is the cause of so much *groping* and *fruitless effort* in this department of education. Why this *aimless, mechanical* "ciphering," that is so prevalent in our schools?

The present work was undertaken, and is now offered to the public, with the hope of contributing something toward the removal of these inveterate evils. Its plan is the following:

1. To lead the pupil to a knowledge of each rule by *induction;* that is, by the examination and solution of a large number of practical examples which involve the principles of the rule.

2. The operation is then *defined*, each principle is *analyzed separately*, and illustrated by other examples.

3. The *general rule* is now deduced, and put in its proper place, both for convenient reference and review; thus combining the *inductive* and *synthetic* modes of instruction.

4. The general rule is followed by *copious examples for practice*, which are drawn from the various departments of business, and are calculated both to call into exercise the different principles of the rule, and to prepare the learner for the active duties of life.

It is believed that much of this *guess-work* in "figuring," and its concomitant habits of *listlessness* and *vacuity* of mind, have arisen from the use, at first, of *abstract* numbers and *intricate* questions, requiring combinations above the capacity of children. Taking his slate and pencil, the pupil sits down to the solution of his problem, but soon finds himself involved in an impenetrable *maze*. He anxiously asks for light, and is directed "to learn the rule." He does it to the letter, but his mind is still in the dark. By *puzzling* and repeated *trials*, he perhaps finds that certain multiplications and divisions produce the answer in the book; but as to the *reasons* of the process, he is *totally ignorant*. To require a pupil to *learn* and *understand* the rule, before he is permitted to see its principles illustrated by simple practical examples, places him in the condition of the boy, whose mother charged him never to go into the *water* till he had *learned* to swim.

These embarrassments are believed to be unnecessary, and are attempted to be removed in the following manner:

1. The examples at the commencement of each rule are all *practical*, and are *adapted* to illustrate the particular principle under consideration. Every teacher can bear testimony, that children reason upon *practical* questions with far greater *facility* and *accuracy* than they do upon *abstract* numbers.

2. The numbers contained in the examples are at first small, so that the learner can solve the question *mentally*, and *understand* the reason of each step in the operation.

3. As the pupil becomes familiar with the more simple combinations, the numbers gradually increase, till the slate becomes necessary for the solution, and its proper use is then explained.

4. Frequent *mental* exercises are interwoven with exercises upon the *slate*, for the purpose of strengthening the habit of *analyzing* and *reasoning*, and thus enable the learner to comprehend and solve the *more intricate* problems.

5. In the arrangement of subjects it has been a cardinal point to follow the *natural order* of the science. No principle is used in the explanation of another, until it

has itself been demonstrated or explained. Common fractions, therefore, are placed immediately after division, for two reasons. *First*, they arise from division, and are in fact *unexecuted* division. *Second*, in Reduction, Compound Addition, &c. it is frequently necessary to use fractions; consequently fractions must be understood, before it is possible to understand the Compound rules.

For the same reason, Federal Money, which is based upon the *decimal notation*, is placed after Decimal Fractions. Interest, Insurance, Commission, Stocks, Duties, &c., are also placed after Percentage, upon whose principles they are based.

6. In preparing the tables of Weights and Measures, particular pains have been taken to ascertain those that are in *present use* in our country, and to give the *legal standard* of each, as adopted by the General Government.* It is well known that a great difference of weights and measures formerly existed in different parts of the country. More than ten years have elapsed since the Government wisely undertook to remedy these evils, by adopting *uniform standards* for the custom-houses and other purposes; and yet not a single author of arithmetic, so far as we know, has given these standards to the public.

7. The subject of *Analysis* is deemed so essential to a thorough knowledge of arithmetic and to business calculations, that a whole section is devoted to its development and application. The principles of *Cancelation* have been illustrated, and its most important applications pointed out, in their proper places. The Square and

---

* In the year 1836, Congress directed the Secretary of the Treasury to cause to be delivered to the Governor of each State in the Union, or to such person as he should appoint, a complete set of all the Weights and Measures adopted as standards, for the use of the States respectively; to the end that a *uniform standard* of Weights and Measures may be established throughout the United States. Most of the States have already received them; and may we not hope that every member of this great Union will promptly and cordially unite in the accomplishment of an object so conducive both to individual and public good.

Cube Roots are illustrated by geometrical figures and cubical blocks.

Such is a brief outline of the present work. It is not designed to be a book of *puzzles*, or mathematical *anomalies;* but to present the elements of practical arithmetic, in a *lucid* and *systematic* manner. It embraces, in a word, all the principles and rules which the business man ever has occasion to use, and is particularly adapted to precede the study of Algebra and the higher branches of mathematics.

With what success the plan has been executed remains for teachers and practical educators to decide. If it should be found to *shorten the road* to a thorough knowledge of arithmetic in any degree, its highest aims will be accomplished.

J. B. THOMSON.

New Haven, Oct. 3, 1845.

SUGGESTIONS

ON THE

MODE OF TEACHING ARITHMETIC.

I. QUALIFICATIONS.—The chief qualifications requisite in teaching Arithmetic, as well as other branches are the following:

1. A thorough knowledge of the subject.
2. A love for the employment.
3. An aptitude to teach. These are *indispensable to success.*

II. CLASSIFICATION.—*Arithmetic*, as well as reading, grammar, &c., should be taught in *classes.*

1. This method saves much time, and thus enables the teacher to devote more attention to *oral illustrations.*

2. The action of mind upon mind, is a *powerful stimulant* to exertion, and can not fail to create a *zest* for the study.

3. The mode of analyzing and reasoning of one scholar, will often *suggest new ideas* to the others in the class.

4. In the classification, those should be put together who possess as nearly equal capacities and attainments as possible. If any of the class learn quicker than others, they should be allowed to take up an extra study, or be furnished with additional examples to solve, so that the whole class may advance together.

5. The number in a class, if practicable, should not be less than six, nor over twelve or fifteen. If the number is less, the recitation is apt to be deficient in animation; if greater, the turn to recite does not come round sufficiently often to keep up the interest.

III. APPARATUS.—The *Black-board* and *Numerical Frame* are as indispensable to the teacher, as tables and cutlery are to the housekeeper. Not a recitation passes without use for the black-board. If a principle is to be demonstrated or an operation explained, it should be done upon the *black-board*, so that all may see and understand it at once.

To illustrate the increase of numbers, the process of adding, subtracting, multiplying, dividing, &c., the *Numerical Frame* furnishes one of the most simple and convenient methods ever invented.*

IV. RECITATIONS.—The *first* object in a recitation, is to secure the *attention* of the class. This is done chiefly by throwing *life* and *variety* into the exercise. Children loathe dullness, while animation and variety are their delight.

2. The teacher should not be too much confined to his text-book, nor depend upon it wholly for illustrations.

---

* Every one who ciphers, will of course have a slate. Indeed, it is desirable that every scholar in school, even to the very youngest, should be furnished with a small slate, so that when the little fellows have learned their lessons, they may busy themselves in writing and drawing various familiar objects. *Idleness* in school is the parent of *mischief*, and *employment* is the best antidote against *disobedience.*

Geometrical diagrams and *solids* are also highly useful in illustrating many points in arithmetic, and no school should be without them.

3. Every example should be *analyzed;* the "why and wherefore" of every step in the solution should be required, till each member of the class becomes perfectly familiar with the process of reasoning and analysis.

4. To ascertain whether each pupil has the right answer to all the examples, it is an excellent method to name a question, then call upon some one to give the answer, and before deciding whether it is right or wrong, ask how many in the class agree with it. The answer they give by raising their hand, will show at once how many are right. The explanation of the process may now be made.

Another method is to let the class exchange slates with each other, and when an answer is decided to be right or wrong, let every one mark it accordingly. After the slates are returned to their owners, each one will correct his errors.

V. THOROUGHNESS.—The motto of every teacher should be *thoroughness.* Without it, the great ends of the study are *defeated.*

1. In securing this object, much advantage is derived from *frequent reviews.*

2. Not a recitation should pass without *practical exercises* upon the black-board or slates, besides the lesson assigned.

3. After the class have solved the examples under a rule, each one should be required tc give an *accurate account* of its principles with the *reason* for each step, either in his own language or that of the author.

4. *Mental Exercises* in arithmetic, either by classes or the whole school together, are *exceedingly useful* in making ready and accurate arithmeticians, and should be *frequently* practised.

VI. SELF-RELIANCE.—The *habit* of *self-reliance* in study, is confessedly *invaluable.* Its power is proverbial; I had almost said, *omnipotent.* "Where there is a *will,* there is a *way.*"

1. To acquire this habit, the pupil, like a child learning to walk, must be taught to *depend upon himself.* Hence,

2. When assistance in solving an example is required, it should be given *indirectly;* not by taking the slate and performing the example for him, but by explaining the *meaning* of the question, or illustrating the *principle* on which the operation depends, by supposing a more familiar case. Thus the pupil will be able to solve the question himself, and his eye will sparkle with the consciousness of victory.

3. He must learn to perform examples *independent* of the answer, without seeing or knowing what it is. Without this attainment the pupil receives but little or no *discipline* from the study, and is *unfit* to be trusted with business calculations. What though he comes to the recitation with an occasional wrong answer; it were better to solve one question *understandingly* and *alone,* than to copy a *score* of answers from the book. What would the study of mental arithmetic be worth, if the pupil had the answers before him? What is a young man good for in the *counting-room,* who has never learned to perform arithmetical operations alone, but is obliged to look to the *answer* to know what figure to place in the quotient, or what number to place for the third term in proportion, as is too often the case in school ciphering?

CONTENTS.

SECTION I.

SECTION II.

SECTION III.

SECTION IV.

SECTION V.

SECTION VI.

SECTION VII.

SECTION VIII.

SECTION IX.

SECTION X.

SECTION XI.

SECTION XII.

SECTION XIII.

SECTION XIV.

SECTION XV.

ARITHMETIC.

SECTION I.

NOTATION AND NUMERATION.

ART. **1.** Any single thing, as a peach, a rose, a book, is called a *unit*, or *one*; if another single thing is put with it, the collection is called *two*; if another still, it is called *three*; if another, *four*; if another, *five*, &c.

The terms, *one, two, three, &c.*, by which we express *how many single things* or *units* are under consideration, are the *names of numbers*. Hence,

2. NUMBER signifies a *unit*, or a *collection of units.*

OBS. Numbers have various properties and relations, and are applied to various calculations in the practical concerns of life. These properties and applications are formed into a system, called *Arithmetic*. Hence,

3. ARITHMETIC *is the science of numbers.*

Numbers are expressed by *words*, by *letters*, and by *figures*.

Note.—The questions on the observations may be omitted, by beginners, till review, if deemed advisable by the Teacher.

QUEST.—1. What is a single thing called? If another is put with it, what is the collection called? If another, what? What are the terms one, two, three, &c.? 2. What does number signify? *Obs.* To what are numbers applied? 3. What is Arithmetic? How are numbers expressed?

NOTATION.

4. *The art of expressing numbers by letters or figures, is called* NOTATION. There are two methods of notation in use, the *Roman* and the *Arabic.*

5. The Roman method employs seven capital letters, viz: I, V, X, L, C, D, M. When standing alone, the letter I denotes *one;* V, *five;* X, *ten;* L, *fifty;* C, *one hundred;* D, *five hundred;* M, *one thousand.* To express the intervening numbers from one to a thousand, or any number larger than a thousand, we resort to repetitions and various combinations of these letters. The method of doing this will be easily learned from the following

TABLE.

I	denotes	one.	XXX	denote	thirty.
II	"	two.	XL	"	forty.
III	"	three.	L	"	fifty.
IV	"	four.	LX	"	sixty.
V	"	five.	LXX	"	seventy.
VI	"	six.	LXXX	"	eighty.
VII	"	seven.	XC	"	ninety.
VIII	"	eight.	C	"	one hundred.
IX	"	nine.	CI	"	one hundred and one.
X	"	ten.	CX	"	one hundred and ten.
XI	"	eleven.	CC	"	two hundred.
XII	"	twelve.	CCC	"	three hundred.
XIII	"	thirteen.	CCCC	"	four hundred.
XIV	"	fourteen.	D	"	five hundred.
XV	"	fifteen.	DC	"	six hundred.
XVI	"	sixteen.	DCC	"	seven hundred.
XVII	"	seventeen.	DCCC	"	eight hundred.
XVIII	"	eighteen.	DCCCC	"	nine hundred.
XIX	"	nineteen.	M	"	one thousand.
XX	"	twenty.	MM	"	two thousand.
XXI	"	twenty-one.	MDCCCXLV,	one thousand eight	
XXII	"	twenty-two, &c.		hundred and forty-five.	

QUEST.—4. What is notation? How many methods are there in use? What are they? 5. What does the Roman method employ? What does each of these letters denote when standing alone? How are the intervening numbers from one to a thousand expressed? How denote Two? Four? Six? Eight? Nine? Fourteen? Sixteen? Nineteen? Twenty-four? Twenty-eight? What does XL denote? LX? XC? CX?

N. B. Questions on this table should be varied, and continued by the teacher till the class becomes perfectly familiar with it.

OBS. 1. The learner will perceive from the Table above, that every time a letter is repeated, its *value* is repeated. Thus I, standing alone, denotes *one;* II, *two ones* or *two,* &c. So X denotes *ten;* XX, *twenty,* &c.

2. When two letters of different value are joined together, if the less is placed before the greater, the value of the greater is *diminished;* if placed after the greater, the value of the greater is *increased.* Thus, V denotes five; but IV denotes only four; and VI, six. So X denotes ten; IX, nine; XI, eleven.

3. A line or bar (—) placed over a letter, increases its value a *thousand times.* Thus, V denotes five, $\overline{\text{V}}$ denotes five thousand; X, ten; $\overline{\text{X}}$, ten thousand.

4. This method of expressing numbers was invented by the Romans; hence it is called the Roman Notation. It is now seldom used, except to denote chapters, sections, and other divisions of books and discourses.

6. The common method of expressing numbers is by the *Arabic Notation.* The Arabic method employs the following *ten characters* or *figures,* viz:

1	2	3	4	5	6	7	8	9	0
one,	two,	three,	four,	five,	six,	seven,	eight,	nine,	zero.

The first nine are called *significant* figures, because each one always has a value, or denotes some number. They are also called *digits,* from the Latin word *digitus,* which signifies a finger.

The last one is called a *cipher,* or *naught,* because when standing *alone* it has *no value,* or signifies *nothing.*

OBS. It must not be inferred, however, that the cipher is *useless;* for when placed on the right of any of the significant figures, it increases their value. It may therefore be regarded as an *auxiliary* digit, whose office, it will be seen hereafter, is as important as that of any other figure in the system.

Note.—The pupil must be able to distinguish and to write these characters, before he can make any progress in Arithmetic.

7. It will be seen that *nine* is the greatest number that

QUEST.—*Obs.* What is the effect of repeating a letter? If a letter is placed before another of greater value, what is the effect? If placed after, what? When a letter has a line placed over it, how is its value affected? Why is this method of notation called Roman? To what use is it chiefly applied? 6. How are numbers commonly expressed? How many characters does this method employ? What are their names? What are the first nine called? Why? What else are they called? What is the last one called? Why? *Obs.* Is the cipher useless? What may it be regarded?

can be expressed by *any single* figure. All numbers *larger* than nine are expressed by combining together two or more of the ten characters just explained. To express ten for example, we combine the 1 and 0, thus 10; eleven is expressed by two 1s, thus 11; twelve, thus 12; two tens, or twenty, thus 20; one hundred, thus 100, &c.

The numbers from one to a thousand are expressed in the following manner:

1, one.
2, two.
3, three.
4, four.
5, five.
6, six.
7, seven.
8, eight.
9, nine.
10, ten.
11, eleven.
12, twelve.
13, thirteen.
14, fourteen.
15, fifteen.
16 sixteen.
17, seventeen.
18, eighteen.
19, nineteen.
20, twenty.
21, twenty-one, &c.
30, thirty.
31, thirty-one, &c.
40, forty.
41, forty-one, &c.
50, fifty.
51, fifty-one, &c.
60, sixty.
61, sixty-one, &c.
70, seventy.
71, seventy-one, &c.
80, eighty.
81, eighty-one, &c.
90, ninety.
91, ninety-one, &c.
100, one hundred.
101, one hundred and one.
102, one hundred and two.
103, one hundred and three.
110, one hundred and ten.
111, one hundred and eleven.
112, one hundred and twelve.
120, one hundred and twenty.
130, one hundred and thirty.
140, one hundred and forty.
150, one hundred and fifty.
160, one hundred and sixty.
170, one hundred and seventy.
180, one hundred and eighty.
190, one hundred and ninety
200, two hundred.
300, three hundred.
400, four hundred.
500, five hundred.
600, six hundred.
700, seven hundred.
800, eight hundred.
900, nine hundred.
990, nine hundred and ninety.
991, nine hundred and ninety-one.
992, nine hundred and ninety-two.
998, nine hundred & ninety-eight.
999, nine hundred & ninety-nine.
1000, one thousand.

QUEST.—7. What is the greatest number that can be expressed by one figure? How are larger numbers expressed? How express ten? Eleven? Twelve? Twenty? What is the greatest number that can be expressed by two figures? How express a hundred? One hundred and ten? One hundred and forty-five? Five hundred and sixty-eight? What is the greatest number that can be expressed by three figures? How express a thousand?

Note.—Questions on the foregoing table should be continued till the class becomes familiar with the mode of expressing any number from 1 to 1000. They may be answered orally; but the best way is to let the pupil write the figures denoting the number upon the black-board, and at the same time pronounce the answer audibly.

OBS. 1. The terms *thirteen*, *fourteen*, *fifteen*, &c., are obviously derived from three and ten, four and ten, five and ten, which by contraction become thirteen, fourteen, fifteen, &c., and are therefore significant of the numbers which they denote. The terms *eleven* and *twelve*, are generally regarded as primitive words; at all events, there is no perceptible analogy between them and the numbers which they represent. Had the terms *oneteen* and *twoteen* been adopted in their stead, the names would then have been significant of the numbers one and ten, two and ten; and their etymology would have been similar to that of the succeeding terms.

2. The terms *twenty*, *thirty*, *forty*, &c., were formed from two tens, three tens, four tens, which were contracted into twenty, thirty, forty, &c.

3. The terms *twenty-one*, *twenty-two*, *twenty-three*, &c., are compounded of twenty and one, twenty and two, &c. All the other numbers as far as ninety-nine are formed in a similar manner.

4. The terms *hundred* and *thousand* are primitive words, and bear no analogy to the numbers which they denote. The numbers between a hundred and a thousand are expressed by a repetition of the number below a hundred. Thus we say, one hundred and one, one hundred and two, one hundred and three, &c.

8. It will be perceived from the foregoing table, that the figures standing in different places have different values. Thus the digits, 1, 2, 3, &c., standing alone or in the right hand place, respectively denote *units* or *ones*. But when they stand in the second place, they express *tens;* thus the 1 in 10, 12, 15, &c., expresses *ten*, or *ten ones;* that is, its value is *ten times* as much as when it stands in the first or right hand place, and it is called a *unit* of the *second order.* So the other digits, 2, 3, 4, &c., standing

---

QUEST.—*Obs.* From what is the term thirteen formed? Fourteen? Sixteen? Eighteen? What is said of the terms eleven and twelve? How are the terms twenty, thirty, &c., formed? What is said of the terms hundred, and thousand? How are the numbers between a hundred and a thousand expressed? 8. Does the same figure always express the same value? What does each of the digits, 1, 2, 3, &c., denote, when standing in the right hand place? What does the figure 1 denote when it stands in the second place? What is its value then? What do the other figures denote when standing in the second place?

in the second place, denote *two tens*, *three tens*, *four tens*, &c.

When standing in the third place, they express *hundreds:* thus the 1 in 100, 102, 123, &c., denotes a *hundred*, or *ten tens;* that is, its value is *ten times* as much as when it stands in the second place, and it is called a *unit* of the *third order.* In like manner, 2, 3, 4, &c., standing in the third place, denote *two hundred*, *three hundred*, *four hundred*, &c.

When a digit occupies the fourth place, it expresses *thousands:* thus the 1 in 1000, 1845, &c., denotes a *thousand*, or *ten hundreds;* that is, its value is *ten times* as much as when it stands in the third place, and it is called a *unit* of the *fourth order.* Thus,

It will be seen that ten units make one ten, ten tens make one hundred, and ten hundreds make one thousand; that is, *ten* in an *inferior* order are equal to *one* in the next *superior* order. Hence, we may infer universally, that

9. *Numbers increase from right to left in a tenfold ratio; that is, each removal of a figure one place towards the left, increases its value ten times.*

10. The different values which the same figures have, are called *simple* and *local* values.

The *simple* value of a figure is the value which it expresses when it stands alone, or in the right hand place. The simple value of a figure, therefore, is the number which its name denotes. (Art. 6.)

The *local* value of a figure is the *increased* value which

QUEST.—What is a figure called when it occupies the third place? What is its value then? What is it called when in the fourth place? What is its value? What do the other figures denote when standing in the fourth place? How many units are required to make one ten? How many tens make a hundred? How many hundreds make a thousand? Generally, how many of an inferior order are required to make one of the next superior order? 9. What is the general law by which numbers increase? What is the effect upon the value of a figure to remove it one place towards the left? 10. What are the different values of the same figure called? What is the simple value of a figure? What the local value? Upon what does the local value of a figure depend? *Obs.* Why is this system of notation called Arabic? What else is it sometimes called? Why?

it expresses by having other figures placed on its right. Hence, the local value of a figure depends on its locality, or the place which it occupies in relation to other numbers with which it is connected. (Art. 8.)

OBS. 1. This system of notation is called *Arabic*, because it is supposed to have been invented by the Arabs.

2. It is also called the *decimal system*, because numbers increase in a tenfold ratio. The term *decimal* is derived from the Latin word *decem*, which signifies ten.

11. *The art of reading numbers when expressed by figures, is called* NUMERATION.

The pupil has already become acquainted with the names of numbers, from one to a thousand. He will now easily learn to read and express the higher numbers in common use, from the following scheme, called the

NUMERATION TABLE.

Hundreds of Quadrillions.	Tens of Quadrillions.	*Quadrillions.*	Hundreds of Trillions.	Tens of Trillions.	*Trillions.*	Hundreds of Billions.	Tens of Billions.	*Billions.*	Hundreds of Millions.	Tens of Millions.	*Millions.*	Hundreds of Thousands.	Tens of Thousands.	*Thousands.*	Hundreds.	Tens.	*Units.*
5	6	8,	3	4	2,	9	7	5,	8	9	7,	6	4	5,	4	3	2.
Period VI. Quadrillions.			Period V. Trillions.			Period IV. Billions.			Period III. Millions.			Period II. Thousands.			Period I. Units.		

12. The different orders of numbers are divided into *periods* of three figures each, *beginning* at the *right hand*. The first, which is occupied by units, tens and hundreds,

---

QUEST.—11. What is numeration? Repeat the Numeration Table, beginning at the right hand. What is the first place on the right called? The second place? The third? Fourth? Fifth? Sixth? Seventh? Eighth? Ninth? Tenth, &c.? 12. How are the orders of numbers divided? What is the first period called? By what is it occupied? What is the second called? By what occupied? What is the third called? By what occupied? What is the fourth called? By what occupied? What is the fifth called? By what occupied?

is called *units'* period; the second is occupied by thousands, tens of thousands and hundreds of thousands, and is called *thousands'* period, &c.

The figures in the table are read thus: Five hundred and sixty-eight *quadrillions*, three hundred and forty-two *trillions*, nine hundred and seventy-five *billions*, eight hundred and ninety-seven *millions*, six hundred and forty-five *thousand*, four hundred and thirty-two.

13. To read numbers which are expressed by figures.

Point them off into periods of three figures each; then, beginning at the left hand, read the figures of each period in the same manner as those of the right hand period are read, and at the end of each period pronounce its name.

OBS. 1. The learner must be careful, in *pointing off* figures, always to begin at the *right* hand; and in *reading* them, to begin at the *left* hand.

2. Since the figures in the first or right hand period always denote units, the name of the period is not pronounced. Hence, in reading figures, when no period is mentioned, it is always understood to be the right hand, or units' period.

EXERCISES IN NUMERATION.

Note.—At first the pupil should be required to apply to each figure the name of the place which it occupies. Thus, beginning at the right hand, he should say, "Units, tens, hundreds," &c., and point at the same time to the figure standing in the place which he mentions. It will be a profitable exercise for young scholars to write the examples upon their slates or paper, then point them off into periods, and read them.

QUEST.—13. How do you read numbers expressed by figures? *Obs.* Where begin to point them off? Where to read them? Do you pronounce the name of the right hand period? When no period is named, what is understood? 14. In the French method of numeration, how many figures are there in a period? How many in the English method? Which method is preferable?

Read the following numbers:

Ex. 1.	127	11.	75407	21.	5604700
2.	172	12.	125242	22.	2020105
3.	721	13.	240251	23.	45001003
4.	520	14.	407203	24.	30407045
5.	603	15.	300200	25.	145560800
6.	4506	16.	1255673	26.	8900401
7.	7045	17.	5704086	27.	250708590
8.	8700	18.	207047	28.	803068003
9.	25008	19.	2605401	29.	2175240670
10.	40625	20.	4040680	30.	7240305060

31.	45290100300	36.	13657240129698
32.	160000050000	37.	98609006006906
33.	7005003007	38.	80079401697000
34.	101279200361	39.	167540000000465
35.	1143206000675	40.	504069470300400

14. The method of dividing numbers into periods of *three* figures, is the *French* Numeration. The *English* divide numbers into periods of *six* figures. The French method is the more simple and convenient. It is generally used throughout the continent of Europe, as well as in America, and has been recently adopted by some English authors.

EXERCISES IN NOTATION.

Write the following numbers in figures:

1. Twenty-seven. *Ans.* 27.
2. Seventy-two. *Ans.* 72.
3. One hundred and twenty-five.
4. Three hundred and fifty-two.
5. Two hundred and four. *Ans.* 204.
6. One thousand and forty-two. *Ans.* 1042.
7. Thirty thousand nine hundred and seven. *Ans.* 30907.

OBS. It will be observed, that in the 5th example no tens are mentioned, in the 6th no hundreds, and that these places in the answers are filled by ciphers. In all cases, when any intervening order is omitted in the given example, the place of that order in the answer must be filled by a cipher. Hence,

15. To express numbers by figures.

Begin at the left hand, and write in each order the figure which denotes the given number in that order.

If any intervening orders are omitted in the proposed number, write ciphers in their places.

8. Forty-six thousand and four hundred.
9. Ninety-two thousand, one hundred and eight.
10. Sixty-eight thousand and seventy.
11. One hundred and twenty-four thousand, six hundred and thirty.
12. Two hundred thousand, one hundred and sixty.
13. Four hundred and five thousand, and forty-five.
14. Three hundred and forty thousand.
15. Nine hundred thousand, seven hundred and twenty.
16. One million, and seven hundred thousand.
17. Thirty-six millions, twenty thousand, one hundred and fifty.
18. One hundred millions, and forty-five.
19. Mercury is thirty-seven millions of miles from the sun.
20. Venus, sixty-nine millions.
21. The Earth, ninety-five millions.
22. Mars, one hundred and forty-five millions.
23. Jupiter, four hundred and ninety-four millions.
24. Saturn, nine hundred and seven millions.
25. Herschel, one billion, eight hundred and ten millions.
26. Seven billions, nine hundred millions, and forty thousand.
27. Sixty billions, seven millions, and four hundred.
28. One hundred and thirteen billions, six hundred and fifty thousand.
29. Four hundred and six billions, eighty millions, and seven hundred.
30. Twenty-five trillions, and ten thousand.

QUEST.—15. How are numbers expressed by figures? If any intervening order is omitted in the example, how is its place supplied?

SECTION II.

ADDITION.

MENTAL EXERCISES.

ART. **16.** Ex. 1. George bought a slate for 9 cents, a sponge for 6 cents, and a pencil for 1 cent: how many cents did he pay for all?

OBS. To solve this example, we must add together the number of cents which he paid for the several articles. Thus, 9 cents and 6 cents are 15 cents, and one cent more makes 16 cents. *Ans.* He paid 16 cents.

2. Henry gave 8 cents for a writing-book, 6 cents for an inkstand, and 4 cents for some quills: how many cents did he give for all?

3. Sarah obtained 4 credit marks yesterday, 3 the day before, and 5 to-day: how many credit marks has she in all?

4. John had 6 peaches, and his mother gave him 10 more: how many peaches had he then?

5. Harriet has 7 pins; she has given away 4, and lost 2: how many pins had she at first?

6. If a quart of cherries is worth 5 cents, a pound of figs 9 cents, and a lemon 4 cents, how much are they all worth?

7. Joseph paid 6 cents for some raisins, 7 cents for a top, and 3 cents for some fish-hooks: how many cents did he pay for all?

8. Mary has 9 white roses and 8 red ones: how many roses has she in all?

9. A beggar met four men, one of whom gave him 3 shillings, another 2, another 1, and the last 5 shillings: how many shillings did the beggar receive?

10. A farmer sold 4 bushels of apples to one customer, 6 to another, 5 to a third, and 2 to a fourth: how many bushels did he sell to all?

ADDITION TABLE.

2 and	3 and	4 and	5 and	6 and	7 and	8 and	9 and
1 are 3	1 are 4	1 are 5	1 are 6	1 are 7	1 are 8	1 are 9	1 are 10
2 " 4	2 " 5	2 " 6	2 " 7	2 " 8	2 " 9	2 " 10	2 " 11
3 " 5	3 " 6	3 " 7	3 " 8	3 " 9	3 " 10	3 " 11	3 " 12
4 " 6	4 " 7	4 " 8	4 " 9	4 " 10	4 " 11	4 " 12	4 " 13
5 " 7	5 " 8	5 " 9	5 " 10	5 " 11	5 " 12	5 " 13	5 " 14
6 " 8	6 " 9	6 " 10	6 " 11	6 " 12	6 " 13	6 " 14	6 " 15
7 " 9	7 " 10	7 " 11	7 " 12	7 " 13	7 " 14	7 " 15	7 " 16
8 " 10	8 " 11	8 " 12	8 " 13	8 " 14	8 " 15	8 " 16	8 " 17
9 " 11	9 " 12	9 " 13	9 " 14	9 " 15	9 " 16	9 " 17	9 " 18
10 " 12	10 " 13	10 " 14	10 " 15	10 " 16	10 " 17	10 " 18	10 " 19

Note.—It is an interesting and profitable exercise for young pupils to recite tables in concert. But it will not do to depend upon this method alone. It is indispensable for every scholar who desires to be *accurate* either in *arithmetic* or *business*, to have the common arithmetical tables *distinctly* and *indelibly* fixed in his mind. Hence, after a table has been repeated by the class in concert, or individually, the teacher should ask many promiscuous questions, to prevent its being recited *mechanically*, from a knowledge of the regular increase of numbers.

Ex. 11. How many are 12 and 10? 22 and 10? 32 and 10? 42 and 10? 52 and 10? 62 and 10? 72 and 10? 82 and 10? 92 and 10?

12. How many are 24 and 10? 36 and 10? 48 and 10? 53 and 10? 67 and 10? 91 and 10? 86 and 10? 78 and 10? 69 and 10? 97 and 10?

13. How many are 19 and 4? 29 and 4? 39 and 4? 79 and 4? 59 and 4? 89 and 4? 99 and 4? 69 and 4? 49 and 4?

14. How many are 17 and 8? 27 and 8? 47 and 8? 67 and 8? 57 and 8? 97 and 8? 87 and 8?

15. How many are 16 and 7? 26 and 7? 56 and 7? 86 and 7? 76 and 7? 96 and 7?

16. How many are 14 and 6? 24 and 6? 84 and 6? 74 and 6? 54 and 6? 64 and 6? 94 and 6?

17. Add 2 to itself till the sum is a hundred.

OBS. This and the next four examples may be recited in concert. Thus, 2 and 2 are four, and 2 are 6, and 2 are 8, &c.

18. Add 3 in the same manner, till the sum is a hundred and two.

19. Add 5 in the same manner, till the sum is a hundred and ten.

20. Add 4 in the same manner, till the sum is a hundred and twelve.

21. Add 10 in the same manner, till the sum is a hundred and twenty.

22. A man bought a sheep for 3 dollars, a cow for 21 dollars, and a calf for 5 dollars: how much did he pay for the whole.

23. A shopkeeper sold a dress to a lady for 15 dollars, a muff for 10 dollars, and a bonnet for 6 dollars: what was the amount of her bill?

24. A drover bought 16 sheep of one farmer, 9 of another, 10 of another, and 6 of another: how many sheep did he buy?

25. Harry gave 31 cents for his arithmetic, 10 cents for a writing-book, 8 cents for a ruler, and 6 cents for a lead pencil: how many cents did he pay for all?

26. What is the sum of 10 and 12 and 5 and 4?

27. William bought a pair of boots for 26 shillings, and a cap for 9 shillings: how many shillings did he give for both?

28. Susan bought a comb for 17 cents, a purse for 8 cents, and a spool of cotton for 5 cents: how much did she pay for all?

29. A farmer sold a ton of hay for 18 dollars, a cow for 10 dollars, and a cord of wood for 3 dollars: how much did he receive for all?

30. A merchant sold 15 barrels of flour to one man, 5 to another, and 7 to another: how many barrels of flour did he sell?

31. In a certain school there are 60 boys, and 30 girls: how many scholars does that school contain?

Analysis.—60 is 6 tens, and 30 is 3 tens; (Art. 7. Obs. 2;) 6 tens and 3 tens are 9 tens, and 9 tens are 90.

Ans. 90 scholars.

32. A mechanic sold a wagon for 30, and a sleigh for 20 dollars: how much did he get for both?

33. 40 is how many tens? 60? 20? 30? 70? 80? 50? 90? 100?

34. 6 tens are how many? 8 tens? 9 tens? 10 tens? 11 tens? 12 tens? 13 tens? 14 tens? 15 tens? 16 tens? 17 tens? 18 tens? 19 tens? 20 tens?

35. 7 tens and 2 tens are how many? *Ans.* 9 tens, or 90.

36. 8 tens and 3 tens are how many? 5 tens and 8 tens? 7 tens and 8 tens? 6 tens and 9 tens? 9 tens and 8 tens? 10 tens and 6 tens?

37. In a certain orchard there are 80 apple-trees, and 40 peach-trees: how many trees does the orchard contain?

38. A traveler rode 90 miles in the cars, and 60 miles in stages: how many miles did he travel?

39. A man gave 60 dollars for his horse, 30 dollars for his harness, and 20 dollars for his cart: how much did he pay for all?

40. A man bought a horse for 98 dollars, and a wagon for 65 dollars: how much did he give for both?

Analysis.—98 is composed of 9 tens and 8 units, and 65 is composed of 6 tens and 5 units. (Art. 7. Obs. 3.) 9 tens and 6 tens are 15 tens, or 1 hundred and 5 tens; 8 units and 5 units are 13 units, or 1 ten and 3 units; now 1 ten added to 5 tens, makes 6 tens or 60, and 3 units are 63, which, joined with the hundred, makes 163.

Ans. He paid 163 dollars.

41. How many are 63 and 24? *Ans.* 87.

42. How many are 68 and 25?

43. How many are 56 and 23 and 5?

44. How many are 83 and 72 and 4 and 6?

45. How many are 72 and 25 and 10 and 2?

46. Bought a pound of tea for 60 cents, an ounce of pepper for 8 cents, and a quart of molasses for 10 cents: what does my bill amount to?

47. The price of a geography is 55 cents, and the price of a grammar is 42 cents: what is the cost of both?

48. Paid 7 dollars for a barrel of flour, 17 dollars for a ton of hay, and 30 dollars for a cow: what is the cost of all?

49. In January there are 31 days, and in February 28 days: how many days are there in both months?

50. A man, having three sons, gave 50 dollars to the oldest, 40 dollars to the second, and 30 dollars to the youngest: how many dollars did he give to the three?

17. The learner will perceive that the solution of each of the preceding examples, consists in finding a *single number* which will exactly express the value of the *several given* numbers united together.

18. *The process of uniting two or more numbers together, so as to form one single number, is called* ADDITION.

The *answer*, or the number thus found, is called the *sum* or *amount.*

OBS. When the numbers to be added are all of the *same denomination*, as all dollars, or all pounds, &c., the operation is called *Simple Addition.*

19. *Signs.*—Addition is often represented by the sign (+), which is called *plus.* It consists of two lines, one horizontal, the other perpendicular, forming a cross, and shows that the numbers between which it is placed, are to be added together. Thus the expression 6+8, signifies that 6 is to be added to 8. It is read, "6 plus 8," or "6 added to 8."

Note.—*Plus* is a Latin word, originally signifying "more," hence "added to."

20. The *equality* between two numbers, or sets of numbers, is expressed by two parallel lines (=), called *the sign of equality.* It shows that the numbers between which it is placed are equal to each other. Thus the expression 5+3=8, denotes that 5 added to 3 are equal to 8. It is read, "5 plus 3 equal 8," or "the sum of 5 plus 3 is equal to 8." So 7+5=8+4=12.

Q.—18. What is addition? What is the answer called? *Obs.* When the numbers to be added are all of the same denomination, what is the operation called? 19. What is the sign of addition called? Of what does it consist? What does it show? *Note.* What is the meaning of the word plus? 20. How is the equality between two numbers represented? What does the sign of equality show? How is the expression 5+3=8, read? How, 7+5=8+4=12?

EXERCISES FOR THE SLATE.

21. Examples in which the numbers to be added are *small*, should be solved *mentally*; but when the numbers are *large*, the operation may be facilitated by setting them down upon a slate, or black-board. The manner of doing this will now be explained.

OBS. Pupils not unfrequently seem to infer, that when they take up the slate and pencil, they can lay aside *thinking*; that the *hands* are to solve the question without the aid of the *intellect*. Hence operations upon the slate are often a merely mechanical effort, as listless and mindless as the talking of a parrot, or the trudging of a dray-horse. This is a sad mistake. It is sure to render the study of arithmetic irksome, and to destroy the progress of the learner.

It is not the object in using the slate to supersede *thinking* and *reasoning*, but to assist the memory in retaining the numbers and the several steps of the operation, while the intellect is carrying on the process of thinking and reasoning.

The hands simply write down the figures or the result of the operation, but it is the *mind*, and the *mind only*, that performs the addition and all other arithmetical calculations, whether we use the slate or not. Hence, whoever wishes to become a proficient in arithmetic, must never allow his mind to become *inactive* when using his slate, nor pass a single solution without understanding the *reason* of the several steps.

Ex. 1. A man bought a pound of tea for 63 cents, and a pound of coffee for 24 cents: how much did he pay for both?

Directions.—Write the numbers under each other, so that *units* may stand under *units*, *tens* under *tens*, and draw a line beneath them. Then, beginning at the right hand or units, add each column *separately* in the following manner: 4 units and 3 units are 7 units.—

Operation.

tens	units	
6	3	price of tea.
2	4	" of coffee.
8	7	cts. price of both.

QUEST.—21. How should examples, in which the numbers to be added are small, be solved? When they are large, how may the operation be facilitated? *Obs.* Is the slate designed to supersede thinking and reasoning? What is its use? How are all arithmetical calculations performed? What direction is given to those who wish to become proficients in arithmetic?

Write the 7 in units' place, under the column added. 2 tens and 6 tens are 8 tens. Write the 8 in tens' place. The amount is 87 cents.

Note.—The learner will perceive that the operation upon the slate is essentially the same as the mental solution of the same question; (Art. 16. Ex. 41 ;) and that both give the same result.

2. A butcher purchased two droves of sheep, the first containing 436, and the second 243: how many sheep did both droves contain?

Write the numbers under each other, and proceed as before. Thus, 3 units and 6 units are 9 units; 4 tens and 3 tens are 7 tens; 2 hundreds and 4 hundreds are 6 hundreds. The amount is 679.

Operation.

436 First drove.
243 Second "
———
679 *Ans.*

22. It will be perceived, from these examples, that *units* are added to *units*, *tens* to *tens*, and *hundreds* to *hundreds;* that is, figures of the *same order* are added to each other. This is the only way numbers can be added. For, figures standing in different orders or columns, express different values; (Art. 8;) consequently, if united together in a *single* sum, the amount can neither be of one order nor another. Thus, 3 units and 3 tens will neither make *six units*, nor *six tens*, any more than 3 oranges and 3 apples will make 6 apples, or 6 oranges. In like manner it is plain that 4 tens and 4 hundreds will neither make 8 tens, nor 8 hundreds.

OBS. In writing numbers to be added, great care should be taken to place *units* under *units*, *tens* under *tens*, &c., in order to prevent mistakes which would otherwise be liable to occur from adding *different* orders to each other.

3. A man found two purses of money, one containing 425 dollars, the other 361 dollars: how many dollars did both purses contain?

QUEST.—In the 1st example how do you write the numbers for addition? Which column do you add first? Which next? *Note.* Does the operation upon the slate differ from the mental solution of the same question? 22. Can figures standing in different orders be added to each other? Why not? Illustrate by an example. *Obs.* What is the object in writing units under units, &c.?

4. What is the sum of 3261 and 5428?
5. What is the sum of 45436 and 12321?
6. What is the sum of 420261 and 231204?
7. What is the sum of 3021040 and 5630721?
8. What is the sum of 730043000 and 268900483?

Write the following examples upon the slate, and find the sum of each:

9.	10.	11.	12.
221	4212	62022	82202310
345	3120	5103	3060231
422	341	21640	617403

23. When the *sum* of a column does not exceed 9, it must be written, as we have seen, under the column added. But when the sum of a column exceeds 9, it requires two or more figures to express it; (Art. 7;) consequently, it cannot all be written under the column added. What then must be done? We will now illustrate this case.

13. A man paid 98 dollars for a horse, and 65 dollars for a wagon: how much did he pay for both?

Operation.

98 price of horse.
65 " of wagon.
———
163 *Amount.*

Directions.—Write the numbers, and begin at the right hand, as before. Thus 5 units and 8 units are 13 units. Now 13 is 1 ten and 3 units, and requires two figures to express it; (Art. 7;) consequently it cannot be written under the column of units. Hence we write the 3 units in the units' place, and reserving the 1 ten or left hand figure in the mind, add it with the tens in the next column. Thus 1 ten (which was reserved) and 6 tens are 7 tens, and 9 are 16 tens, which are equal to 1 hundred and 6 tens. Write the 6 tens under the column added, and the 1 hundred in the place of hundreds. The amount is 163 dollars.

QUEST.—23. When the sum of a column does not exceed 9, where is it written? Can the whole sum be written under the column when it exceeds 9? Why not? In the 13th example, what is the sum of the units' column? How do you dispose of it? What do you do with the sum of the next column?

OBS. It will be perceived that the operation upon the slate is substantially the same as the mental solution of the same question. (Art. 16. Ex. 40.) In each case, we add the orders separately; in each, finding the sum of the unit's column to be 13, or 1 ten and 3 units, we add the 1 ten to the number of tens which is contained in the example; and in each we obtain the same result.

14. A gentleman bought a span of horses for 645 dollars, a carriage for 467 dollars, and a set of harness for 158 dollars: how much did he give for the whole establishment?

Operation.

645 price of horses.
467 " carriage.
158 " harness.
——
1270 dollars. *Ans.*

Proceed as before. Thus 8 units and 7 units are 15 units, or we simply say, 8 and 7 are 15, and 5 are 20. Set the 0 under the column added, and, reserving the 2, add it with the next column. 2 (which was reserved) and 5 are 7, and 6 are 13, and 4 are 17. Set the 7 under the column added, and add the 1 with the next column. 1 (which was reserved) and 1 are 2, and 4 are 6, and 6 are 12. Set the 2 under the column added, and since there is no other column to be added, write the 1 in the next place on the left. The amount is 1270 dollars.

24. The process of *reserving the tens or left hand figure*, when the sum of a column exceeds 9, and *adding* it mentally to the next column, is called *carrying tens.*

25. When the sum of the column exceeds 9, set the *units* or *right hand figure* under the column added, and carry the *tens* or *left hand figure* to the next column. In adding the *last* column on the left, it will be noticed we set down the *whole* sum. This is done for the obvious reason that there are no figures in the next column to which the left hand figure can be added, and is in fact *carrying* it to the next order.

QUEST.—*Obs.* Does the operation upon the slate differ essentially from the mental solution of the same example? In what respects do they coincide? 24. What is the process of reserving the tens and adding them to the next column, called? 25. When the sum of any column exceeds 9, what is to be done with it? When the sum is 20, what do you set down, and what do you carry? If 27, what?

ILLUSTRATION OF THE PRINCIPLE OF CARRYING.

26. To illustrate the *principle* of *carrying*, let us take the thirteenth example, and as we add the columns, write down the whole sum of each in a separate line. The sum of the units' column is 13 units, or 1 ten and 3 units; the sum of the tens' column is 15 tens, or 1 hundred and 5 tens. Now adding these results together as they stand, i. e. adding units to units, tens to tens, &c., the amount is 163, the same as before. Thus, it will be seen that the 1 ten or left hand figure in the sum of the first column, is added to the sum of the next column or the 15 tens, in the same manner as it was in the solution above.

Operation.

```
 98 price of horse.
 65   "   "  wagon.
 ——
 13  sum of units.
15*   "   "  tens.
———
163  Amount.
```

Again, the *principle* of *carrying* may be illustrated by separating the numbers to be added into the *parts* or *orders* of which they are composed. Thus,

```
98 is composed of 9 tens or 90, and 8 units.
65       "        6 tens or 60, and 5 units.
                            ——        —
                            150 and  13.

        Adding the sum of the units (13)    13
            to the sum of the tens, (150)   150
                                            ———
                         the amount is      163
```

Take also the fourteenth example:

```
645 is composed of 600,  40 and 5 units.
467       "        400,  60 and 7 units.
158       "        100,  50 and 8 units.
————               ————  ———     ——
1270 Amount.       1100  150 and 20
```

QUEST.—If the sum is 36, what? If 70, what? What do you do with the sum of the left hand column? Why? Does this differ from carrying?

Adding these results, units to units, tens to tens, &c.,	1100	sum of hundreds.
	150	" of tens.
	20	" of units.
we have	1270	*Amount.*

Here it will also be noticed, that when the sum of any column exceeds 9, the *tens* or *left hand* figure is added, in every instance, to the *same column* or *order* to which it is carried in the solution.

27. From these illustrations it will be seen, that the *process of carrying tens* is, in effect, simply adding the tens to tens, the hundreds to hundreds, &c., which are contained in the given example; or adding figures of the same order together, which is the only way they can be added. (Art. 22.) For, if the sum of any column exceeds 9, and thus requires two or more figures to express it, (Art. 7,) the right hand figure denotes units of the same order as the column added, and the left hand figure denotes units of the next higher order; (Art. 8;) consequently, it is of the same order as the next column to which it is *carried.* The result will obviously be the same, whether we add the tens in their proper place, as we proceed in the operation, or reserve them till we have added the respective columns, and then add them to the same orders. The *former* method is the more *convenient* and *expeditious*, and is therefore adopted in practice.

15. What is the sum of 473 and 987? *Ans.* 1460.

16.	17.	18.	19.
4674	67375	84056	405673
6206	87649	5721	720021
4321	6048	41630	369115
8569	452	163	505181

QUEST.—27. What, in effect, is the process of carrying the tens to the next column? How does this appear? Does it make any difference with the result, when the tens are added to the next column? When are they commonly added? Why?

28. PROOF.—*Beginning at the top, add each column downwards, and if the second result is the same as the first, the work is supposed to be right.*

OBS. The object of beginning at the top and adding downwards, is that the figures may be taken in a different order from that in which they were added before; otherwise, if a mistake has been made the first time adding, we should be liable to fall into the same again. But the order being reversed, the presumption is, that any mistake which may have been made will thus be detected; for it can hardly be supposed that two mistakes exactly equal will occur.

20. Find the sum of 256, 763, and 894, and prove the operation.

21. Find the sum of 8054, 5730, and 3056, and prove the operation.

22. Find the sum of 74502, 83000, and 62581, and prove the operation.

23. Find the sum of 68056, 31067, 680, and 200, and prove the operation.

24. Find the sum of 50563, 8276, 75009, 31, and 856, and prove the operation.

25. Find the sum of 65031, 2900, 35221, and 870, and prove the operation.

29. From the preceding illustrations and principles we derive the following

GENERAL RULE FOR ADDITION.

I. *Write the numbers to be added under each other, so that units may stand under units, tens under tens, &c.* (Art. 21, Ex. 1.)

II. *Begin at the right hand, and add each column separately. When the sum of a column does not exceed* 9, *write it under the column; but if the sum of a column exceeds* 9, *write the units' figure under the column added, and carry the tens to the next column.* (Arts. 23, 25.)

QUEST.—28. How is addition proved? *Obs.* Why add the columns downwards, instead of upwards? 29. What is the general rule for addition?

III. *Proceed in this manner through all the orders, and finally set down the whole sum of the last or the left hand column.* (Art. 25.)

EXAMPLES FOR PRACTICE.

1. A man bought a quantity of flour for 38 dollars, a ton of hay for 14 dollars, and a firkin of butter for 12 dollars. How much did he give for the whole?

2. A grocer bought three boxes of honey; the first contained 22 pounds, the second 15, and the third 9 pounds. How many pounds were there in all?

3. A man being asked his age, answered that it was equal to the united ages of his three children, the oldest of whom was 18, the second 16, and the third 14 years old. What was his age?

4. A man bought 5 hogsheads of molasses for 238 dollars, and sold it so as to gain 75 dollars. How much did he sell it for?

5. A lady purchased materials for 3 dresses; for the first she paid 15 dollars, for the second, 9 dollars, and for the third, 7 dollars. How much did she pay for them all?

6. A boy bought a cap for 12 shillings, a pair of gloves for 6 shillings, a pair of boots for 16 shillings, and a book for 6 shillings. How much did he give for the whole?

7. A gentleman owns 3 houses; for the first he receives a rent of 150 dollars, for the second 175, and for the third 225 dollars. What is the sum of all his rents?

8. A shopkeeper commenced business with 1530 dollars; after trading some time, he found he had gained 950 dollars. How much had he then?

9. A man bought a horse for 87 dollars, a carriage for 75 dollars, and a harness for 28 dollars. How much did he give for the whole?

10. What number of dollars are there in four purses; the first containing 25 dollars, the second 73, the third 84, and the fourth 96 dollars?

11. A poor man having lost his house by fire, to help him repair his loss, one man gave him 25 dollars, another 15, another 10, another 5, and another 3. How much did he receive from all?

12. In a certain school there were three classes in arithmetic; the first class contained 8 scholars, the second 11, and the third 14. How many scholars were studying arithmetic?

13. A merchant, on closing his business for the day, found he had received 23 dollars from one customer, 57 from another, 31 from another, and 25 from various others. How much did he receive that day?

14. A laborer, in pursuit of employment, walked 7 miles the first day, 10 the second, 12 the third, 15 the fourth, and 20 the fifth day. How far had he then walked?

15. A man, owning a large farm, gave to one of his sons 112 acres, to another 123, to the third 147, and had 200 acres left. How large was his farm at first?

16. A man bought a barrel of oil for 30 dollars, and sold it so as to gain 15 dollars. How much did he sell it for?

17. A lad bought a geography for 50 cents, a grammar for 25 cents, an arithmetic for 13 cents, and a slate for 10 cents. How much did he give for them all?

18. A gentleman purchased a carpet for 38 dollars, a dozen chairs for 36 dollars, a bureau for 15 dollars, and a table for 12 dollars. What did his bill amount to?

19. A merchant had 4 notes; one for 157 dollars, another for 368, another for 576, and another for 1687 dollars. What was the whole amount of his notes?

20. A gentleman bought a cloak for 56 dollars, a coat for 25 dollars, a vest for 9 dollars, a hat for 7 dollars, and a pair of boots for 5 dollars. What did he give for the whole?

21. A fashionable lady purchased a cashmere shawl for 469 dollars, a watch for 237 dollars, a pocket handkerchief for 87 dollars, and a bonnet for 53 dollars. What was the amount of her bill?

22. A farmer had 375 sheep and 168 lambs in one pasture, in another 379 sheep and 197 lambs. How many sheep had he? How many lambs? How many sheep and lambs together?

23. Four men entered into partnership; one furnished 2878 dollars, another 1784 dollars, a third 1265 dollars, and the fourth 894 dollars. What was the amount of their stock?

24. A man sold three house lots; for one he received 975 dollars, for another 763 dollars, and for the third 586 dollars. What did the whole amount to?

25. A gentleman purchased a store for 4500 dollars, and paid 75 dollars for repairs, and 150 dollars for having it enlarged. For how much must he sell it in order to gain 175 dollars?

26. A gentleman paid 75 dollars for one piece of cloth, 67 dollars for another, 54 dollars for another, and 48 dollars for another. How much did he pay for all?

27. A certain orchard contains 56 apple-trees, 19 peach-trees, 23 plum-trees, and 15 cherry-trees. How many trees are there in the orchard?

28. The distance from New York to Albany is 150 miles, from Albany to Utica 93 miles, from Utica to Rochester 158 miles, and from Rochester to Buffalo 75 miles. How far is it from New York to Buffalo?

29. A man being asked his age, said he was 17 years old when he left the academy, he spent 4 years in college, 3 years in a law school, practiced law 15 years, was a member of congress 18 years, and it was 16 years since he retired from business. How old was he?

30. A shopkeeper having a note due, paid 184 dollars at one time, at another 268 dollars, at another 379 dollars, at another 467 dollars, and there were 350 dollars still unpaid. What was the amount of his note?

31. A gentleman owns a house and lot worth 10800 dollars, a store worth 5450 dollars, a house-lot worth 3700 dollars, and has 15000 dollars in personal property. What is the whole amount of his property?

32. A man left his estate to his wife, his three sons, and two daughters; to his wife he gave 10350 dollars, to his sons 5450 dollars apiece, and his daughters 3500 dollars apiece. How large was his estate?

33. A merchant, on looking over his accounts, found he owed one man 750 dollars, another 648, another 597, another 486, another 379, and another 287 dollars. What was the amount of his debts?

34. A man bought a span of horses for 275 dollars, a

carriage for 150 dollars, and a harness for 87 dollars. How much did he give for the whole?

35. A man bought 268 bushels of wheat for 287 dollars, 187 bushels of corn for 98 dollars, and 156 bushels of oats for 128 dollars. How many bushels of grain did he buy; and how much did he give for the whole?

36. A man wishing to stock his farm, paid 197 dollars for a span of horses, 86 dollars for a yoke of oxen, 175 dollars for cows, and 169 dollars for sheep. How much did he give for the whole?

37. A butcher sold to one customer 157 pounds of meat, to another 159, to another 149, to another 97, and to another 68. How much did he sell to all?

38. A carpenter received 879 dollars for one job, for another 786, for another 693, for another 587, for another 476, and for another 368 dollars. How much did he receive in all?

39. A grocer bought 375 dollars worth of sugar, 287 dollars worth of molasses, 168 dollars worth of tea, 158 dollars worth of coffee, and 137 dollars worth of spices. What was the amount of his bill?

40. A merchant bought calico to the amount of 568 dollars, silks to the amount of 479 dollars, and broadcloths to the amount of 784 dollars. He sold them so as to gain 134 dollars on the calico, 178 dollars on the silks, and 242 dollars on the broadcloths. How much did he sell them for; and what was the amount of his gains?

41. A merchant pays 560 dollars a year for store rent, 386 dollars to one clerk, 267 to another, and 369 dollars for various other expenses. What does it cost him a year to carry on his business?

42. A man receives 568 dollars rent for one store, 479 for another, and 276 for another. How much does he receive for them all?

43. The distance from Boston to Springfield is 98 miles, from Springfield to Pittsfield is 53 miles, from Pittsfield to Albany is 49 miles, from Albany to Auburn is 173 miles, and from Auburn to Buffalo is 152 miles. How far is it from Boston to Buffalo?

44. A man bought a quantity of oil for 2649 dollars, and candles for 1367 dollars; he afterwards sold them so as to gain 568 dollars on the oil, and 346 dollars on the candles. How much did he receive for the whole?

45. In 1840, the state of Maine contained 501793 inhabitants; New Hampshire, 284574; Vermont, 291948; Massachusetts, 737699; Connecticut, 309978; and Rhode Island, 108830. What was the population of New England?

46. In 1840, the state of New York contained 2428921 inhabitants; New Jersey, 373306; Pennsylvania, 1724-033; and Delaware, 78085. What was the population of the Middle States?

47. In 1840, the state of Maryland contained 470019* inhabitants; Virginia, 1239797; North Carolina, 753419; South Carolina, 594398; Georgia, 691392; Alabama, 590756; Mississippi, 375651; and Louisiana, 352411. What was the population of the Southern States?

48. In 1840, the state of Tennessee contained 829210 inhabitants; Kentucky, 779828; Ohio, 1519467; Michigan 212267; Indiana, 685866; Illinois, 476183; Missouri, 383702; and Arkansas, 97574. What was the population of the Western States?

49. In 1840, the territory of Florida contained 54477 inhabitants; Wisconsin, 30945; Iowa, 43112; and the District of Columbia, 43712; on board vessels of war, 6100. What was the population of the Territories and naval service of the United States?

50. What was the whole population of the United States in 1840?

---

* According to the Official Revision.

SECTION III.

SUBTRACTION.

MENTAL EXERCISES.

ART. **30.** Ex. 1. Henry having 7 peaches, gave 4 to his sister: how many had he left?

OBS. To solve this question, consider what number added to 4 makes 7. Now from addition we know that 4 and 3 make 7; that is, 7 is composed of the numbers 4 and 3. It is evident, therefore, if one of these numbers be taken from 7, the other number will be left. Hence, 4 peaches from 7 peaches leave 3 peaches. *Ans.* 3 peaches.

2. James had 7 cents, and spent three of them: how many had he left?

3. Jack has 6 marbles: how many more must he get to make 10?

4. A farmer having 9 cows, sold 5 of them: how many had he left?

5. A pound of raisins costs 11 cents, and a pound of sugar 8 cents: what is the difference in their prices?

6. In a stage coach there were 10 passengers, 6 of whom got out at a hotel: how many remained in the coach?

7. Dick bought a knife for 12 cents, and having but 7 cents in his pocket, agreed to pay the rest to-morrow: how much does he owe for it?

8. John gathered 8 quarts of chestnuts: how many more must he gather to make 14 quarts?

9. The cost of a cap is 13 shillings, and the cost of a comforter is 3 shillings: what is the difference in their cost?

10. Susan is 15 years old, and Harriet is only 9: what is the difference in their ages?

SUBTRACTION TABLE.

2 from	3 from	4 from	5 from	6 from	7 from	8 from	9 from
2 leaves 0	3 lea. 0	4 lea. 0	5 lea. 0	6 lea. 0	7 lea. 0	8 lea. 0	9 lea. 0
3 " 1	4 " 1	5 " 1	6 " 1	7 " 1	8 " 1	9 " 1	10 " 1
4 " 2	5 " 2	6 " 2	7 " 2	8 " 2	9 " 2	10 " 2	11 " 2
5 " 3	6 " 3	7 " 3	8 " 3	9 " 3	10 " 3	11 " 3	12 " 3
6 " 4	7 " 4	8 " 4	9 " 4	10 " 4	11 " 4	12 " 4	13 " 4
7 " 5	8 " 5	9 " 5	10 " 5	11 " 5	12 " 5	13 " 5	14 " 5
8 " 6	9 " 6	10 " 6	11 " 6	12 " 6	13 " 6	14 " 6	15 " 6
9 " 7	10 " 7	11 " 7	12 " 7	13 " 7	14 " 7	15 " 7	16 " 7
10 " 8	11 " 8	12 " 8	13 " 8	14 " 8	15 " 8	16 " 8	17 " 8
11 " 9	12 " 9	13 " 9	14 " 9	15 " 9	16 " 9	17 " 9	18 " 9
12 " 10	13 " 10	14 " 10	15 " 10	16 " 10	17 " 10	18 " 10	19 " 10

OBS. This Table is the reverse of Addition Table. Hence, if the pupil has thoroughly learned that, it will cost him but little time or trouble to learn this. (See observations under Addition Table.)

11. 4 from 7 leaves how many? 4 from 9? 4 from 12? 4 from 8? 4 from 11? 4 from 13?

12. 6 from 8 leaves how many? 6 from 10? 6 from 13? 6 from 11? 6 from 15? 6 from 12? 6 from 16?

13. 7 from 9 leaves how many? 7 from 11? 7 from 14? 7 from 15? 7 from 16? 7 from 13? 7 from 17?

14. 8 from 11? 8 from 13? 8 from 16? 8 from 12? 8 from 15? 8 from 17? 8 from 14? 8 from 18?

15. 9 from 12? 9 from 14? 9 from 11? 9 from 13? 9 from 17? 9 from 15? 9 from 18? 9 from 19?

16. 2 from 4 leaves how many? 2 from 14? 2 from 24? 2 from 34? 2 from 44? 2 from 54? 2 from 64? 2 from 74? 2 from 84? 2 from 94?

17. 3 from 6? 3 from 16? 3 from 26? 3 from 36? 3 from 46? 3 from 56? 3 from 66? 3 from 76? 3 from 86? 3 from 96?

18. 4 from 9? 4 from 29? 4 from 39? 4 from 49? 4 from 59? 4 from 69? 4 from 79? 4 from 89? 4 from 99?

19. 6 from 15? 6 from 25? 6 from 35? 6 from 45? 6 from 55? 6 from 65? 6 from 75? 6 from 95?

20. 8 from 14? 8 from 24? 8 from 34? 8 from 44? 8 from 54? 8 from 64? 8 from 74? 8 from 84? 8 from 94?

21. A gentleman bought a coat for 15 dollars, and a hat for 6 dollars: how much more did his coat cost than his hat?

22. A farmer having sold 6 cords of wood for 18 dollars, took a barrel of flour at 6 dollars towards his pay and the rest in cash: how much money did he receive?

23. A lady bought a shawl for 15 dollars, and handed the shopkeeper a 20 dollar bill: how much change ought she to receive back?

24. A man having 25 watermelons in his garden, some wicked boys stole 9 of them: how many had he left?

25. James is 14 years old, and his sister is 19: what is the difference in their ages?

26. A merchant had a piece of calico which contained 33 yards; on measuring the remnant he finds he has but 7 yards left: how many yards has he sold?

27. A hogshead of cider contains 63 gallons: after drawing out 9 gallons, how many will be left?

28. Henry had 48 silver dollars, and gave 8 to the orphan asylum: how many dollars did he have left?

29. A man bought a piece of cloth containing 39 yards, and sold 6 yards of it: how many yards had he left?

30. George gave 75 cents for a pair of skates, and sold them for 9 cents less than he gave: how much did he get for his skates?

31. William had 67 cents; he spent 5 for chestnuts and 2 for apples: how many cents has he left?

32. A man sold a load of wood for 18 shillings; he laid out 4 shillings for tea and 6 for sugar: how many shillings had he to carry home?

33. Sarah having 85 cents, gave 10 cents to the Sabbath School Society, 8 to the Bible Society, and spent 6 for candy: how many cents had she left?

34. If I pay 27 dollars for a cow and sell it for 18 dollars, how much do I lose by the bargain?

35. Richard had 45 marbles; he lost 7 and gave away 5: how many had he left?

36. A man having 56 dollars in his pocket, bought a hat for 5 dollars, a coat for 10, and a pair of boots for 4: how much money had he left?

37. If I owe a merchant 50 dollars and pay him 20 dollars, how many dollars shall I then owe him?

Ans. 30 dollars.

Suggestion.—It is advisable for beginners to analyze the numbers in this question, as in Art. 16, Ex. 31, and then take 2 tens from 5 tens.

38. A farmer having 80 sheep, sold all but 30: how many did he sell?

39. A man having 90 acres of land, gave 50 acres to his son: how many acres has he left?

40. George had 70 cents and spent 30: how many had he left?

41. In a certain orchard there are 100 trees, 60 of them are apple-trees and the rest are peach-trees: how many peach-trees are there?

42. A grocer bought 150 eggs, and afterwards found that 20 of them were rotten: how many sound ones were there?

43. In the Centre School there are 150 scholars, 60 of whom are girls: how many boys are there?

44. A man bought a horse for 90 dollars, and sold it immediately for 130 dollars: how much did he make by his bargain?

45. A man owing me 200 dollars, turned me out a horse worth 80 dollars, and is to pay the balance in cash: how much money must he pay me?

46. A boy going to market with 80 cents, bought 20 cents worth of cheese, and 30 cents worth of butter: how much change had he left?

47. 35 from 42 leaves how many? 63 from 75?

48. 26 from 40 leaves how many? 35 from 45?

49. 65 from 85, how many? 82 from 94, how many?

50. 8 from 17, how many? 13 from 26, how many? 6 from 25, how many? 8 from 94, how many? 5 from 68, how many? 17 from 34, how many? 7 from 43, how many? 6 from 72, how many? 9 from 75, how many? 7 from 86, how many?

31. It will be observed that all the preceding examples of this section, though expressed in a variety of ways, involve the *same* principle; that the object aimed at in each of them, is to find the *difference* between two numbers; consequently, they are all performed in the same manner. The operation consists in taking a *less* number from a *greater*, and is called *subtraction*. Hence,

32. SUBTRACTION *is the process of finding the difference between two numbers.*

The *difference*, or the *answer* to the question, is called the *remainder*.

OBS. 1. The number to be subtracted is often called the *subtrahend*, and the number from which it is subtracted, the *minuend*. These terms, however, are calculated to embarrass, rather than assist the learner, and are properly falling into disuse.

2. Subtraction, it will be perceived, is the *reverse* of addition. Addition unites two or more numbers into one single number; subtraction, on the other hand, separates a number into two parts.

3. When the given numbers are of the *same denomination*, the operation is called *Simple Subtraction*. (Art. 18. Obs.)

33. Subtraction is often represented by a short horizontal line (—), which is called *minus*. When placed between two numbers, this sign shows that the number after it is to be subtracted from the one before it. Thus the expression 8—5, signifies that 5 is to be subtracted from 8; and is read, " 8 minus 5," or " 8 less 5."

Note.—The term *minus*, is a Latin word signifying *less*.

EXERCISES FOR THE SLATE.

34. When we wish to find the difference between two *small* numbers, it is the most convenient way to perform the subtraction in the mind. But when the numbers are *large*, it is difficult to retain them in the mind, and carry on the operation at the same time. By setting them down upon a slate or black-board, however, the process of subtracting large numbers is rendered short and simple. (Art. 21.)

Q.—What is subtraction? What is the answer called? *Obs.* What is the number to be subtracted sometimes called? That from which it is subtracted? Of what is subtraction the reverse? When the given numbers are of the same denomination, what is the operation called? 33. What is the sign of subtraction called? Of what does it consist? What does it show? How is the expression 8—5, read? *Note.* What is the meaning of the term minus? 34. What is the most convenient way of finding the difference between two small numbers? What between two large ones?

Ex. 1. Suppose a man gave 475 dollars for a span of horses, and 352 dollars for a carriage: how much more did he pay for his horses than for his carriage?

Directions.—Write the less number under the greater, so that units may stand under units, tens under tens, &c. Now, beginning with the units, proceed thus: 2 units from 5 units leave 3 units; write the 3 in units' place, under the figure subtracted. 5 tens from 7 tens leave 2 tens; set the 2 in tens' place. 3 hundreds from 4 hundreds leave 1 hundred; write the 1 in hundreds' place. The remainder is 123 dollars.

Operation.

	hund.	tens	units	
Horses,	4	7	5	Dolls.
Carriage,	3	5	2	Dolls.
Rem.	1	2	3	Dolls.

OBS. It is important for the learner to observe, that we subtract *units* from *units*, *tens* from *tens*, &c.; that is, we subtract figures of the *same* order from each other. This is done for the same reason that we *add* figures of the *same* order to each other. (Art. 22.) Hence, in writing numbers for subtraction, great care should be taken to set units under units, &c., in order to prevent the mistake of subtracting *different* orders from each other.

2. A merchant bought 268 barrels of flour; and on examination, found that only 123 barrels were fit for use: how many were damaged? *Ans.* 145.

Suggestion.—Write the less number under the greater, &c., and proceed as above.

3. A traveler having 576 dollars, was robbed of 344 dollars: how many dollars had he left?
4. What is the difference between 648 and 235?
5. What is the difference between 876 and 523?
6. What is the difference between 759 and 341?
7. What is the difference between 4567 and 1235?
8. What is the difference between 8643 and 5412?

QUEST.—In the 1st example how do you write the numbers for subtraction? Where begin to subtract? *Obs.* What orders do you subtract from each other? Why not subtract different orders from each other? Why place units under units, &c., in subtraction?

	9.	10.	11.	12.
From	68476	765274	563181	3286732
Take	36124	152140	32040	135011

35. When the figures in the lower number are all *smaller* than those directly over them, each lower figure, as we have seen in the preceding examples, must be subtracted from that above it, and the remainder must be placed under the figure subtracted.

But it often happens that a figure in the lower number is *larger* than that above it, and consequently cannot be taken from it.

13. It is required to find the difference between 75 and 48.

Operation.
75
48
27 *Rem.*

It is plain that we cannot take 8 units from 5 units, for 8 is larger than 5. What then shall we do? Since 75 is composed of 7 tens and 5 units, we can take 1 ten from the 7 tens, and adding it mentally to the 5 units, it will make 15 units. Then subtracting the 8 units from 15 units, will leave 7 units; write the 7 under the units' column. As we took 1 ten from the 7 tens, we have but 6 tens left; and 4 tens from 6 tens leave 2 tens: write the 2 under the tens' column. The whole remainder, therefore, is 2 tens and 7 units, or 27.

36. The process of taking one from a higher order in the upper number, and adding it to the figure from which the subtraction is to be made, is called *borrowing ten*, and is the reverse of *carrying ten*. (Art. 24.)

OBS. The 1 taken from a higher order, is always equal to 10 in the next lower order to which it is added. (Art. 8.)

37. The principle of *borrowing* may be illustrated by the following *analytic* solution of the last example.

QUEST.—35. When the figures in the lower number are each smaller than those over them, how proceed? Where do you place the remainder? Is a figure in the lower number ever larger than that above it? 36. What is meant by borrowing 10? What is the 1 taken from the higher order equal to?

75=60+15
48=40+ 8
Rem.=20+ 7, or 27.

Taking 1 ten from 7 tens, and uniting it with the 5 units, we have 60 plus 15 for the upper number. And we simply separate the lower number into the tens and units of which it is composed. Now subtracting, as in the last article, 8 from 15 leaves 7: 40 from 60 leaves 20. Thus the remainder is 20+7, or 27, the same as before.

OBS. It is manifest that this process of *borrowing ten*, does not change the value of the upper number; for, it consists simply in transposing a part of one order to another order in the same number, which can no more *diminish* or *increase* the number, than it will diminish or increase the amount of money a man has, if he takes a part from one pocket and puts it into another. It is advisable for the pupil to analyze several examples as above, until the process of borrowing becomes familiar.

14. From 6042
Take 2367
Rem. 3675

Since 7 units cannot be taken from 2 units, we borrow 10, which added to the 2, will make 12: then 7 units from 12 units leave 5. Now having borrowed 1 of the 4 tens, it becomes 3 tens; and 6 from 3 is impossible: hence we must borrow again. But the next figure in the upper number, i. e. the figure in the hundreds' place, is a 0, and consequently has nothing to lend. We must therefore borrow 1 from the next order still, i. e. from thousands, and adding it to the 0, it will make 10 hundreds. Then, borrowing 1 of the 10 hundreds and adding it to the 3 tens, it will make 13 tens, and 6 from 13 leaves 7. Diminishing the 10 hundreds by 1, (which we borrowed,) it becomes 9, and 3 from 9 leaves 6. Again, diminishing the 6 thousands by 1, (which we borrowed,) it becomes 5, and 2 from 5 leaves 3. The answer is 3675.

37. *a.* There is another method of *borrowing*, or rather of *paying*, which the learner will often find more con-

---

QUEST.—How illustrate the principle of borrowing upon the blackboard? *Obs.* Is the value of the upper number increased by borrowing? Is it diminished? How does this appear? 37. *a.* When we borrow 10, what other way is there to compensate for it?

venient in practice than the preceding, and less liable to lead him into mistakes, especially, when the figure in the next higher order is a *cipher*.

When we borrow 10, that is, when we add 10 to the upper figure, instead of considering the next figure in the upper number to be diminished by 1, the result will manifestly be the same, if we simply add 1 to the next figure in the lower number.

Thus, in the last example, instead of diminishing the 4 tens in the upper number by 1, we may add 1 to the 6 tens in the lower number, which will make 7; and 7 from 14 leaves 7, the same as 6 from 13. Again, adding 1 to the 3 hundreds (to compensate for the 10 we borrowed) makes 4 hundreds; and 4 from 10 leaves 6, the same as 3 from 9. Finally, adding 1 to the 2 (because we borrowed) makes 3; and 3 from 6 leaves 3. The remainder is 3675, the same as before.

15. From 574
Take 326

Rem. 248

6 from 4 is impossible: add 10 to the 4, and it will make 14; then 6 from 14 leaves 8. Adding 1 to the 2 makes 3, and 3 from 7 leaves 4. 3 from 5 leaves 2. *Ans.* 248.

OBS. This method of *borrowing* depends on the self-evident principle, that if any two numbers are *equally* increased, their difference will not be *altered*. That the two given numbers are equally increased by this process, is evident from the fact that the 1 added to the lower number, is of the next superior order to the 10 added to the upper number, and will compensate for it; for 1 in a superior order, is equal to 10 in an inferior order. (Art. 8.) Hence,

38. When a figure in the lower number is larger than that above it, borrow 10, i. e. add 10 to the upper figure, and from the number thus produced, subtract the lower figure: to compensate this, add 1 to the next figure in the lower number; or diminish the next figure in the upper number by 1, and proceed as before.

	16.	17.	18.
From	78562	645630	70430256
Take	24380	520723	4326107

QUEST.—*Obs.* Upon what does the second method of borrowing depend? How does it appear that you increase the given numbers equally?

39. PROOF.—*Add the remainder to the smaller number; and if the sum is equal to the larger number, the work is right.*

19. A man bought a horse for 175 dollars, and sold it for 127 dollars: how much did he lose by his bargain?

Operation.		*Proof.*	
Paid	175 dolls.	127	Smaller No.
Rec'd	127 dolls.	48	Remainder.
Lost	48 dolls.	175	Larger No.

Since the sum of the smaller number and remainder is equal to the larger number, the operation is correct.

OBS. This method of proof depends upon the obvious principle, that if the *difference* between two numbers be added to the *less*, the *sum* must be equal to the *greater*.

20. From 8796 subtract 2675, and prove the operation?

21. From 6210896 subtract 3456809, and prove the operation.

22. From 1000000 subtract 67583, and prove the operation.

23. From 7834501 subtract 1000000, and prove the operation.

24. From 68436907 subtract 59476012, and prove the operation.

25. From 8006754231 subtract 7975663417, and prove the operation.

40. From the preceding illustrations and principles we derive the following

GENERAL RULE FOR SUBTRACTION.

I. *Write the less number under the greater, so that units may stand under units, tens under tens,* &c.

II. *Beginning at the right hand, subtract each figure in the lower number from the figure above it, and set the remainder directly under the figure subtracted.* (Art 35.)

QUEST.—38. How then do you proceed, when a figure in the lower number is larger than the one over it? Why do you add 1 to the next figure in the lower line? 39. How is subtraction proved? *Obs.* Upon what principle does the proof of subtraction depend? 40. What is the general rule for subtraction?

III. *When a figure in the lower number is larger than that above it, add* 10 *to the upper figure; then subtract as before, and add* 1 *to the next figure in the lower number.* (Arts. 37, 38.)

EXAMPLES FOR PRACTICE.

1. A man bought a piece of cloth containing 37 yards, and sold 24 yards of it. How much had he left?

2. A merchant had on hand a quantity of flour, for which he asked 245 dollars; but for ready money he made a deduction of 24 dollars. How much did he receive for his flour?

3. In a certain Academy there were 357 scholars, 168 of whom were young ladies. How many young gentlemen were there?

4. A farmer raised 4879 bushels of wheat, and sold 3876 bushels. How much had he left?

5. A man purchased a farm for 4687 dollars, but the times becoming hard he was obliged to sell it for 896 dollars less than he gave for it. How much did he sell it for?

6. A merchant bought 2268 dollars worth of goods, which, in consequence of getting damaged, he sold for 848 dollars less than cost. How much did he sell them for?

7. A merchant sold a lot of silks for 561 dollars, which was 179 dollars more than the cost of them. How much did he give for them?

8. A man bought an estate for 8796 dollars, and sold it again for 9875 dollars. How much did he gain by his bargain?

9. A farmer raised 1389 bushels of wheat one year, and 1763 the next. How much more did he raise the second year than the first?

10. A man bought a house and lot for 5687 dollars. The house was worth 3698 dollars: how much was the lot worth?

11. Suppose a gentleman's income is 3268 dollars a year, and his expenses are 2789 dollars. How much does he save in a year?

12. The United States declared their independence in 1776: how many years is it since?

13. Two brothers commenced business at the same time; one gained 3678 dollars in five years, the other gained 2387 dollars in the same time. How much more did one gain than the other?

14. The distance from Boston to Springfield is 98 miles, and from Boston to Pittsfield it is 151 miles. How far is it from Springfield to Pittsfield?

15. From New York to Utica it is 243 miles, and from New York to Albany it is 150 miles. How far is it from Albany to Utica?

16. America was discovered by Columbus in 1492: how many years is it since?

17. Dr. Franklin died A. D. 1790, and was 84 years old when he died: in what year was he born?

18. General Washington was born A. D. 1732, and died in 1799: how old was he when he died?

19. The first settlement in New England was made at Plymouth in the year 1620: how many years is it since?

20. A ship sailed having on board a cargo valued at 100000 dollars, but being overtaken by a storm, 27680 dollars worth of goods were thrown overboard. How much of the cargo was saved?

21. The population of Massachusetts in 1840, was 737699, and that of Connecticut was 309978. How many more inhabitants were there in Massachusetts than in Connecticut?

22. In 1840, the population of Massachusetts was 737699, and in 1820 it was 523287. How much did the population increase during this period?

23. In 1840, the population of the state of New York was 2428921, and in 1820 it was 1372812. How much did the population increase during that period?

24. In 1840, the population of the New England States was 2234822, and that of the State of New York was 2428921. How many more inhabitants were there in the State of New York than in New England?

25. In 1800, the population of the United States was 5305925, and in 1840 it was 17069453. How much did it increase in forty years?

26. A farmer having 389 acres of land, sold to one man 126 acres, and to another 163. How many acres had he left?

27. A gentleman having 1768 dollars deposited in the bank, gave a check for 175 dollars to one man, to another for 238 dollars, and to another for 369 dollars. How much remained on deposit?

28. A man bought a horse for 87 dollars, a carriage for 75 dollars, and a harness for 16 dollars, and sold them all together for 200 dollars. How much did he gain by the bargain?

29. A man bought a quantity of sugar for 25 dollars, a quantity of molasses for 27 dollars, and a quantity of raisins for 29 dollars, for which he paid a hundred dollar-bill. How much change ought he to receive back?

30. An orchard contained 120 apple-trees, 47 peach-trees, and 28 pear-trees. Of the apple-trees 26 were cut down for a Railroad to pass through, 18 of the peach-trees died, and 5 of the pear-trees were blown down. How many trees were left in the orchard?

31. A gentleman had 2700 dollars which he wished to distribute among his three sons. To the oldest he gave 825 dollars, to the second 785 dollars, and the remainder to the youngest. How much did the youngest son receive?

32. A man owing 5648 dollars, paid at one time 536 dollars, at another 378 dollars, and at another 896 dollars. How much did he then owe?

33. A man having 7689 dollars, invested 689 dollars in Railroad stock, 500 dollars in a woolen factory, and 1250 dollars in bank stock. How much had he left?

34. A man bought a quantity of oil for 1763 dollars, and a lot of candles for 598 dollars. He afterwards sold them both for 2684 dollars. How much did he gain by the bargain?

35. A man owning 3789 acres of land, gave to one son 869 acres, and to another 987 acres. How much land had he left?

36. A ship of war sailing with 650 men, lost in one battle 29 men, in another 37, and by sickness 19 more. How many were still living?

37. A merchant owes one man 2684 dollars, another 1786 dollars, another 987 dollars. The whole amount of his property is 4684 dollars. How much more does he owe than he is worth?

38. A man bought three farms: for the first he gave 4673 dollars, for the second 5674 dollars, and for the third 9287 dollars. He sold them all for 37687 dollars. How much did he gain by the bargain?

39. A man bought 86 dollars worth of wheat, 48 dollars worth of butter, and a fine horse worth 148 dollars. He gave his note for 128 dollars, and paid the rest in cash. How much money did he pay?

40. A gentleman left a fortune of 18864 dollars to be divided between his two sons and one daughter; to one son he gave 6389 dollars, to the other 6984 dollars. How much did the daughter receive?

41. A man owing 8648 dollars, paid at one time 486, at another 684, at another 729 dollars. How much did he still owe?

42. Suppose a man gains by one speculation 867 dollars, by another 687; another time he gains 563 dollars, and then loses 479; still another time he gains 435 dollars, and loses 378. How much more has he gained than lost?

43. A man borrowed of a friend 684 dollars at one time, 786 at another, 874 at another, and 976 at another. He has paid 568 dollars. How much does he still owe?

44. If a man's income is 4586 dollars a year, and he spends 384 dollars for clothing, 568 for house rent, 784 for provisions, 568 for servants, and 369 for traveling, how much will he have left at the end of the year?

45. A merchant bought a quantity of sugar for 8978 dollars, paid 374 dollars freight, and then sold it for 9684 dollars. How much did he gain by the trade?

46. A merchant had in his storehouse 6384 bushels of wheat, 3752 bushels of corn, 4564 bushels of oats, and 1384 bushels of rye: it was broken open and 3564 bushels of grain taken out. How many bushels remained?

47. A man bought a quantity of beef for 5493 dollars,

a quantity of coffee for 261 dollars, and a quantity of sugar for 157 dollars; in exchange he gave 3687 dollars worth of flour, 568 dollars worth of oats, and 165 dollars worth of potatoes. How much did he then owe?

48. A gentleman has real estate valued at 3879 dollars, and personal property amounting to 9857 dollars. He owes one man 1350 dollars, and another 2687 dollars. How much would he have left if he should pay his debts?

49. A man had property to the amount of 30000, and lost by fire a store worth 5000 dollars, and goods to the amount of 3578 dollars. How much property had he left?

50. A man died leaving an estate of 175000 dollars. He gave to his wife 25000 dollars, to his three sons 32000 apiece, to his two daughters, 23000 dollars each, and the rest he gave to a literary institution. How much did the institution receive?

SECTION IV.

MULTIPLICATION.

MENTAL EXERCISES.

ART. **41.** Ex. 1. What will 3 lead pencils cost, at 4 cents apiece?

Solution.—Three pencils will evidently cost three times as much as one pencil. Now if 1 pencil costs 4 cents, 3 pencils will cost 3 times 4 cents; and 3 times 4 cents are 12 cents. *Ans.* 12 cents.

Note.—It is highly important for the pupil to give the reason in full for the solution of every example.

2. What will 2 yards of cloth cost, at 8 dollars a yard?
3. At 6 cents apiece, what will 4 oranges cost?
4. What cost 5 pounds of ginger, at 7 cents a pound?

5. If 1 pair of gloves cost 6 shillings, what will 6 pair cost?

6. At 9 cents a pound, what will 4 pounds of butter come to?

7. What will 7 barrels of flour cost, at 4 dollars a barrel?

8. In 1 bushel there are 4 pecks: how many pecks are there in 6 bushels?

9. What cost 8 pair of boots, at 6 dollars a pair?

10. At 9 shillings apiece, what will 5 caps cost?

11. What cost 6 pounds of sugar, at 10 cents a pound?

12. What cost 9 inkstands, at 8 cents apiece?

MULTIPLICATION TABLE.

2 times	3 times	4 times	5 times	6 times	7 times
1 are 2	1 are 3	1 are 4	1 are 5	1 are 6	1 are 7
2 " 4	2 " 6	2 " 8	2 " 10	2 " 12	2 " 14
3 " 6	3 " 9	3 " 12	3 " 15	3 " 18	3 " 21
4 " 8	4 " 12	4 " 16	4 " 20	4 " 24	4 " 28
5 " 10	5 " 15	5 " 20	5 " 25	5 " 30	5 " 35
6 " 12	6 " 18	6 " 24	6 " 30	6 " 36	6 " 42
7 " 14	7 " 21	7 " 28	7 " 35	7 " 42	7 " 49
8 " 16	8 " 24	8 " 32	8 " 40	8 " 48	8 " 56
9 " 18	9 " 27	9 " 36	9 " 45	9 " 54	9 " 63
10 " 20	10 " 30	10 " 40	10 " 50	10 " 60	10 " 70
11 " 22	11 " 33	11 " 44	11 " 55	11 " 66	11 " 77
12 " 24	12 " 36	12 " 48	12 " 60	12 " 72	12 " 84

8 times	9 times	10 times	11 times	12 times
1 are 8	1 are 9	1 are 10	1 are 11	1 are 12
2 " 16	2 " 18	2 " 20	2 " 22	2 " 24
3 " 24	3 " 27	3 " 30	3 " 33	3 " 36
4 " 32	4 " 36	4 " 40	4 " 44	4 " 48
5 " 40	5 " 45	5 " 50	5 " 55	5 " 60
6 " 48	6 " 54	6 " 60	6 " 66	6 " 72
7 " 56	7 " 63	7 " 70	7 " 77	7 " 84
8 " 64	8 " 72	8 " 80	8 " 88	8 " 96
9 " 72	9 " 81	9 " 90	9 " 99	9 " 108
10 " 80	10 " 90	10 " 100	10 " 110	10 " 120
11 " 88	11 " 99	11 " 110	11 " 121	11 " 132
12 " 96	12 " 108	12 " 120	12 " 132	12 " 144

OBS. The pupil will find assistance in learning this table, by observing the following particulars.

1. The several results of multiplying by 10 are formed by simply adding a cipher to the figure that is to be multiplied. Thus, 10 times 2 are 20; 10 times 3 are 30, &c.

2. The results of multiplying by 5, terminate in 5 and 0, alternately. Thus, 5 times 1 are 5; 5 times 2 are 10; 5 times 3 are 15, &c.

3. The first nine results of multiplying by 11 are formed by repeating the figure to be multiplied. Thus, 11 times 2 are 22; 11 times 3 are 33, &c.

4. In the successive results of multiplying by 9, the right hand figure regularly decreases by 1, and the left hand figure regularly increases by 1. Thus, 9 times 2 are 18; 9 times 3 are 27; 9 times 4 are 36, &c.

13. At 2 dollars a cord, what will 12 cords of wood cost? 10 cords? 9 cords? 8 cords? 7 cords? 6 cords? 5 cords? 4 cords? 3 cords?

14. In one yard there are 3 feet: how many feet are there in 12 yards? in 11 yards? 10 yards? 9 yards? 8 yards? 7 yards? 6 yards? 5 yards? 4 yards?

15. In 1 gallon there are 4 quarts: how many quarts in 12 gallons? in 11 gallons? 10 gallons? 9 gallons? 8 gallons? 7 gallons? 6 gallons? 5 gallons? 4 gallons?

16. If you buy 5 marbles for a cent, how many can you buy for 12 cents? for 11 cents? 10 cents? 9 cents? 8 cents? 7 cents? 6 cents? 5 cents? 4 cents?

17. In New England a dollar contains 6 shillings: how many shillings do 12 dollars contain? 11 dolls.? 10 dolls.? 9 dolls.? 8 dolls.? 7 dolls.? 6 dolls.? 5 dolls.?

18. If 7 pounds of sugar cost a dollar, how many pounds can you buy for 12 dollars? for 11 dollars? 10 dolls.? 9 dolls.? 8 dolls.? 7 dolls.? 6 dolls.? 5 dolls.?

19. In New York a dollar contains 8 shillings: how many shillings do 12 dollars contain? 11 dolls.? 10 dolls.? 9 dolls.? 8 dolls.? 7 dolls.? 6 dolls.? 5 dolls.?

20. At 9 cents a quart, what will 12 quarts of blackberries cost? 11 quarts? 10 quarts? 9 quarts? 8 quarts? 7 quarts? 6 quarts? 5 quarts?

21. What will be the cost of 12 yards of silk at 10 shillings per yard? of 11 yards? 10 yards? 9 yards? 8 yards? 7 yards? 6 yards? 5 yards?

22. What cost 8 cords of wood, at 5 dollars per cord?

23. If 7 yards of cloth make a cloak, how many yards will it take to make 8 cloaks?

24. What cost 9 pounds of ginger, at 8 cents a pound?

25. At 12 dollars apiece, what will 10 cows cost?

26. What cost 10 barrels of cider, at 9 shillings a barrel?

27. What will 11 pair of shoes come to, at 10 shillings a pair?

28. If 8 men can do a job of work in 9 days, how long will it take 1 man to do it?

29. If a barrel of beer will last 7 persons 8 weeks, how long will it last 1 person?

30. How much will 3 cows cost, at 14 dollars apiece?

Analysis.—14 is composed of 1 ten and 4 units, or 10 and 4. Now, 3 times 10 are 30, and 3 times 4 are 12; but 12 added to 30 make 42. Hence, 3 times 14 dollars are 42 dollars. *Ans.* 42 dollars.

31. What cost 5 tons of hay, at 13 dollars per ton?

32. What cost 4 hogsheads of molasses, at 15 dollars per hogshead?

33. How much can a man earn in 6 months, at 15 dollars per month?

34. A butcher bought 6 sheep, at 17 shillings apiece: how many shillings did they come to?

35. If a scholar performs 18 examples in 1 day, how many can he perform in 5 days?

36. In 1 pound there are 16 ounces: how many ounces are there in 8 pounds?

37. How far will a man walk in 5 days, if he walks 20 miles per day?

38. If 19 men can build a house in 4 days, how long would it take one man to do it?

39. If a shoemaker packs 16 pair of boots in 1 box, how many pair can he pack in 7 boxes?

40. If 1 acre of land produces 23 bushels of wheat, how many bushels will 4 acres produce?

Analysis.—23 is composed of 2 tens and 3 units, or 20 and 3. Now 4 times 20 are 80; 4 times 3 are 12; and 12 and 80 are 92. *Ans.* 92 bushels.

41. A merchant bought 4 pieces of silk, each piece having 24 yards: how many yards did they all contain?

42. What will 6 sleighs cost, at 25 dollars apiece?

43. What cost 4 reading books, at 42 cents apiece?

44. In 1 guinea there are 21 shillings: how many shillings are there in 5 guineas?

45. In 1 hogshead there are 63 gallons: how many gallons are there in 4 hogsheads?

46. What cost 32 pounds of sugar, at 8 cents per pound?

47. What cost 85 reams of paper, at 3 dollars per ream?

48. What cost 90 hats, at 4 dollars apiece?

49. In 1 week there are 7 days: how many days are there in 70 weeks?

50. In 1 hour there are 60 minutes: how many minutes are there in 9 hours?

42. Let us now attend to the nature of the preceding operations in this section. Take, for instance, the first example. Since 1 pencil costs 4 cents, 3 pencils will cost 3 times 4 cents. Now 3 times 4 cents is the same as 4 cents added to itself 3 times; and 4 cents + 4 cents + 4 cents are 12 cents.

Again, in the second example: since 1 yard of cloth costs 6 dollars, 4 yards will cost 4 times 6 dollars: and 4 times 6 dollars is the same as 6 dollars added to itself 4 times; and 6 dollars + 6 dollars + 6 dollars + 6 dollars are 24 dollars.

43. *This repeated addition of a number or quantity to itself, is called* MULTIPLICATION.

The number to be *repeated* or *multiplied*, is called the *multiplicand.*

The number by which we *multiply*, or which shows how many times the multiplicand is to be repeated, is called the *multiplier.*

The number *produced*, or the *answer* to the question, is called the *product.*

QUEST.—43. What is multiplication? What is the number to be repeated called? What the number by which we multiply? What does the multiplier show? What is the number produced called? When we say, 6 times 12 are 72, which is the multiplicand? Which the multiplier? Which the product?

Thus, when we say, 6 times 12 are 72, 12 is the multiplicand, 6 the multiplier, and 72 the product.

OBS. When the multiplicand denotes things of *one denomination only,* the operation is called *Simple Multiplication.*

44. The multiplier and multiplicand together are often called *factors*, because they make or produce the product.

Note.—The term *factor*, is derived from a Latin word which signifies an *agent,* a *doer*, or *producer.*

45. Multiplying by 1, is taking the multiplicand *once:* thus, 4 multiplied by 1=4.

Multiplying by 2, is taking the multiplicand *twice:* thus, 2 times 4, or 4+4=8.

Multiplying by 3, is taking the multiplicand *three times:* thus 3 times 4, or 4+4+4=12, &c. Hence,

Multiplying by any whole number, is taking the multiplicand as many times, as there are units in the multiplier.

Note—The application of this principle to *fractional* multipliers, will be illustrated under fractions.

OBS. 1. From the definition of multiplication, it is manifest that the *product* is of the *same kind* or *denomination* as the multiplicand; for, *repeating* a number or quantity does not alter its nature. Thus, if the multiplicand is an *abstract number;* that is, a number which does not express money, yards, pounds, bushels, or have reference to any particular object, the product will be an *abstract number;* if the multiplicand is *money*, the product will be money; if *weight*, the product will be weight; if *measure*, measure, &c.

2. Every *multiplier* is to be considered an *abstract number.* In familiar language it is sometimes said, that the price multiplied by the *weight* will give the value of an article; and it is often asked how much 25 cents multiplied by 25 cents will produce. But these are abbreviated expressions, and are liable to convey an erroneous idea, or rather no idea at all. If taken literally, they are absurd; for multiplication is *repeating* a number or quantity a certain *number of times.* Now to say that the price is repeated as many times as the given

---

QUEST.—When we say, 6 times 9 are 54, what is the 6 called? The 9? The 54? 44. What are the multiplicand and multiplier together called? Why? *Note.* What does the term factor signify? 45. What is it to multiply by 1? By 2? By 3? What is it to multiply by any whole number? Of what denomination is the product? How does this appear? What must every multiplier be considered? Can you multiply by a given weight, a measure, or a sum of money?

quantity is *heavy*, or that 25 cents are repeated 25 *cents times*, is nonsense. But we can multiply the price of 1 pound by a *number* equal to the number of pounds in the *weight* of the given article, and the product will be the value of the article. We can also multiply 5 cents by the *number* 5; that is, repeat 5 cents 5 times, and the product is 25 cents. Construed in this manner, the multiplier becomes an abstract number, and the expressions have a consistent meaning.

46. *Multiplication* is often denoted by two oblique lines crossing each other (×), called *the sign of multiplication.* It shows that the numbers between which it is placed, are to be multiplied together. Thus the expression 9×6, signifies that 9 and 6 are to be multiplied together, and is read, "9 multiplied by 6," or simply, "9 into 6."

OBS. The product will be the same, whether we multiply 9 by 6, or 6 by 9; for, by the table, 6 times 9 are 54, also 9 times 6 are 54. So 6×4=4×6; 5×3=3×5; 8×7=7×8, &c.

To illustrate this point; suppose there is a certain orchard which contains 4 rows of trees, and each row has 6 trees. Let the number of rows be represented by the number of horizontal rows of stars in the margin, and the number of trees in each row by the number of stars in a row. Now it is evident, that the whole number of trees in the orchard is equal either to the number of stars in a *horizontal* row repeated *four times*, or to the number of stars in a *perpendicular* row repeated *six times;* that is, equal to 6×4, or 4×6. Hence,

* * * * * *
* * * * * *
* * * * * *
* * * * * *

47. *The product of any two numbers will be the same, whichever factor is taken for the multiplier.*

EXERCISES FOR THE SLATE.

Ex. 1. What will 3 house-lots cost, at 231 dollars each?

Suggestion.—If 1 house-lot costs 231 dollars, 3 lots will cost 3 times 231 dollars; that is, three lots will cost 231+231+231, or 693 dollars.

QUEST.—46. How is multiplication sometimes denoted? What does the sign of multiplication show? How is the expression 9×6, read? How 6×7=42? 47. Does it make any difference in the product, which factor is made the multiplier? How illustrate this?

Having written the numbers upon the slate, as in the margin, we proceed thus: 3 times 1 unit are 3 units. Set the 3 in units' place under the multiplier. 3 times 3 tens are 9 tens; set the 9 in tens' place. 3 times 2 hundreds are 6 hundreds; set the 6 in hundreds' place. The product is 693 dollars.

Operation.

	231	Multiplicand.
	3	Multiplier.
Dolls.	693	*Product.*

2. What will 4 horses cost, at 120 dollars apiece?

Suggestion.—Write the less number under the greater, and proceed as before. *Ans.* 480 dolls.

3. What is the product of 312 multiplied by 3? *Ans.* 936.

4. What is the product of 121 multiplied by 4? *Ans.* 484.

5. In 1 mile there are 320 rods: how many rods are there in 3 miles?

6. If a man travels 110 miles in 1 day, how far can he travel in 8 days?

	7.	8.	9.	10.
Multiplicand,	3032	22120	101101	3012302
Multiplier,	3	4	5	3

11. What will 6 stage-coaches cost, at 783 dollars apiece?

Proceeding as before, 6 times 3 units are 18 units, or simply say, 6 times 3 are 18. Now 18 requires two figures to express it; hence, we set the 8 under the figure multiplied, and reserving the 1, carry it to the next product, as in addition. (Art. 25.) 6 times 8 are 48, and 1 (to carry) makes 49. Set the 9 under the figure multiplied, and carry the 4 to the next product, as before. 6 times 7 are 42, and 4 (to carry) make 46. Since there are no more figures to be multiplied, set down the 46 in full. The product is 4698 dollars. Hence,

Operation.

	783	
	6	
Ans.	4698	dolls.

49. When the multiplier contains but *one* figure.

Write the multiplier under the multiplicand; then, beginning at the right hand, multiply each figure of the multiplicand by the multiplier separately. If the product of any figure of the multiplicand into the multiplier does not exceed 9, *set it in its proper place under the figure multiplied; but if it does exceed* 9, *write the units' figure under the figure multiplied, and carry the tens to the next product on the left, as in addition.* (Art. 25.)

50. The principle of *carrying the tens* in multiplication is the same as in addition, and may be illustrated in a similar manner. (Art. 26.)

Take, for instance, the last example, and set the product of each figure in a separate line.

Thus, 783
6
18 units,
48* tens,
42** hunds.
4698 *Prod.*

Or, separate the multiplicand into the orders of which it is composed: thus, 783=700+80+3

Now 700×6=4200 hund.
80×6= 480 tens.
3×6= 18 units.

Adding these results, we have 4698 *Product.*

In this *analytic solution* it will be seen that the tens' figure in each product which exceeds 9, is added to the next product on the left, the same as in the common method of solving this and similar examples. The only difference between the two operations is, that in one case we add the tens as we proceed in the multiplication; in the other we reserve them till each figure is multiplied, and then add them to the same orders as before: consequently, the result must be the same in both. (Art. 27.)

QUEST.—49. How do you write the numbers for multiplication? Where begin to multiply? When the product of a figure in the multiplicand does not exceed 9, where is it written? When it exceeds 9, what is to be done with it? 50. How illustrate the principle of carrying in multiplication?

51. From this and the preceding illustrations, the learner will perceive, that units multiplied by units produce units; tens into units produce tens; hundreds into units produce hundreds, &c. Hence,

When the multiplier is units, the product will always be of the same order as the figure multiplied.

12. What cost 83 pounds of opium, at 8 dollars per pound?

13. At 9 shillings per day, how much can a man earn in 213 days?

14. If 1 sofa costs 78 dollars, how much will 8 sofas cost?

15. What cost 879 barrels of flour, at 7 dollars a barrel?

16. At 8 shillings apiece, what will a drove of 650 lambs come to?

	17.	18.	19.	20.
Multiply	8006	76030	10906	4608790
By	5	8	7	9

21. What will 26 horses cost, at 113 dollars apiece?

Suggestion.—Reasoning as before, if 1 horse costs 113 dollars, 26 horses will cost 26 times as much.

Operation.

	113	Multiplicand.
	26	Multiplier.
	678	cost of 6 horses.
	226*	cost of 20 "
Ans.	2938	cost of 26 "

Since it is not convenient to multiply by 26 at once, we first multiply by the 6 units, then by the 2 tens and add the two results together.—Thus 6 times 3 are 18; set down the 8 and carry the 1, as above. 6 times 1 are 6, and 1 to carry makes 7. 6 times 1 are 6. Next, multiply by the 2 tens thus: 20 times 3 units are 60 units or 6 tens; or we may simply say, 2 times 3 are 6. Now the 6 must denote tens; for units into tens, or what is

---

QUEST.—51. What do units multiplied into units produce? Tens into units? Of what order is the product universally, when the multiplier is units?

the same thing, (Art. 47,) tens into units, produces tens: consequently the 6 must be written in tens' place in the product; that is, under the figure 2 by which we are multiplying. 20 times 1 ten are 20 tens or 200; or simply say, 2 times 1 are 2: and since the 2 denotes hundreds, as we have just seen, set it on the left of the 6 in hundreds' place. 20 times 1 hundred are 20 hundred or 2000; or simply say, 2 times 1 are 2: and since the 2 denotes thousands, set it in the thousands' place on the left of the last figure in the product. Finally, adding these two results together as they stand, units to units, tens to tens, &c., we have 2938 dollars, which is the whole product required.

Note.—The several products of the multiplicand into the separate figures of the multiplier, are called *partial* products. Hence,

52. When the multiplier contains *more* than *one* figure.

Multiply each figure of the multiplicand by each figure of the multiplier separately, and write each partial product in a separate line, placing the first figure of each line directly under that by which you multiply; finally, add the several partial products together, and the sum will be the true product or answer required.

53. PROOF.—*Multiply the multiplier by the multiplicand, and if the product thus obtained is the same as the other product, the work is supposed to be right.*

OBS. 1. This method of proof depends upon the principle, that the product of any two numbers is the same, whichever is taken for the multiplier. (Art. 47.)

2. When the multiplier is small, we may add the multiplicand to itself as many times as there are units in the multiplier, and if the sum is equal to the product, the work is right. Thus 78×3=234. *Proof.*—78+78+78=234, which is the same as the product.

3. Multiplication may also be proved by *division*, and *by casting out the nines;* but neither of these methods can be explained here

---

QUEST.—*Note.* What is meant by partial products? 52. How do you proceed when the multiplier contains more than one figure? How should the partial products be written? Where write the first figure of each line? What do you finally do with the partial products? 53. How is multiplication proved? *Obs.* On what principle does this method of proof depend? When the multiplier is small, how may we prove it?

without anticipating principles belonging to division, with which the learner is supposed as yet to be unacquainted.

22. What will 45 cows cost, at 27 dollars a head?

Operation.	*Proof.*
45 Multiplicand,	27
27 Multiplier,	45
315	135
90	108
1215 *Product.*	1215 *Product.*

23. What cost 63 hats, at 36 shillings apiece?

24. How much corn can a man raise on 87 acres, at 45 bushels per acre?

25. How many pounds of sugar will 75 boxes contain, if each box holds 256 pounds?

26. What cost 278 hogsheads of molasses, at 23 dollars per hogshead?

27. What is the product of 347 multiplied by 256?

Suggestion.—Proceed in the same manner as when the multiplier contains but *two* figures, remembering to place the *right hand figure of each partial product directly under the figure by which you multiply.*

Operation.

```
  347
  256
 2082
1735
694
88832 Ans.
```

28. What is the product of 569 into 308?

After multiplying by the 8 units, we must next multiply by the 3 hundreds, since there are no tens in the multiplier, and place the first figure of this partial product directly under the figure 3 by which we are multiplying.

Operation.

```
    569
    308
   4552
 1707
 175252 Ans.
```

29. What is the product of 67025 into 4005?

Ans. 268435125.

30. What is the product of 841072 into 603?

54. From the preceding illustrations and principles we derive the following

GENERAL RULE FOR MULTIPLICATION.

I. *Write the multiplier under the multiplicand, units under units, tens under tens, &c.*

II. When the multiplier contains but *one* figure.

Begin with the units, and multiply each figure of the multiplicand by the multiplier, setting down the result and carrying as in addition. (Art. 49.)

III. When the multiplier contains *more* than one figure.

Multiply each figure of the multiplicand by each figure of the multiplier separately, beginning at the right hand, and write the partial products in separate lines, placing the first figure of each line directly under the figure by which you multiply. (Art. 52.)

Finally, add the several partial products together, and the sum will be the whole product.

OBS. It is immaterial as to the result which of the factors is taken for the multiplier. (Art. 47.) But it is more convenient and therefore customary to place the *larger* for the multiplicand and the *smaller* for the multiplier. Thus, it is easier to multiply 254672381 by 7, than it is to multiply 7 by 254672381, but the product will be the same.

EXAMPLES FOR PRACTICE.

1. What will 465 hats cost, at 6 dollars apiece?
2. What will 638 sheep cost, at 4 dollars a head?
3. What will 1360 yards of cloth cost, at 7 dollars a yard?
4. What cost 169 bushels of potatoes, at 4 shillings per bushel?
5. What cost 279 barrels of salt, at 9 shillings a barrel?
6. At 12 dollars a suit, how much will it cost to furnish 1161 soldiers with a suit of clothes apiece?
7. What cost 1565 acres of wild land, at 7 dollars per acre?

QUEST.—54. What is the general rule for multiplication? *Obs.* Which number is usually taken for the multiplicand?

8. What will 758 baskets of peaches cost, at 5 dollars per basket?

9. What cost 25650 pounds of opium, at 6 dollars a pound?

10. How much can a man earn in 12 months, at 15 dollars per month?

11. What will 23 loads of hay come to, at 18 dollars a load?

12. What will 45 cows come to, at 21 dollars apiece?

13. What will 56 hogsheads of molasses cost, at 32 dollars a hogshead?

14. What cost 128 firkins of butter, at 13 dollars a firkin?

15. What cost 97 kegs of tobacco, at 26 dollars per keg?

16. What cost 110 barrels of pork, at 19 dollars per barrel?

17. How much will 235 sheep come to, at 21 shillings a head?

18. How many bushels of corn will grow on 83 acres, at the average rate of 37 bushels to an acre?

19. In one bushel there are 32 quarts: how many quarts are there in 92 bushels?

20. What will a drove of 463 cattle come to, at 48 dollars per head?

21. How much will 78 thousand of boards cost, at 19 dollars per thousand?

22. What cost 243 chests of tea, at 37 dollars per chest?

23. A man bought 168 horses, at 63 dollars apiece: what did they come to?

24. What cost 256 barrels of beef, at 16 dollars a barrel?

25. If 376 men can build a fortification in 95 days, how long would it take 1 man to build it?

26. Allowing 365 days to a year, how many days has a man lived who is 45 years old?

27. If a garrison consume 725 pounds of beef in one day, how many pounds will they consume in 125 days?

28. How many pounds will the same garrison consume in 243 days?

29. How far will a ship sail in 365 days, at 215 miles per day?

30. What costs 678 tons of Railroad iron, at 115 dollars per ton?

CONTRACTIONS IN MULTIPLICATION.

55. The general rule is adequate to the solution of all examples that occur in multiplication. In many instances, however, by the exercise of judgment in applying the preceding principles, the operation may be very much abridged.

CASE I.—*When the multiplier is a composite number.*

Ex. 1. What will 14 hats cost, at 8 dollars apiece?

Analysis.—Since 14 is twice as much as 7; that is, 14=7×2, it is manifest that 14 hats will cost twice as much as 7 hats.

Operation.

```
        8
        7
       ——
       56 cost of 7 hats.
        2
      ———
Dolls. 112 cost of 14 hats.
```

Instead of multiplying by 14, we may first find the cost of 7 hats, and then multiply that product by 2, which will give the cost of 14 hats. In other words, we may first multiply by the factor 7, and that product by 2, the other factor of 14.

Proof.—14×8=112, the same as before.

2. What will 27 horses cost, at 85 dollars apiece?

Suggestion.—Find the factors of 27; that is, find two numbers, which being multiplied together, produce 27, and multiply first by one of these factors, and the product thus arising by the other.

OBS. 1. Any number which may be produced by multiplying two or more numbers together, is called a *composite* number, and the factors, which being multiplied together, produce the composite number, are sometimes called the *component parts* of the number. Thus, 14, 27, 32, &c., are composite numbers, and the factors 7 and 2, 9 and 3, 8 and 4, are their component parts.

2. The process of finding the factors of which a given number is composed, is called *resolving the number into factors.*

56. Some numbers may be resolved into *more* than *two* factors; and also into *different sets* of factors. Thus, the factors of 24 are 3, 2, 2 and 2; or 4, 3 and 2; or 6, 2 and 2; or 8 and 3; or 6 and 4; or 12 and 2.

OBS. We have seen that the product of any *two* numbers is the same, whichever factor is taken for the multiplier. (Art. 47.) In like manner, the product of any *three* or *more* factors is the same, in whatever order they are multiplied. For, the product of two factors may be considered as one number, and this may be taken either for the multiplicand, or the multiplier. Again, the product of three factors may be considered as one number, and be taken for the multiplicand, or the multiplier, &c. Thus, $24=3\times2\times2\times2=6\times2\times2=12\times2=6\times4=4\times2\times3=8\times3$.

3. What will 24 hogsheads of molasses cost, at 37 dollars per hogshead? *Ans.* 888 dollars.

Suggestion.—Resolve 24 into any *two* or *more* factors, and proceed as before. Hence,

57. To multiply by a composite number.

Resolve the multiplier into two or more factors; multiply the multiplicand by one of these factors, and this product by another factor, and so on till you have multiplied by all the factors. The last product will be the product required.

OBS. The *factors* into which a number may be *resolved*, must not be confounded with the *parts* into which it may be *separated*. (Art. 26.) The *former* have reference to multiplication, the latter to addition; that is, *factors* must be *multiplied* together, but *parts* must be *added* together to produce the given number. Thus, 56 may be resolved into two *factors*, 8 and 7; it may be separated into two *parts*, 5 tens or 50, and 6. Now $8\times7=56$, and $50+6=56$.

4. What will 36 cows cost, at 19 dollars a head?

QUEST.—*Obs.* What is a composite number? What are the factors which produce it, sometimes called? What is meant by resolving a number into factors? 56. Are numbers ever composed of more than two factors? What are the factors of 24? 32? 36? 40? 42? 60? 64? 72? 108? *Obs.* When three or more factors are to be multiplied together, does it make any difference in what order they are taken? 57. When the multiplier is a composite number, how do you proceed? *Obs.* What is the difference between the factors into which a number may be resolved, and the parts into which it may be separated?

5. What cost 45 acres of land, at 110 dollars per acre?

6. At 36 shillings per week, how much will it cost a person to board 52 weeks?

7. If a man travels at the rate of 42 miles a day, how far can he travel in 205 days?

8. At the rate of 56 bushels per acre, how much corn can be raised on 460 acres of land?

9. What cost 672 yards of broadcloth, at 24 shillings per yard?

10. What cost 1265 yoke of oxen, at 72 dollars per yoke?

CASE II.—*When the multiplier is* 1 *with ciphers annexed to it.*

58. It is a fundamental principle of notation, that each removal of a figure one place towards the left, increases its value *ten times;* (Art. 9;) consequently, annexing a *cipher* to a number will increase its value *ten times*, or *multiply* it by 10; annexing *two* ciphers, will increase its value a *hundred times*, or multiply it by 100; annexing *three* ciphers will increase it a *thousand times*, or multiply it by 1000, &c.; for each cipher annexed, removes each figure in the number one place towards the left. Thus, 12 with a cipher annexed, becomes 120, and is the same as 12×10; 12 with *two* ciphers annexed, becomes 1200, and is the same as 12×100; 12 with *three* ciphers annexed, becomes 12000, and is the same as 12×1000, &c. Hence,

59. To multiply by 10, 100, 1000, &c.

Annex as many ciphers to the multiplicand as there are ciphers in the multiplier, and the number thus formed will be the product required.

Note.—To *annex* means to place *after*, or at the *right hand.*

11. What will 10 drums of figs weigh, at 28 pounds a drum? *Ans.* 280 pounds.

QUEST.—58. What effect does it have to remove a figure one place towards the left hand? Two places? 59. How do you proceed when the multiplier is 10, 100, 1000, &c.? *Note.* What is the meaning of the term annex?

12. How many pages are there in 100 books, each book having 352 pages?
13. Multiply 476 by 1000.
14. Multiply 53486 by 10000.
15. Multiply 12046708 by 100000.
16. Multiply 26900785 by 1000000.
17. Multiply 89063457 by 10000000.
18. Multiply 9460305068 by 100000.
19. Multiply 78312065073 by 10000.

CASE III.—*When the multiplier has ciphers on the right.*

20. What will 20 acres of land cost, at 32 dollars per acre?

Note.—Any number with ciphers on its right hand, is obviously a composite number; the significant figure or figures being one factor, and 1 with the given ciphers annexed to it, the other factor. Thus 20 may be resolved into the factors 2 and 10. We may therefore first multiply by 2 and then by 10, by annexing a cipher as above.

Solution.—32×2=64, and 64×10=640 dolls. *Ans.*

21. If the expenses of an army are 2000 dollars per day, what will it cost to support the same army 365 days?

Operation.
365
2
730000

2000 may be resolved into the factors 2 and 1000. Then 2 times 365 are 730; now adding three ciphers to this product, multiplies it by 1000, (Art. 59,) and we have 730000 dollars for the answer. Hence,

60. When there are ciphers on the right hand of the multiplier.

Multiply the multiplicand by the significant figures of the multiplier, and to this product annex as many ciphers as are found on the right hand of the multiplier.

OBS. It will be perceived that this case combines the principles of the two preceding cases; for, the multiplier is a composite number, and one of its factors is 1 with *ciphers annexed* to it.

QUEST.—60. When there are ciphers on the right of the multiplier, how do you proceed? *Obs.* What principles are combined in this case?

22. How many days are there in 36 months, reckon ing 30 days to a month?

23. If 1 barrel of flour weighs 192 pounds, how much will 200 barrels weigh?

24. Multiply 4376 by 2500.

25. Multiply 50634 by 41000.

26. Multiply 630125 by 620000.

CASE IV.—*When the multiplicand has ciphers on the right.*

27. Multiply 12000 by 31.

Operation.

```
       12000
          31
       -----
          12
         36
      ------
Ans. 372000
```

Suggestion.—12000 is a composite number, the factors of which are 12 and 1000. But the product of two or more numbers is the same in whatever order they are multiplied; (Art. 47;) consequently multiplying the factor 12 by 31, and this product by 1000, will give the same result as 12000×31. Thus, 31 times 12 are 372; then annexing three ciphers, we have 372000, which is the same as 12000×31. Hence,

61. When there are ciphers on the right of the multiplicand.

Multiply the significant figures of the multiplicand by the multiplier, and to the product annex as many ciphers as are found on the right of the multiplicand.

OBS. When the multiplier and multiplicand both have ciphers on the right, multiply the significant figures together, and to their product annex as many ciphers as are found on the right of both factors.

28. Multiply 370000 by 32.

29. Multiply 8120000 by 46.

30. Multiply 56300000 by 64.

31. Multiply 623000000 by 89.

32. Multiply 54000 by 700. *Ans.* 37800000.

QUEST.—61. When there are ciphers on the right hand of the multiplicand, how proceed? *Obs.* How, when there are ciphers on the right both of the multiplier and multiplicand?

33. Multiply 4300 by 600. *Ans.* 2580000.
34. Multiply 563800 by 7200.
35. Multiply 1230000 by 12000.
36. Multiply 310200 by 20000.
37. Multiply 2065000 by 810000.
38. Multiply 2109090 by 510000.

SECTION V.

DIVISION.

MENTAL EXERCISES.

ART. **63.** Ex. 1. How many oranges, at 3 cents apiece, can you buy for 12 cents?

Suggestion.—If 3 cents buy one orange, 12 cents will buy as many oranges as there are 3 cents in 12 cents; that is, as many as 3 is contained times in 12. Now 3 is contained in 12, 4 times. *Ans.* 4 oranges.

2. How many lemons, at 4 cents apiece, can you buy for 20 cents?

Suggestion.—To find how many times 4 cents are contained in 20 cents, think how many times 4 make 20, or what number multiplied by 4, produces 20.

3. At 3 dollars per yard, how many yards of cloth can be bought for 15 dollars?

4. How many hats, at 5 dollars apiece, can you buy for 30 dollars?

5. How many barrels of flour will 36 bushels of wheat make, allowing 4 bushels to one barrel?

6. If you pay 6 cents a mile for riding in a stage, how far can you ride for 48 cents?

7. If a pound of sugar cost 7 cents, how many pounds can you buy for 56 cents.

8. How many slates, at 8 cents apiece, can you buy for 40 cents?

9. Four quarts make one gallon: how many gallons are there in 48 quarts?

10. At 7 dollars a ton, how many tons of coal can be bought for 63 dollars?

DIVISION TABLE.

2 in 2, once		3 in 3, once		4 in 4, once		5 in 5, once		6 in 6, once		7 in 7, once		8 in 8, once		9 in 9, once	
4,	2	6,	2	8,	2	10,	2	12,	2	14,	2	16,	2	18,	2
6,	3	9,	3	12,	3	15,	3	18,	3	21,	3	24,	3	27,	3
8,	4	12,	4	16,	4	20,	4	24,	4	28,	4	32,	4	36,	4
10,	5	15,	5	20,	5	25,	5	30,	5	35,	5	40,	5	45,	5
12,	6	18,	6	24,	6	30,	6	36,	6	42,	6	48,	6	54,	6
14,	7	21,	7	28,	7	35,	7	42,	7	49,	7	56,	7	63,	7
16,	8	24,	8	32,	8	40,	8	48,	8	56,	8	64,	8	72,	8
18,	9	27,	9	36,	9	45,	9	54,	9	63,	9	72,	9	81,	9

11. How many pair of boots, at 2 dollars a pair, can be bought for 24 dollars? for 22? 20? 18? 16? 14? 12? 10?

12. How many barrels of cider, at 3 dollars a barrel, can you buy for 36 dollars? for 30? 27? 24? 21? 18? 15? 12?

13. How many quarts of milk, at 4 cents a quart, can you buy for 48 cents? for 44? 40? 36? 32? 28? 24? 20? 16?

14. At 5 cents an ounce, how many ounces of wafers can you buy for 60 cents? for 55? 50? 45? 40? 35? 30? 25?

15. At 6 shillings a pair, how many pair of gloves can be bought for 60 shillings? for 54? 48? 42? 36? 30? 24? 18?

16. How many pounds of butter, at 7 cents a pound, can be purchased for 63 cents? 56? 49? 42? 35? 28? 21? 14?

17. How many cloaks will 72 yards of cloth make, allowing 8 yards to a cloak? how many 64? 56? 48? 40? 32? 24?

18. How many cows, at 9 dollars apiece, can be

bought for 81 dollars? for 72? 63? 54? 45? 36? 27? 18? 9?

19. How many times is 4 contained in 36? 48? 40?

20. How many times is 8 contained in 40? 56? 48? 64? 72?

21. In 25, how many times 4, and how many over?

Ans. 6 times and 1 over.

22. In 34, how many times 5, and how many over? In 43? 45? 37? 28? 39?

23. In 23, how many times 3, and how many over? How many times 4? 2? 10? 6?

24. In 24, how many times 7, and how many over? 6? 5? 9? 12? 2?

25. In 36, how many times 6? 7? 3? 8? 12? 5? 9?

26. In 32, how many times 6? 4? 3? 16?

27. How many hats, at 6 dollars apiece, can be bought for 60 dollars?

28. How many tons of hay, at 9 dollars per ton, can you buy for 81 dollars.

29. If you travel 7 miles an hour, how long will it take to travel 70 miles?

30. If you pay 10 cents apiece for slates, how many can you buy for 95 cents, and how many cents over?

31. George bought 12 oranges, which he wishes to divide equally between his 2 brothers: how many can he give to each?

Suggestion.—Since there are 12 oranges to be divided equally between 2 boys, each boy must receive 1 orange as often as 2 oranges are contained in 12 oranges; that is, each must receive as many oranges as 2 is contained times in 12. But 2 is contained in 12, 6 times; for 6 times 2 make 12.

Ans. 6 oranges.

32. Henry has 15 apples, which he wishes to divide equally among 3 of his companions: how many can he give to each?

33. A gentleman sent 20 peaches to be divided equally among 4 boys: how many did each boy receive?

34. A dairy-woman having 30 pounds of butter, wish-

es to pack it in 5 boxes, so that each box shall have an equal number of pounds: how many pounds must she put in each box?

35. I have 21 acres of land, which I wish to fence into 7 equal lots: how many acres must I put into each lot?

36. A boy having 28 marbles, wished to divide them into 4 equal piles: how many must he put in a pile?

37. I have 40 peach-trees, which I wish to set out in 5 equal rows: how many must I set in a row?

38. There were 45 scholars in a certain school, and the teacher divided them into 5 equal classes: how many did he put in a class?

39. If 50 dollars were divided equally among 10 men, how many dollars would each man receive?

40. A company of 8 boys buying a boat for 32 dollars, agreed to share the expense equally: how much must each one pay?

41. In a certain orchard there are 54 apple-trees, and 6 trees in each row: how many rows are there in the orchard?

42. If 63 quills are divided equally among 7 pupils, how many will each receive?

43. If you divide 36 into 4 equal parts, how many will there be in a part?

44. If you divide 56 into 8 equal parts, how many will each part contain?

45. If you divide 48 into 6 equal parts, how many will each part contain?

46. A gentleman distributed 40 dollars equally among 8 beggars: how many dollars did he give to each?

47. A company of 6 boys found a pocket-book, and on returning it to its owner, he handed them 60 dollars to be shared equally among them: what was each one's share?

48. A merchant received 72 dollars for 6 coats of equal value: how much was that apiece?

49. A man paid 81 cents for the use of a horse and buggy to ride 9 miles: how much was that a mile?

50. If you divide 90 dollars into 10 equal parts, how many dollars will there be in each part?

OBS. The object in each of the last twenty questions, is to divide a given number into several *equal* parts, and ascertain the *value* of these parts; but the method of solving them is precisely the *same* as that of the preceding ones.

64. The process by which the foregoing examples are solved, is called DIVISION.

It consists in finding how many times one given number is contained in another.

The number to be *divided*, is called the *dividend.*

The number by which we *divide*, is called the *divisor.*

The number *obtained* by division, or the *answer* to the question, is called the *quotient.* It shows how *many times* the dividend contains the divisor. Hence, it may be said

65. *Division is finding a quotient, which multiplied into the divisor, will produce the dividend.*

Note.—The term *quotient* is derived from the Latin word *quoties,* which signifies *how often,* or *how many times.*

66. The number which is sometimes *left* after division, is called the *remainder.* Thus, in the twenty-first example, when we say 4 is contained in 25, 6 times and 1 over, 4 is the divisor, 25 the dividend, 6 the quotient, and 1 the remainder.

OBS. 1. The remainder is always *less* than the divisor; for if it were equal to, or greater than the divisor, the divisor could be contained *once more* in the dividend.

2. The remainder is also of the same denomination as the dividend; for it is a part of it.

67. Division is denoted in two ways:

QUEST.—64. In what does division consist? What is the number to be divided, called? The number by which we divide? What is the number obtained, called? What does the quotient show? 65. What then may division be said to be? 66. What is the number called which is sometimes left after division? When we say 4 is in 25, 6 times and 1 over, what is the 4 called? The 25? The 6? The 1? When we say 6 is in 45, 7 times and 3 over, which is the divisor? The dividend? The quotient? The remainder. *Obs.* Is the remainder greater or less than the divisor? Why? Of what denomination is it? Why? 67. How many ways is division denoted?

First, by a horizontal line between two dots (÷), called the *sign of division*, which shows that the number *before* it, is to be divided by the number *after* it. Thus the expression 24÷6, signifies that 24 is to be divided by 6.

Second, division is often expressed by placing the divisor *under* the dividend with a short line between them. Thus the expression $\frac{35}{7}$, shows that 35 is to be divided by 7, and is equivalent to 35÷7.

OBS. It will be perceived that division is similar in principle to subtraction, and may be performed by it. For instance, to find how many times 3 is contained in 12, as in the first example, subtract 3 (the divisor) continually from 12 (the dividend) until the latter is exhausted; then counting these repeated subtractions, we shall have the true quotient. Thus, 3 from 12 leaves 9; 3 from 9 leaves 6; 3 from 6 leaves 3; 3 from 3 leaves 0. Now by counting, we find that 3 can be taken from 12, 4 times; or that 3 is contained in 12, 4 times. Hence,

67. *a. Division is sometimes defined to be a short way of performing repeated subtractions of the same number.*

OBS. 1. It will also be observed that division is the *reverse* of multiplication. Multiplication is the *repeated addition* of the same number; division is the *repeated subtraction*, of the same number. The *product* of the one answers to the *dividend* of the other: but the latter is always *given*, while the former is *required*.

2. When the dividend denotes things of *one denomination only*, the operation is called *Simple Division*.

EXERCISES FOR THE SLATE.

Ex. 1. How many barrels of cider, at 2 dollars a barrel, can you buy for 648 dollars?

Suggestion.—Since 2 dollars will buy 1 barrel, 648 dollars will buy as many barrels as 2 is contained times in 648.

QUEST.—What is the first? What does this sign show? What is the second way of denoting division? *Obs.* To what rule is division similar in principle? How is division sometimes defined? Of what is division the reverse? How does this appear? When the dividend denotes things of one denomination only, what is the operation called?

Operation.	
Divisor.	Dividend.
2)	6 4 8
Quot.	3 2 4

Having written the numbers upon the slate, as in the margin, we proceed thus: 2 is contained in 6, 3 times. Now as the 6 denotes hundreds, the 3 must also be hundreds. We therefore write it in hundreds' place; that is, under the figure which we are dividing. 2 in 4, 2 times. Since the 4 is tens, the 2 must also be tens, and we write it in tens' place. 2 in 8, 4 times. The 8 is units; hence the 4 must be units, and we write it in units' place. The answer is 324 barrels.

2. Divide 63 by 7. *Ans.* 9.

3. Divide 56 by 8. 4. Divide 42 by 7.

5. Divide 54 by 9. 6. Divide 72 by 8.

7. How many hats, at 2 dollars apiece, can be bought for 468 dollars? *Ans.* 234 hats.

8. How many sheep, at 3 dollars a head, can be bought for 369 dollars?

9. A man wishes to divide 248 acres of land equally between his two sons: how many acres will each receive?

10. How many times is 4 contained in 488?

68. Hence, when the divisor contains but *one* figure,

Write the divisor on the left hand of the dividend with a curve line between them; then, beginning at the left hand, divide each figure of the dividend by the divisor, and set each quotient figure directly under the figure from which it arose.

11. A farmer bought 96 dollars worth of dry goods, and agreed to pay in wood at 3 dollars a cord: how many cords will it take to pay his bill? *Ans.* 32 cords.

12. In 963 feet, how many yards are there, allowing 3 feet to a yard?

13. Divide 63936 by 3. 14. Divide 48848 by 4.

15. Divide 55555 by 5. 16. Divide 2486286 by 2.

69. When the divisor is not contained in the *first*

QUEST.—68. How do you write the numbers for division? Where begin to divide? Where place each quotient figure? 69. When the divisor is not contained in the first figure of the dividend, what must be done?

figure of the dividend, we must find how many times it is contained in the *first two* figures.

17. How many hats, at 3 dollars apiece, can be bought for 249 dollars?

Operation.
3)249
Quot. 83

Since the divisor 3, is not contained in 2 the first figure of the dividend, we say, 3 is in 24, 8 times, and write the 8 under the 4. 3 in 9, 3 times. *Ans.* 83 hats.

18. Divide 124 by 4.
19. Divide 366 by 6.
20. Divide 255 by 5.
21. Divide 1248 by 4.
22. Divide 24693 by 3.
23. Divide 4266 by 6.
24. Divide 35555 by 5.
25. Divide 5677 by 7.
26. Divide 64888 by 8.
27. Divide 8199 by 9.

70. After dividing any figure of the dividend, if there is a *remainder*, prefix it mentally to the next figure of the dividend, and then divide this number as before.

Note.—To *prefix* means to place *before*, or at the *left hand.*

28. A man bought 741 acres of land, which he divided equally among his 3 sons: how many acres did each receive?

Operation.
3)741
Ans. 247 acres.

When we divide 7 by 3, there is 1 remainder. This we prefix mentally to the next figure of the dividend. We then say, 3 in 14, 4 times, and 2 over. Prefixing the remainder 2 to the next figure, as before, we say, 3 in 21, 7 times.

29. If a man travel at the rate of 5 miles an hour, how long will it take him to travel 345 miles? *Ans.* 69 hours.

30. If 192 pounds of flour were equally divided among 4 persons, how many pounds would each receive?

31. Divide 45690 by 6.
32. Divide 52584 by 8.
33. Divide 81670 by 5.
34. Divide 28296 by 9.

35. When flour is 6 dollars a barrel, how much can be bought for 642 dollars?

QUEST.—70. If there is a remainder after dividing a figure of the dividend, what must be done with it? *Note.* What does the word prefix mean? *Obs.* When the divisor is not contained in a figure of the dividend, what must be done?

OBS. In this example the divisor is not contained once in the tens' figure of the dividend; we must therefore write a cipher in the quotient, and prefix the 4 to the next figure of the dividend, as if it were a remainder. We then say, 6 in 42, 7 times, and place the 7 under the 2.

Operation.
6)642
Ans. 107 barrels.

36. Divide 36060 by 6. 37. Divide 49000 by 7.

38. Divide 45900 by 9. 39. Divide 568000 by 8.

40. Allowing 5 yards of cloth for a suit of clothes, how many suits can be made from 1525 yards?

Ans. 305 suits.

41. A company of 3 men agree to pay a bill of 321 dollars: how many dollars must each man pay?

42. Divide 14350 by 7. 43. Divide 30420 by 6.

44. Divide 25105 by 5. 45. Divide 643240 by 8.

46. A merchant wished to divide 49 oranges equally among 4 boys: how many must he give to each?

Operation.
4)49
Ans. 12–1 remainder.

After giving them 12 apiece, it will be seen that there is one remainder, or 1 orange left, which is not divided. Now it is plain that the whole dividend must be divided, in order to render the division complete. But 4 is not contained in 1; hence the division must be represented by writing the 4 under the 1, thus $\frac{1}{4}$, (Art. 67,) and in order to complete the quotient, the $\frac{1}{4}$ must be annexed to the 12. The true quotient, therefore, is 12 and 1 divided by 4, and should be written thus, $12\frac{1}{4}$. Hence,

71. *When there is a remainder, after dividing the last figure of the dividend, it should always be written over the divisor and annexed to the quotient.*

47. A shoemaker has 375 pair of boots, which he wishes to pack in 6 boxes: how many pair can he put into a box?

Ans. $62\frac{3}{6}$.

48. A baker wishes to lay out 756 dollars in flour: how much can he buy, when the price is 5 dollars a barrel?

QUEST.—71. When there is a remainder after dividing the last figure of the dividend, what must be done with it?

49. How many yearlings, at 9 dollars a head, can be bought for 468 dollars?

50. How many acres of land, at 6 dollars an acre, can I buy for 973 dollars?

72. The preceding method of dividing is called *Short Division.* From the illustrations and principles now explained, we derive the following

RULE FOR SHORT DIVISION.

I. *Write the divisor on the left hand of the dividend with a curve line between them. Then beginning at the left hand, divide successively each figure of the dividend by the divisor, and place each quotient figure directly under the figure divided.* (Art. 68.)

II. *If there is a remainder after dividing any figure, prefix it to the next figure of the dividend and divide this number as before; and if the divisor is not contained in any figure of the dividend, place a cipher in the quotient, and prefix this figure to the next one of the dividend, as if it were a remainder.* (Arts. 69, 70. Obs.)

III. *When a remainder occurs after dividing the last figure, write it over the divisor and annex it to the quotient.* (Art. 71.)

73. PROOF.—*Multiply the divisor by the quotient, to the product add the remainder, and if the sum is equal to the dividend, the work is right.*

51. Divide 6973 by 6.

Solution.

6)6973
Quot. 1162 and 1 *rem.*

Proof. 1162 quotient.
6 divisor.
6972
Add the rem. 1
The result is 6973, the div.

QUEST.—72. What is the rule for short division? 73. How is division proved? *Obs.* How does it appear that the product of the divisor and quotient should be equal to the dividend? What other way of proving division is mentioned?

OBS. 1. Since the quotient shows how many times the divisor is contained in the dividend, (Art. 64,) it follows, that if the divisor is repeated as many times as there are units in the quotient, it must produce the dividend.

2. Division may also be proved by subtracting the remainder, if any, from the dividend, then dividing the result by the quotient.

PROOF OF MULTIPLICATION BY DIVISION.

74. *Divide the product by one of the factors, and if the quotient thus arising is equal to the other factor, the work is right.*

OBS. This method of proof depends on this obvious principle, viz: if the divisor and quotient, multiplied together, produce the dividend, the product of the two numbers, divided by one of those numbers, must give the other number.

LONG DIVISION.

Ex. 1. A father bought 741 acres of land, which he divided equally among his 3 sons: how many acres did each receive?

Note.—This example has been solved by short division. (Art. 70. Ex. 28.) We have introduced it here for the purpose of illustrating a different mode of dividing.

Operation.

```
Divisor. Divid. Quot.
   3)  741  (247
       6
       --
       14
       12
       --
        21
        21
        --
```

Having written the divisor on the left of the dividend as before, we find 3 is contained in 7, 2 times, and place the 2 on the right of the dividend, with a curve line between them. We next multiply the divisor by this quotient figure—2 times 3 are 6—and, placing the product under the 7, the figure divided, subtract it therefrom. We now bring down the next figure of the dividend, and placing it on the right of the remainder 1, we have 14. And 3 is in 14, 4 times. Set the 4

QUEST.—74. How is multiplication proved by division? *Obs.* Upon what principle does this proof depend? How are the numbers written for long division? Where begin to divide? Where is the quotient placed?

on the right hand of the last quotient figure, and multiply the divisor by it: 4 times 3 are 12. Write the product under 14, and subtract as before. Finally, bringing down the last figure of the dividend to the right of the last remainder, we have 21; and 3 is in 21, 7 times. Set the 7 in the quotient, then multiply and subtract as before. The quotient is 247, the same as in short division.

75. This method of dividing is called *Long Division.* It is the *same* in *principle* as Short Division. The only difference between them is, that in *Long Division* the result of each step in the operation is written down, while in *Short Division* we carry on the process in the mind, and simply write the quotient.

Note.—To prevent mistakes, it is advisable to put a dot under each figure of the dividend, when it is *brought down.*

The following questions are designed to be performed by long division, and each operation should be proved.

2. How many times is 2 contained in 578? *Ans.* 289.
3. How many times is 5 contained in 7560? *Ans.* 1512.
4. How many times is 4 contained in 126332? *Ans.* 31583.
5. How many times is 6 contained in 763251?
6. How many times is 3 contained in 4026942?
7. How many times is 8 contained in 2612488?
8. How many times is 5 contained in 1682840?
9. How many times is 7 contained in 45063284?
10. How many times is 9 contained in 650031507?
11. Divide 2234 by 21.

Operation.

```
21)2234(106 8/21. Ans.
   21
   ---
    134
    126
    ---
      8 rem.
```

21 is contained in 22 *once.*—Write the 1 in the quotient. Then multiplying and subtracting, the remainder is 1. Bringing down the next figure, we have 13 to be divided by 21. But 21 is not contained in 13, therefore we put a

QUEST.—75. What is the difference between long and short division?

cipher in the quotient, (Art. 70. Obs,) and bring down the next figure. Then, 21 in 134, 6 times, and 8 remainder. Write the 8 over the divisor, and annex it to the quotient. (Art. 71.)

76. After the first quotient figure is obtained, for *each figure of the dividend which is brought down*, either *a significant figure* or a *cipher* must be put in the *quotient*.

12. Divide 345 by 15. *Ans.* 23.
13. Divide 5378 by 25. *Ans.* 215$\frac{3}{25}$.
14. Divide 7840 by 32.
15. Divide 59690 by 45.
16. Divide 81229 by 67.
17. Divide 99435 by 81.
18. How many times is 131 contained in 18602? *Ans.* 142.

OBS. When the divisor is not contained in the first *two* figures of the dividend, find how many times it is contained in the first *three*; and, generally, find how many times it is contained in the *fewest* figures which will contain it, and proceed as before.

19. How many times is 93 contained in 100469?
20. How many times is 156 contained in 140672?

77. From the preceding principles we derive the following

RULE FOR LONG DIVISION.

Begin on the left of the dividend, find how many times the divisor is contained in the fewest figures that will contain it, and place the quotient figure on the right of the dividend with a curve line between them. Then multiply the divisor by this figure and subtract the product from the figures divided; to the right of the remainder bring down the next figure of the dividend and divide this number as before. Proceed in this manner till all the figures of the dividend are divided.

When there is a remainder after dividing the last figure, write it over the divisor and annex it to the quotient, as in short division. (Art. 71.)

QUEST.—76. What is placed in the quotient, on bringing down each figure of the dividend? *Obs.* When the divisor is not contained in the first two figures of the dividend, what is to be done? 77. What is the rule for long division?

OBS. When the divisor contains but *one* figure, the operation by *Short Division* is the most expeditious, and should therefore be practiced; but when the divisor contains *two* or *more* figures, it will generally be the most convenient to divide by *Long Division.*

EXAMPLES FOR PRACTICE.

1. If a man travel at the rate of 8 miles an hour, how long will it take him to travel 192 miles?

2. How many yards of broadcloth, at 9 dollars a yard, can be bought for 324 dollars?

3. A farmer bought a lot of young cattle, at 11 dollars per head, and paid 473 dollars for them: how many did he buy?

4. How many tons of coal, at 7 dollars a ton, can be bought for 756 dollars?

5. At 12 dollars a month, how long will it take a man to earn 156 dollars?

6. In one day there are 24 hours: how many days are there in 480 hours?

7. A man traveled 215 miles in 21 hours: how many miles did he travel per hour?

8. At 16 dollars a ton, how many tons of hay can be bought for 176 dollars?

9. How many casks of wine, at 25 dollars a cask, can be bought for 275 dollars?

10. The ship George Washington was 25 days in crossing the Atlantic Ocean, a distance of 3000 miles. How many miles did the ship sail per day?

11. The steamer Great Western crossed it in 15 days. How many miles did she sail per day?

12. The steamer Caledonia crossed it in 12 days. How many miles did she sail per day?

13. If a man can earn 32 dollars a month, how long will it take him to earn 420 dollars?

14. If 63 gallons make a hogshead, how many hogsheads will 1260 gallons make?

15. If a ship can sail 264 miles per day, how far can she sail in an hour?

QUEST.—*Obs.* When should short division be used? When long division?

16. How many times 12 in 172, and how many over?
17. How many times 15 in 630, and how many over?
18. How many times 22 in 865, and how many over?
19. 1236 is how many times 17, and how many over?
20. 7652 is how many times 13, and how many over?
21. 3061 is how many times 125, and how many over?
22. 1861 is how many times 231, and how many over?
23. 8 times 256 is how many times 9?
24. 12 times 157 is how many times 7?
25. 15 times 2251 is how many times 12?
26. 19 times 136 is how many times 75?
27. 63 times 102 is how many times 37?
28. 78 times 276 is how many times 136?
29. 115 times 321 is how many times 95?
30. 144 times 137 is how many times 312?

CONTRACTIONS IN DIVISION.

77. *a.* The operations in division, as well as in multiplication, may often be shortened by a careful attention to the application of the preceding principles.

Case I.—*When the divisor is a composite number.*

Ex. 1. A gentleman divided 168 oranges equally among 14 grandchildren who belonged to 2 families, each family containing 7 children: how many oranges did he give to each child?

Suggestion.—First find how many each family received, then how many each child received.

If 2 families receive 168 oranges, 1 family will receive as many oranges, as 2 is contained times in 168, viz: 84. But there are 7 children in each family. If then 7 children receive 84 oranges, 1 child will receive as many, as 7 is contained times in 84, viz: 12. He therefore gave 12 oranges to each child.

Operation.

```
2)168
 7)84
   12 Ans.
```

Note.—This operation is exactly the reverse of that in Ex. 1. Art. 55. The divisor 14 being a composite number, we divide first by one of its factors, and the quotient thus found by the other. The final result would have been the same, if we had divided by 7 first, then by 2. Hence,

78. To divide by a *composite* number.

Divide the dividend by one of the factors of the divisor, and the quotient thus obtained by the other factor. The last quotient will be the answer required.

To find the *true remainder*, should there be any.

Multiply the last remainder by the first divisor, and to the product add the first remainder.

OBS. 1. If the divisor can be resolved into more than two factors, we may divide by them successively, as above.

2. To find the true remainder when more than two factors are employed, multiply each remainder by *all the preceding divisors*, and to the sum of the products add the first remainder.

2. Divide 465 by 35.

Operation.

7)465

5)66–3 rem.

13–1 rem.

1 last remainder.
7 first divisor.
7 product.
3 first rem. added.
10 true rem. *Ans.* $13\frac{10}{35}$.

3. A teacher having 36 scholars arranged in 4 equal classes, wishes to distribute 216 pears among them equally: how many can he give to each scholar?

4. How many cows, at 27 dollars a head, can be bought for 945 dollars?

5. How many times is 64 contained in 453?

6. How many times is 72 contained in 237?

CASE II.—*When the divisor is* 1 *with ciphers annexed to it.*

79. It has been shown that *annexing* a cipher to a number, *increases* its value *ten times*, or *multiplies* it by 10. (Art. 58.) Reversing this process; that is, *removing* a cipher from the right hand of a number, will evidently *diminish* its value *ten times*, or *divide* it by 10; for,

QUEST.—78. How proceed when the divisor is a composite number? How find the true remainder? *Obs.* How proceed when the divisor can be resolved into more than two factors? How find the remainder in this case? 79. What is the effect of annexing a cipher to a number? What is the effect of removing a cipher from the right of a number?

each figure in the number is thus restored to its original place, and consequently to its original value. Thus, annexing a cipher to 12, it becomes 120, which is the same as 12×10. On the other hand, removing the cipher from 120, it becomes 12, which is the same as 120÷10.

In the same manner it may be shown, that removing *two* ciphers from the right of a number, divides it by 100; removing *three*, divides it by 1000; removing *four*, divides it by 10000, &c. Hence,

80. To divide by 10, 100, 1000, &c.

Cut off as many figures from the right hand of the dividend as there are ciphers in the divisor. The remaining figures of the dividend will be the quotient, and those cut off the remainder.

7. How many times is 10 contained in 120?

Ans. 12.

8. In one dime there are 10 cents: how many dimes are there in 100 cents? In 250 cents? In 380 cents?

9. In one dollar there are 100 cents: how many dollars are there in 6500 cents? In 76500 cents? In 432000 cents?

10. Divide 675000 by 10000.

Ans. 67 and 5000 rem.

11. Divide 44360791 by 1000000.

12. Divide 82367180309 by 10000000.

CASE III.—*When the divisor has ciphers on the right.*

13. How many acres of land, at 20 dollars per acre, can you buy for 645 dollars?

Analysis.—The divisor 20 is a composite number, the factors of which are 2 and 10. (Art. 55. Obs. 1.) We may, therefore, divide first by one factor, and the quotient thence arising by the other. (Art. 78.) Now cutting off the right hand figure of the dividend, divides it by 10; (Art. 80;) consequently, dividing the remaining

---

QUEST.—80. How proceed when the divisor is 10, 100, 1000, &c.

figures of the dividend by 2, the other factor of the divisor, will give the true quotient.

Operation.

2|0)64|5

—

32–5 rem.

Cut off the cipher on the right of the divisor; also cut off the right hand figure of the dividend; then divide the 64 by 2. The 5 which we cut off, is the remainder. *Ans.* $32\frac{5}{20}$ acres. Hence,

81. When there are ciphers on the right hand of the divisor.

Cut off the ciphers, also cut off as many figures from the right of the dividend. Then divide the other figures of the dividend by the remaining figures of the divisor, and annex the figures cut off from the dividend to the remainder.

14. How many horses, at 80 dollars apiece, can you buy for 640 dollars?

15. How many barrels will 6800 pounds of beef make, allowing 200 pounds to the barrel?

16. How many regiments of 4000 each, can be formed from 840000?

17. Divide 143900 by 2100.

18. Divide 4314670 by 24000.

81. *a.* The four preceding rules, viz: *Addition, Subtraction, Multiplication,* and *Division,* are usually called the FUNDAMENTAL RULES of Arithmetic, because they are the *foundation* or *basis* of all arithmetical calculations.

GENERAL PRINCIPLES IN DIVISION.

82. From the nature of division, it is evident, that the *value* of the *quotient* depends both on the *divisor* and the *dividend.*

If a given divisor is contained in a given dividend a

QUEST.—81. When there are ciphers on the right of the divisor, how proceed? What is to be done with figures cut off from the dividend? 81. *a.* What are the four preceding rules called? Why? 82. Upon what does the value of the quotient depend?

certain number of times, the same divisor will obviously be contained,

In *double* that dividend, *twice* as many times;

In *three times* that dividend, *thrice* as many times, &c.

Thus, 4 is contained in 12, 3 times; in 2 times 12 or 24, 4 is contained 6 times; (i. e. twice 3 times;) in 3 times 12 or 36, 4 is contained 9 times; (i. e. thrice 3 times;) &c. Hence,

83. *If the divisor remains the same, multiplying the dividend by any number, is in effect multiplying the quotient by that number.*

Again, if a given divisor is contained in a given dividend a certain number of times, the same divisor is contained,

In *half* that dividend, *half* as many times;

In a *third* of that dividend, a *third* as many times, &c.

Thus, 4 is contained in 24, 6 times; in 24÷2 or 12, (half of 24,) 4 is contained 3 times; (i. e. half of 6 times;) in 24÷3 or 8, (a third of 24,) 4 is contained 2 times; (i. e. a third of 6 times;) &c. Hence,

84. *If the divisor remains the same, dividing the dividend by any number, is in effect dividing the quotient by that number.*

If a given divisor is contained in a given dividend a certain number of times, then, in the same dividend,

Twice that divisor is contained only *half* as many times;

Three times that divisor, a *third* as many times, &c.

Thus, 2 is contained in 12, 6 times; 2 times 2 or 4, is contained in 12, 3 times; (i. e. half of 6 times;) 3 times 2 or 6, is contained in 12, 2 times; (i. e. a third of 6 times;) &c. Hence,

85. *If the dividend remains the same, multiplying the divisor by any number, is in effect dividing the quotient by that number.*

QUEST.—83. If the divisor remains the same, what effect has it on the quotient to multiply the dividend? 84. What is the effect of dividing the dividend by any given number? 85. If the dividend remains the same, what is the effect of multiplying the divisor by any given number?

If a given divisor is contained in a given dividend a certain number of times, then, in the same dividend,

Half that divisor is contained *twice* as many times;

A *third* of that divisor, *three times* as many times, &c.

Thus, 6 is contained in 24, 4 times: 6÷2 or 3, (half of 6,) is contained in 24, 8 times; (i. e. twice 4 times;) 6÷3 or 2, (a third of 6,) is contained in 24, 12 times; (i. e. three times 4 times;) &c. Hence,

86. *If the dividend remains the same, dividing the divisor by any number, is in effect multiplying the quotient by that number.*

87. From the preceding articles, it is evident that any given divisor is contained in any given dividend, just as many times, as *twice* that divisor is contained in *twice* that dividend; *three times* that divisor in *three times* that dividend, &c.

Conversely, any given divisor is contained in any given dividend just as many times, as *half* that divisor is contained in *half* that dividend; a *third* of that divisor, in a *third* of that dividend, &c.

Thus, 4 is contained in 12, 3 times;
2 times 4 is contained in 2 times 12, 3 times;
3 times 4 is contained in 3 times 12, 3 times, &c.

Again, 6 is contained in 24, 4 times;
6÷2 is contained in 24÷2, 4 times;
6÷3 is contained in 24÷3, 4 times, &c. Hence

88. *If the divisor and dividend are both multiplied, or both divided by the same number, the quotient will not be altered.*

89. *If any given number is multiplied and the product divided by the same number, its value will not be altered.* Thus, 12×5=60; and 60÷5=12, the given number.

QUEST.—86. What of dividing the divisor? 88. What is the effect upon the quotient if the divisor and dividend are both multiplied or both divided by the same number? 89. What is the effect of multiplying and dividing any given number by the same number?

CANCELATION.*

90. We have seen that division is finding a quotient, which multiplied into the divisor will produce the dividend. (Art. 65.) If, therefore, the dividend is resolved into two such factors that one of them is the divisor, the other factor will, of course, be the quotient. Suppose, for example, 42 is divided by 6. Now the factors of 42 are 6 and 7, the first of which being the divisor, the other must be the quotient. Therefore,

Canceling a factor of any number, divides the number by that factor. Hence,

91. *When the dividend is the product of two or more factors, one of which is the same as the divisor, the division may be performed by* CANCELING *that factor in the divisor and dividend.* (Art. 88.)

Note.—The term *cancel*, means to *erase* or *reject*.

21. Divide the product of 19 into 25 by 19.

Common Method.

```
   19
   25
   ──
   95
  38
  ───
19)475(25 Ans.
   38
   ──
    95
    95
```

By Cancelation.

$\not{1}\not{9})\not{1}\not{9}\times25$

25 *Ans.*

Cancel the factor 19, which is common both to the divisor and dividend, and 25, the other factor of the dividend, is the quotient. (Art. 90.)

22. Divide 85×31 by 85. *Ans.* 31.
23. Divide 76×58 by 58.
24. Divide 75×40 by 40.
25. Divide 63×28 by 7.

Analysis.—28=4×7. We may therefore contract the division by canceling the 7, which is a factor both of the dividend and the divisor. (Arts. 88, 90.)

QUEST.—90. What is the effect of canceling a factor of any number? *Note.* What is meant by the term cancel? 91. When the divisor is a factor of the dividend, how may the division be performed?

* Birk's Arithmetical Collections, London, 1764.

Operation.

$\not{7})63\times4\times\not{7}$

252 *Ans.*

The product of 63×4, the other factors of the dividend, is the answer required.

26. In 32 times 84, how many times 8? *Ans.* 336.
27. In 35 times 95, how many times 7?
28. In 48 times 133, how many times 8?
29. In 96 times 156, how many times 12?
30. Divide 168×2×7 by 7×3.

Operation.

$\not{7}\times3)168\times2\times\not{7}$

3)336

112 *Ans.*

We cancel the factor 7, which is common to the divisor and dividend, then divide the product of 168 into 2 by 3.

31. Divide the product of 8, 6, and 12 by the product of 2, 6, and 8.

Solution.—$\not{2}\times\not{6}\times\not{8})\not{8}\times6\times\not{12}=6$. *Ans.*

Note.—We cancel the factors 2, 6 and 8 in the divisor, and the 12 and 8 in the dividend. Canceling the same or equal factors, both in the divisor and dividend, is dividing them both by the same number, and consequently does not affect the quotient. (Arts. 88, 90.) Hence,

91. *a. When the divisor and dividend have factors common to both, the division may be performed by canceling the common factors, and then dividing those that are left as before.*

32. Divide the product of 7, 9, 15, and 8 by the product of 5, 7, and 8.
33. Divide the product of 6, 3, 7, and 4 by the product of 12 and 6.
34. Divide the product of 2, 28, and 15 by 30.
35. Divide the product of 5, 6, and 56 by 7×8.

92. The method of contracting arithmetical operations, by rejecting equal factors, is called CANCELATION. It applies with great advantage to that class of examples and problems which involve both multiplication and division; that is, when the product of two or more numbers is to be divided by another number, or by the product of two or more numbers.

Note.—Its further developments and application may be seen in reduction of compound fractions to simple ones; in multiplication and division of fractions; in simple and compound proportion, &c., &c.

GREATEST COMMON DIVISOR.

92. *a.* A *Common Divisor* of two or more numbers, is a number which will *divide* them without a remainder. Thus, 2 is a common divisor of 4, 6, 8, 12, 16.

93. The *Greatest Common Divisor* of two or more numbers, is the *greatest* number which will divide them without a remainder. Thus, 6 is the greatest common divisor of 12, 18, and 24.

OBS. 1. One number is said to be a *measure* of another, when the former is *contained* in the latter any number of times without a remainder. Hence, a Com. divisor is often called a *Common Measure.*

2. It will be seen that a *common divisor* of two or more numbers, is simply a factor which is *common* to those numbers, and the *greatest* common divisor is the *greatest* factor common to them. Hence,

94. To find a common divisor of two or more numbers.

Resolve each number into two or more factors, one of which shall be common to all the given numbers.

Ex. 1. Find a common divisor of 8, 10, and 12.

Analysis.—8 may be resolved into the factors 2 and 4; that is, $8=2\times4$; $10=2\times5$; and $12=2\times6$. Now the factor 2 is common to each number and is therefore a common divisor of them.

2. Find a common divisor of 9, 15, 18, and 24.

OBS. The following facts may assist the learner in finding common divisors:

1. Any number ending in 0, or an even number, as 2, 4, 6, &c. may be divided by 2.
2. Any number ending in 5 or 0, may be divided by 5.
3. Any number ending in 0, may be divided by 10.
4. When the two right hand figures are divisible by 4, the whole number may by divided by 4.

3. Find a common divisor of 16, 20, and 36.

QUEST.—92. *a.* What is a common divisor of two or more numbers? 93. What is the greatest common divisor of two or more numbers? *Obs.* When is one number said to be a measure of another? What is a common divisor sometimes called? 94. How do you find a common divisor of two or more numbers?

4. Find a common divisor of 35, 50, 75, and 80.
5. Find a common divisor of 148 and 184.
6. Find a common divisor of 126 and 4653.

95. No two numbers can have a common divisor greater than a unit, unless they have a common factor. Thus, the factors of 8 are 2 and 4; the factors of 15 are 3 and 5; hence, 8 and 15 have no common divisor.

96. To find the *greatest common divisor* of two numbers.

Divide the greater number by the less; then the preceding divisor by the last remainder, and so on, till nothing remains. The last divisor will be the greatest common divisor.

7. What is the greatest common divisor of 70 and 84?

Operation.

```
70)84(1
   70
   14)70(5
      70
```

Dividing 84 by 70, the remainder is 14; then dividing 70 (the preceding divisor) by 14, (the last remainder,) nothing remains. Hence, 14 the last divisor, is the greatest common divisor.

8. What is the greatest common divisor of 63 and 147?

9. What is the greatest common divisor of 91 and 117?

10. What is the greatest common divisor of 247 and 323?

11. What is the greatest common divisor of 285 and 465?

12. What is the greatest common divisor of 2145 and 3471?

97. To find the greatest common divisor of more than two numbers.

First find the greatest common divisor of any two of them; then, that of the common divisor thus obtained and of another given number, and so on through all the given

QUEST.—95. If two numbers have not a common factor, what is true as to a common divisor? 96. How find the greatest common divisor of two numbers? 97. Of more than two?

numbers. The last common divisor found, will be the one required.

13. What is the greatest common divisor of 63, 105, and 140? *Ans.* 7.

Suggestion.—Find the greatest common divisor of 63 and 105, which is 21. Then, that of 21 and 140.

14. What is the greatest common divisor of 16, 24, and 100?

15. What is the greatest common divisor of 492, 744, and 1044?

LEAST COMMON MULTIPLE.

98. One number is said to be a *multiple* of another, when the former can be *divided* by the latter without a remainder. Thus, 4 is a multiple of 2; 10 is a multiple of 5.

OBS. A *multiple* is therefore a composite number, and the number thus contained in it, is always one of its factors.

99. A *common multiple* of two or more numbers, is a number which can be *divided* by *each* of them without a remainder. Thus, 12 is a common multiple of 2, 3, and 4; 15 is a common multiple of 3 and 5.

OBS. A *common* multiple is also a composite number, of which each of the given numbers must be a factor; otherwise it could not be divided by them.

100. The *continued product* of two or more given numbers, will always form a common multiple of those numbers.

The same numbers, therefore, may have an *unlimited number* of common multiples; for, multiplying their continued product by any number, will form a new common multiple. (Art. 99. Obs.)

QUEST.—98. What is a multiple of a number? *Obs.* What kind of a number is a multiple? 99. What is a common multiple? *Obs.* What kind of a number is a common multiple? 100. How may a common multiple of two or more numbers be found? How many common multiples may there be of any given numbers?

101. The *least common multiple* of two or more numbers, is the *least* number which can be divided by each of them without a remainder. Thus, 12 is the least common multiple of 4 and 6, for it is the least number which can be exactly divided by them.

15. Find the least common multiple of 6 and 10.

Analysis.—6=2×3; and 10=2×5. Now it is evident that the number required must contain all the different factors which are in each of the given numbers; otherwise it will not be a *common* multiple of them. (Art. 99. Obs.) The continued product of the factors 2×3×2×5=60, is exactly divisible by 6 and 10, but it will be seen that 60 is *twice* as large as is necessary to be a common multiple of them. We also perceive that the factor 2 is common to both the given numbers; hence it is that the continued product is twice too large. If, therefore, we retain this factor *only once*, the continued product of 2×3×5=30, which is the *smallest* number that is exactly divisible by 6 and 10, and is therefore the *least* common multiple of them.

Operation.

```
2)6 ״ 10
  ──────
  3 ״ 5
2×3×5=30
```

We divide both numbers by 2. This resolves them into factors, and the divisor and quotients contain all the different factors found in each of the given numbers *once*, and *only once*. Then we multiply the divisor and quotients together and the product is 30, which is the least common multiple required. Hence,

102. To find the least common multiple of two or more numbers.

Write the given numbers in a line with two points between them. Divide by the smallest number which will divide any two or more of them without a remainder, and set the quotients and the numbers not divided in a line below. Divide this line and set down the results as before; thus

---

QUEST.—101. What is the least common multiple of two or more numbers? 102. How is the least common multiple of two or more numbers found?

continue the operation till there are no two numbers which can be divided by any number greater than 1. *The continued product of the divisors into the numbers in the last line, will be the least common multiple required.*

16. Find the least common multiple of 6, 8, and 12.

First Operation.

2)6 " 8 " 12
2)3 " 4 " 6
3)3 " 2 " 3
1 " 2 " 1

2×2×3×2=24 *Ans.*

Second Operation.

6)6 " 8 " 12
2)1 " 8 " 2
1 " 4 " 1

Now 6×2×4=48.

OBS. 1. In the first operation, we divide by the *smallest* numbers which will divide any two of the given numbers without a remainder, and the product of the divisors, and the numbers in the last line, is 24, which is the answer required.

In the second operation, we divide by 6, then by 2. But 6 is *not the smallest* number that will exactly divide two of the given numbers, and the continued product of the divisors into the figures in the last line is 48, which is not the least common multiple. Hence,

2. We must divide, in all cases, by the *smallest* number that will divide any two of the given numbers exactly; otherwise, the divisor may contain a factor *common* to it and some one of the quotients, or undivided numbers in the last line, and consequently the continued product of them will be *too large* for the *least* common multiple. Thus in the 2d operation, the 6 and 4 contain a common factor 2, which must be rejected from them, in order that the product of the divisors and quotients may be the least common multiple.

17. Find the least common multiple of 4, 9, and 12.
18. Find the least common multiple of 16, 12, and 24.
19. Find the least common multiple of 15, 9, 6, and 5.
20. Find the least common multiple of 10, 6, 18, 15.
21. Find the least common multiple of 24, 16, 15, 20.
22. Find the least common multiple of 25, 60, 72, 35.
23. Find the least common multiple of 63, 12, 84, 72.
24. Find the least common multiple of 54, 81, 14, 63.
25. Find the least common multiple of 12, 72, 36, 144.

QUEST.—*Obs.* Why do you divide by the smallest number that will divide two or more without a remainder?

SECTION VI.

FRACTIONS.

MENTAL EXERCISES.

ART. **103.** When a number or thing is divided into *two equal* parts, one of these parts is called *one half.* If the number or thing is divided into *three equal* parts, one of the parts is called *one third;* if it is divided into *four equal* parts, one of the parts is called *one fourth*, or *one quarter;* two of the parts, *two fourths;* three, *three fourths;* if divided into *five equal* parts, the parts are called *fifths;* if into *six equal* parts, *sixths;* if into *ten, tenths;* if into a *hundred, hundredths,* &c. That is,

When a number or thing is divided into *equal parts,* the parts always take their *name* from the *number* of *parts* into which the thing or number is divided.

104. The *value* of one of these equal parts manifestly depends upon the number of parts into which the given number or thing is divided. Thus, if an orange is successively divided into 2, 3, 4, 5, 6, &c., equal parts, the thirds will be less than the halves; the fourths, than the thirds; the fifths, than the fourths, &c.

Ex. 1. What is one half of 2 cents? Of 4 cents? 6? 8? 16? 18? 20? 24? 30? 40? 50? 60? 70? 80? 100?

2. What is one third of 6 cents? Of 9? 12? 15?

QUEST.—103. What is meant by one half? How many halves make a whole one? What is meant by one third? How many thirds make a whole one? What is meant by a fourth? 3 fourths? What are fourths sometimes called? How many fourths make a whole one? What is meant by fifths? By sixths? Eighths? How many sevenths make a whole one? How many tenths? What is meant by twentieths? By hundredths? When a number or thing is divided into equal parts, from what do the parts take their name? 104. Upon what does the value of one of these equal parts depend? Which is the greater, a half or a third? A sixth or a fourth? A seventh or a tenth?

OBS. A *half* of any number, it will be perceived, is equal to as many units, as 2 is contained times in that number; a *third* of a number is equal to as many units, as 3 is contained times in the given number; a *fourth* is equal to as many, as 4 is contained in it, &c.

3. What is a third of 12? Of 15? 18? 21? 24? 27? 30? 36? 39? 45? 60?

4. What is a fourth of 8 dollars? Of 12? 16? 20? 24? 28? 32? 36? 40? 44? 48?

5. What is a fifth of 5? 10? 15? 20? 25? 30? 35? 40? 45? 50? 55? 60? 100?

6. What is a sixth of 12? 18? 24? 36? 30? 48? 60? 54? 42? 72?

7. What is a seventh of 14? 28? 35? 21? 42? 56? 49? 63?

8. What is an eighth of 16? 24? 40? 32? 64? 48? 56? 72? 88?

9. What is a ninth of 9? 18? 36? 27? 45? 54? 72? 63? 81? 99?

10. What is a tenth of 20? 40? 60? 50? 30? 100? 90? 120?

11. What part of 2 is 1? *Ans.* One half.

12. What part of 3 is 1? Of 4? 5? 7? 10? 15? 19? 37? 200?

13. What part of 3 is 2?

Suggestion.—Since 1 is 1 third part of 3, 2 must be two times the third part of 3, or two thirds of 3.

14. What part of 5 is 2? is 3? is 4? is 5? is 6? is 8? is 9?

15. What part of 8 is 3? is 7? is 6? is 9? is 8? 12? 15?

16. What part of 17 is 5? 8? 9? 13? 15? 16? 20?

17. What part of 100 is 13? 29? 63? 75? 92?

18. If 1 half an orange cost 2 cents, what will a whole orange cost?

Analysis.—If 1 half of an orange cost 2 cents, 2 halves or a whole orange, will cost twice as much; and 2 times 2 cents are 4 cents. *Ans.* 4 cents.

19. If 1 third of a pie cost 4 cents, what will 2 thirds cost? What will a whole pie cost?

20. If 1 fourth of a pound of ginger cost 3 cents, what will 2 fourths of a pound cost? 3 fourths? What will a whole pound cost?

21. If 1 eighth of a yard of cloth cost 2 shillings, what will 3 eighths cost? 5 eighths? 7 eighths? What will a whole yard cost?

22. If 1 third of a barrel of flour cost 3 dollars, how much will a whole barrel cost? How much will 5 barrels cost? 8 barrels?

23. If 1 sixth of a hogshead of molasses cost 5 dollars, what will be the cost of a hogshead? Of 4 hogsheads? Of 10 hogsheads?

24. If 1 pound of sugar cost 12 cents, what will 1 half a pound cost?

Suggestion.—If 1 pound cost 12 cents, it is plain that 1 half of a pound will cost 1 half of 12 cents; and 1 half of 12 cents is 6 cents. *Ans.* 6 cents.

25. If one yard of ribbon cost 15 cents, how much will 1 third of a yard cost?

26. If one pound of tea cost 4 shillings, how much will 1 fourth of a pound cost? How much will 2 fourths cost?

27. If a ton of hay cost 15 dollars, how much will 1 fifth of a ton cost? How much 2 fifths? 3 fifths?

28. What will 1 tenth of an acre of land cost, at 30 dollars per acre? 2 tenths? 6 tenths?

29. What will 1 eighth of a ton of iron cost, at 48 dollars per ton? 3 eighths? 5 eighths? 7 eighths?

30. If 1 bushel of corn cost 1 half a dollar, what will 2 bushels cost? 4 bushels?

Suggestion.—If 1 bushel cost 1 half a dollar, 2 bushels will cost twice as much. 2 times 1 half are 2 halves, or a whole dollar. 4 bushels will cost 4 times 1 half, or 2 whole dollars.

31. If one man eats 1 half of a loaf of bread at a meal, how many loaves will 3 men eat?

32. How many whole ones are 4 halves equal to? 5 halves? 6 halves? 8 halves? 9 halves?

33. If I burn 1 third of a ton of coal in a week, how much shall I burn in 3 weeks? 4 weeks? 6 weeks? 10 weeks? 12 weeks?

34. How many whole ones in 4 thirds, and how many over? In 6 thirds? 8 thirds? 11 thirds? 14 thirds?

35. If a horse eat 1 fourth of a bushel of oats a day, how many will he eat in 6 days? In 8? In 10? In 12?

36. If a boy can saw 1 eighth of a cord of wood in a day, how much can he saw in 6 days? In 12 days? In 15 days? In 24 days?

37. If 12 oranges were divided equally among 4 boys, what *part* of them would each boy receive; and how many oranges would each have?

Analysis.—1 is 1 fourth of 4; hence, 1 boy must receive 1 fourth part of the oranges. 1 fourth of 12 oranges is 3 oranges.

38. A builder employed 6 men to do a job of work, for which he gave them 24 dollars: what part of the money did 1 man receive? What part did 2 receive? What part did 3 receive? What part did 4 receive? How many dollars did one man receive? How many did two? Three? Four?

39. If 5 yards of cloth cost 40 dollars, what part of 40 dollars will 1 yard cost? 2 yards? 3 yards? 4 yards? How many dollars will 1 yard cost? 2 yards? 3 yards? 4 yards?

40. 2 is 1 third of what number?

Solution.—If 2 is 1 third of a number, 3 thirds or the whole number, must be 3 times as many.

Or thus, 2 is a third of 3 times 2; and 3 times 2 are 6.

41. 4 is 1 fifth of what number? 1 sixth of what number? 1 third? 1 eighth? 1 fourth? 1 seventh?

42. 6 is 1 third of what number? 1 fourth? 1 seventh? 1 tenth? 1 ninth? 1 twelfth?

43. 5 is 1 fourth of what number? 1 sixth? 1 eighth? 1 eleventh? 1 twelfth?

44. 8 is 1 seventh of what number? 1 sixth? 1 tenth? 1 ninth? 1 twelfth?

45. 4 is 2 thirds of what number?

Suggestion.—First find 1 third. Now if 4 is 2 thirds, 1 third is 1 half of 4, which is 2; and 3 thirds is 3 times 2, or 6. *Ans.* 6.

46. 9 is 3 fourths of what number?

47. 8 is 4 fifths of what number?

48. 16 is 4 ninths of what number?

49. 20 is 5 eighths of what number?

50. 32 is 8 twelfths of what number?

105. When a number or thing is divided into *equal parts*, as halves, thirds, fourths, &c., these parts are called FRACTIONS.

A *whole* number is called an *Integer.*

106. Fractions are divided into two classes, *Common* and *Decimal.* (For the illustration of Decimal Fractions, see Section VIII.)

107. *Common Fractions* are expressed by two numbers, one placed over the other, with a line between them. One half is written thus $\frac{1}{2}$; one third, $\frac{1}{3}$; one fourth, $\frac{1}{4}$; nine tenths, $\frac{9}{10}$; thirteen forty-fifths, $\frac{13}{45}$, &c.

The number below the line is called the *denominator*, and shows into *how many parts* the number or thing is divided.

The number above the line is called the *numerator*, and shows *how many parts* are expressed by the fraction. Thus in the fraction $\frac{2}{3}$, the denominator 3, shows that the number is divided into *three* equal parts; the numerator 2, shows that *two* of those parts are expressed by the the fraction.

The denominator and numerator together, are called the *terms* of the fraction.

QUEST.—105. What are fractions? What is an integer? 106. Of how many kinds are fractions? 107. How are common fractions expressed? What is the number below the line called? What does it show? What is the number above the line called? What does it show? What are the denominator and numerator, taken together, called?

OBS. 1. The *term fraction* is of a Latin origin, and signifies *broken*, or *separated* into parts. Hence fractions are sometimes called *broken numbers.*

2. *Common* fractions are often called *vulgar* fractions. This term, however, is very properly falling into disuse.

3. The number below the line is called the *denominator*, because it gives the name or denomination to the fraction; as, halves, thirds, fifths, &c.

The number above the line is called the *numerator*, because it numbers the parts, or shows how many parts are expressed by the fraction.

108. A *proper* fraction is a fraction whose numerator is *less* than its denominator; as, $\frac{1}{2}$, $\frac{2}{3}$, $\frac{4}{5}$.

An *improper* fraction is one whose numerator is *equal to*, or *greater* than its denominator; as, $\frac{3}{3}$, $\frac{7}{2}$.

A *mixed* number is a whole number and a fraction expressed together; as, $4\frac{2}{3}$, $25\frac{1}{12}$.

A *simple* fraction is a fraction which has but one numerator and one denominator, and may be proper, or improper; as, $\frac{3}{5}$, $\frac{9}{4}$.

A *compound* fraction is a fraction of a fraction; as, $\frac{2}{3}$ of $\frac{4}{5}$ of $\frac{7}{8}$.

A *complex* fraction is one which has a fraction in its numerator or denominator, or in both; as, $\frac{2\frac{1}{2}}{5}$, $\frac{4}{5\frac{1}{3}}$, $\frac{2\frac{1}{3}}{8\frac{3}{4}}$.

109. Fractions, it will be seen, both from the definition and the mode of expressing them, arise from *division*, and may be regarded as expressions of *unexecuted* division, the numerator answering to the dividend, and the denominator to the divisor. (Arts. 67, 105.) Hence,

110. The *value* of a fraction is the *quotient* of the numerator divided by the denominator. Thus the value of $\frac{6}{3}$ is *two*; of $\frac{4}{4}$ is *one*; of $\frac{1}{3}$ is *one third*; &c. Hence,

---

QUEST.—*Obs.* What is the meaning of the term fraction? What are common fractions sometimes called? Why is the lower number called the denominator? Why is the upper one called the numerator? 108. What is a proper fraction? An improper fraction? A mixed number? A simple fraction? A compound fraction? A complex fraction? 109. From what do fractions arise? 110. What is the value of a fraction?

111. *If the denominator remains the same, multiplying the numerator by any number, multiplies the value of the fraction by that number.* For, the numerator and denominator answer to the dividend and divisor; therefore, multiplying the numerator is the same as multiplying the dividend. Now multiplying the dividend, we have seen, multiplies the quotient, (Art. 83,) which is the same as the value of the fraction. (Art. 110.) Thus, the value of $\frac{6}{3}=2$. Multiplying the numerator by 3, the fraction becomes $\frac{18}{3}$, whose value is 6, and is the same as 2×3.

112. *Dividing the numerator by any number, divides the value of the fraction by that number.* For, dividing the dividend divides the quotient. (Art. 84.) Thus, $\frac{6}{3}=2$. Now dividing the numerator by 2, the fraction becomes $\frac{3}{3}$, whose value is 1, and is the same as $2\div2$. Hence,

OBS. With a given denominator, the *greater* the *numerator*, the *greater* will be the value of the fraction.

113. *If the numerator remains the same, multiplying the denominator by any number, divides the value of the fraction by that number.* For, multiplying the divisor divides the quotient. (Art. 85.) Thus, $\frac{24}{6}=4$. Now multiplying the denominator by 2, the fraction becomes $\frac{24}{12}$, whose value is 2, and is the same as $4\div2$.

114. *Dividing the denominator by any number, multiplies the value of the fraction by that number.* For, dividing the divisor, multiplies the quotient. (Art. 86.) Thus, $\frac{24}{6}=4$. Now dividing the denominator by 2, the fraction becomes $\frac{24}{3}$, whose value is 8, and is the same as 4×2. Hence,

OBS. With a given numerator, the *greater* the *denominator*, the *less* will be the value of the fraction.

QUEST.—111. What is the effect of multiplying the numerator, while the denominator remains the same? Explain the reason. 112. What is the effect of dividing the numerator? *Obs.* With a given denominator, what is the effect of increasing the numerator? 113. What is the effect of multiplying the denominator? Why? 114. What is the effect of dividing the denominator? Why? *Obs.* With a given numerator, what is the effect of increasing the denominator?

115. It is evident from the preceding articles, that *multiplying the numerator* by any number, has the same effect on the value of the fraction, *as dividing the denominator* by that number. (Arts. 111, 114.)

Dividing the numerator has the same effect, as *multiplying the denominator.* (Arts. 112, 113.)

116. If the numerator and denominator are both *multiplied* or both *divided* by the same number, *the value of the fraction will not be altered.* (Arts. 88, 109.) Thus, $\frac{12}{4}=3$. Now if the numerator and denominator are both multiplied by 2, the fraction becomes $\frac{24}{8}$; whose value is 3. If both terms are divided by 2, the fraction becomes $\frac{6}{2}$, whose value is 3; that is, $\frac{12}{4}=\frac{24}{8}=\frac{6}{2}=3$.

117. Since the value of a fraction is the quotient of the numerator divided by the denominator, it follows, that

If the numerator and denominator are *equal*, the value is a *unit* or *one.* Thus, $\frac{5}{5}=1$, $\frac{7}{7}=1$, &c.

If the numerator is *greater* than the denominator, the value is greater than *one.* Thus, $\frac{4}{2}=2$, $\frac{5}{3}=1\frac{2}{3}$.

If the numerator is *less* than the denominator, the value is less than *one.* Thus, $\frac{1}{3}=1$ *third* of 1, $\frac{4}{5}=4$ fifths of 1.

118. It will be seen from the preceding exercises, that fractions may be *added, subtracted, multiplied,* and *divided,* as well as whole numbers.

OBS. 1. In order to perform these operations, it is often necessary to make certain changes in the terms of the fractions.

QUEST.—115. What may be done to the denominator to produce the same effect on the value of the fraction, as multiplying the numerator by any given number? What, to produce the same effect as dividing the numerator by any given number? 116. What is the effect if the numerator and denominator are both multiplied, or both divided by the same number? 117. When the numerator and denominator are equal, what is the value of the fraction? When the numerator is the larger, what? When smaller, what?

2. It is evident that any changes may be made in the terms of a fraction, which do not alter the quotient of the numerator divided by the denominator; for, if the quotient is not altered, the value remains the same. (Art. 110.) Thus, the terms of the fraction $\frac{4}{2}$ may be changed into $\frac{2}{1}$, $\frac{8}{4}$, $\frac{16}{8}$, &c., without altering its value; for in each case the quotient of the numerator divided by the denominator is 2. Hence, for any given fraction, we may substitute any other fraction, which will give the *same quotient.*

REDUCTION OF FRACTIONS.

119. The process of changing the *terms* of a fraction into others, without altering its value, is called REDUCTION OF FRACTIONS.

EXERCISES FOR THE SLATE.

CASE I.

Ex. 1. Reduce $\frac{6}{12}$ to its lowest terms.

First Operation.

$2)\frac{6}{12}=\frac{3}{6}$: again, $3)\frac{3}{6}=\frac{1}{2}$. *Ans.*

Dividing both terms of the fraction by 2, it becomes $\frac{3}{6}$: again, dividing both by 3, we obtain $\frac{1}{2}$, whose terms are the lowest to which the given fraction can be reduced.

Second Operation.

$6)\frac{6}{12}=\frac{1}{2}$. *Ans.*

If we divide both terms by 6, their greatest common divisor, (Art. 96,) the given fraction will be reduced to its lowest terms by a single division. Hence,

120. To reduce a fraction to its lowest terms.

Divide the numerator and denominator by any number which will divide them both without a remainder; and thus continue the operation, till there is no number greater than 1 *that will divide them exactly.*

Or, divide both the numerator and denominator by their greatest common divisor; and the two quotients thus arising will be the lowest terms to which the given fraction can be reduced. (Art. 96.)

QUEST.—*Obs.* What changes may be made in the terms of a fraction? 119. What is meant by reduction of fractions? 120. How is a fraction reduced to its lowest terms?

OBS. 1. A fraction is said to be reduced to its *lowest terms*, when its numerator and denominator are expressed in the *smallest* numbers possible.

2. The value of a fraction is not altered by reducing it to its lowest terms. (Art. 116.)

3. When the terms of the fraction are small, the former method will generally be found to be the shorter and more convenient; but when the terms are large, it is often difficult to determine whether the fraction is in its simplest form, without finding their greatest common divisor.

2. Reduce $\frac{8}{16}$ to its lowest terms. *Ans.* $\frac{1}{2}$.

3. Reduce $\frac{5}{10}$.
4. Reduce $\frac{6}{9}$.
5. Reduce $\frac{16}{18}$.
6. Reduce $\frac{32}{48}$.
7. Reduce $\frac{48}{72}$.
8. Reduce $\frac{75}{87}$.
9. Reduce $\frac{63}{105}$.
10. Reduce $\frac{55}{121}$.
11. Reduce $\frac{240}{312}$.
12. Reduce $\frac{126}{162}$.
13. Reduce $\frac{272}{425}$.
14. Reduce $\frac{435}{957}$.
15. Reduce $\frac{384}{1152}$.
16. Reduce $\frac{1740}{2900}$.
17. Reduce $\frac{594}{2142}$.
18. Reduce $\frac{6435}{7335}$.

CASE II.

19. Reduce $\frac{17}{5}$ to a whole or mixed number.

Suggestion.—The object in this example, is to find a whole or mixed number, whose value is equal to the given fraction. But the value of a fraction is the quotient of the numerator divided by the denominator. (Art. 110.) Hence,

Operation.

$$5)\underline{17}$$
$$3\frac{2}{5} \text{ Ans.}$$

121. To reduce an improper fraction to a whole, or mixed number.

Divide the numerator by the denominator, and the quotient will be the whole, or mixed number required.

20. Reduce $\frac{29}{3}$ to a whole or mixed number. *Ans.* $9\frac{2}{3}$.

QUEST.—*Obs.* What is meant by lowest terms? Is the value of a fraction altered by reducing it to its lowest terms? 121. How is an improper fraction reduced to a whole or mixed number?

Reduce the following fractions to whole or mixed numbers:

21. Reduce $\frac{24}{6}$.
22. Reduce $\frac{35}{7}$.
23. Reduce $\frac{21}{8}$.
24. Reduce $\frac{45}{6}$.
25. Reduce $\frac{18}{13}$.
26. Reduce $\frac{500}{12}$.
27. Reduce $\frac{750}{25}$.
28. Reduce $\frac{8437}{298}$.
29. Reduce $\frac{845}{30}$.
30. Reduce $\frac{7243}{320}$.

CASE III.

31. Reduce the mixed number $15\frac{3}{4}$ to an improper fraction.

Operation.

$15\frac{3}{4}$
4
$\frac{63}{4}$ *Ans.*

OBS. In 1 there are 4 fourths, and in 15, there are 15 times as many. 4×15=60, and 3 fourths make 63 fourths. Hence,

122. To reduce a mixed number to an improper fraction.

Multiply the whole number by the denominator of the fraction; to the product add the given numerator. The sum placed over the given denominator, will form the improper fraction required.

OBS. 1. Any whole number may be expressed in the form of a fraction without altering its value, by *making* 1 *the denominator.*

2. A whole number may also be reduced to a fraction of any denomination, by *multiplying* the given number by the proposed denominator; the product will be the numerator of the fraction required.

Thus 25 may be expressed by $\frac{25}{1}$, $\frac{100}{4}$, or $\frac{400}{16}$, &c., for $25=\frac{25}{1}=\frac{100}{4}=\frac{400}{16}$, &c. So $12=\frac{12}{1}=\frac{24}{2}=\frac{36}{3}=\frac{60}{5}$; for, the quotient of each of these numerators divided by its denominator, is 12.

32. Reduce $8\frac{1}{3}$ to an improper fraction. *Ans.* $\frac{25}{3}$.

QUEST.—122. How reduce a mixed number to an improper fraction? *Obs.* How express a whole number in the form of a fraction? How reduce a whole number to a fraction of a given denominator?

Reduce the following numbers to improper fractions:

33. Reduce $9\frac{2}{3}$.
34. Reduce $16\frac{5}{8}$.
35. Reduce $23\frac{7}{9}$.
36. Reduce $45\frac{5}{12}$.
37. Reduce $64\frac{8}{15}$.
38. Reduce $56\frac{23}{45}$.
39. Reduce $304\frac{1}{3}$.
40. Reduce $725\frac{1}{5}$.
41. Reduce 45 to fifths.
42. Reduce 72 to eighths.
43. Reduce 830 to sixths.
44. Reduce 743 to fifteenths.

CASE IV.

45. Reduce $\frac{2}{3}$ of $\frac{6}{7}$ to a simple fraction.

Analysis.—$\frac{2}{3}$ of $\frac{6}{7}$ is 2 times as much as 1 third of $\frac{6}{7}$. Now $\frac{1}{3}$ of $\frac{6}{7}$ is $\frac{6}{21}$; for, multiplying the denominator divides the value of the fraction. (Art. 113.) And 2 thirds is 2 times $\frac{6}{21}$, which is equal to $\frac{12}{21}$, or $\frac{4}{7}$. (Art. 120.) The answer is $\frac{4}{7}$.

OBS. This operation consists in simply multiplying the two numerators together and the two denominators. Hence,

123. To reduce compound fractions to simple ones.

Multiply all the numerators together for a new numerator, and all the denominators together for a new denominator.

46. Reduce $\frac{4}{7}$ of $\frac{6}{8}$ of $\frac{2}{9}$ to a simple fraction.

Ans. $\frac{48}{504}$, or $\frac{2}{21}$.

47. Reduce $\frac{1}{3}$ of $\frac{10}{12}$ of $\frac{2}{16}$ to a simple fraction.
48. Reduce $\frac{1}{5}$ of $\frac{2}{7}$ of $\frac{4}{9}$ to a simple fraction.
49. Reduce $\frac{3}{8}$ of $\frac{7}{15}$ of $\frac{11}{13}$ to a simple fraction.
50. Reduce $\frac{1}{2}$ of $\frac{2}{3}$ of $\frac{3}{4}$ of $\frac{5}{7}$ to a simple fraction.

Operation.

$$\frac{1}{\cancel{2}} \text{ of } \frac{\cancel{2}}{\cancel{3}} \text{ of } \frac{\cancel{3}}{4} \text{ of } \frac{5}{7} = \frac{5}{28}$$

Since the product of the numerators is to be divided by the product of the denominators, we may cancel the factors 2 and 3, which are common to both; for, this is dividing the terms of the new fraction by the same number, (Art. 90,) and therefore does not alter its value. (Art. 116.) Multiplying the re-

---

QUEST.—123. How are compound fractions reduced to simple ones?

maining factors together, we have $\frac{5}{28}$, which is the answer required. Hence,

124. To reduce compound fractions to simple ones by CANCELATION.

Cancel all the factors which are common to the numerators and denominators; then multiply the remaining terms together as before. (Art. 123.)

OBS. This method not only shortens the operation of multiplying, but at the same time reduces the answer to its lowest terms. A little practice will give the learner great facility in its application.

51. Reduce $\frac{4}{5}$ of $\frac{10}{12}$ of $\frac{3}{7}$ to a simple fraction.

Operation.

$$\frac{\cancel{4}}{\cancel{5}} \text{ of } \frac{\overset{2}{\cancel{10}}}{\cancel{12}} \text{ of } \frac{\cancel{3}}{7} = \frac{2}{7} \textit{ Ans.}$$

First, we cancel the 4 and 3 in the numerator, then the 12 in the denominator, which is equal to the factors 4 and 3. Finally, we cancel the 5 in the denominator, and the factor 5 in the numerator 10, placing the other factor 2 above. We have 2 left in the numerator, and 7 in the denominator. *Ans.* $\frac{2}{7}$.

52. Reduce $\frac{2}{7}$ of $\frac{6}{9}$ of $\frac{10}{12}$ to a simple fraction.
53. Reduce $\frac{4}{7}$ of $\frac{1}{4}$ of $\frac{7}{14}$ of $\frac{5}{3}$ to a simple fraction.
54. Reduce $\frac{5}{8}$ of $\frac{2}{9}$ of $\frac{4}{5}$ of $\frac{7}{10}$ to a simple fraction.
55. Reduce $\frac{8}{10}$ of $\frac{12}{15}$ of $\frac{20}{32}$ to a simple fraction.
56. Reduce $\frac{6}{10}$ of $\frac{15}{18}$ of $\frac{7}{9}$ of $\frac{9}{20}$ to a simple fraction.
57. Reduce $\frac{2}{9}$ of $\frac{17}{12}$ of $\frac{13}{17}$ of $\frac{9}{26}$ to a simple fraction.
58. Reduce $\frac{3}{12}$ of $\frac{6}{17}$ of $\frac{9}{7}$ of $\frac{2}{3}$ to a simple fraction.

Note.—For the method of reducing *complex* fractions to simple ones, see Art. 143.

CASE V.

Ex. 1. Reduce $\frac{1}{2}$ and $\frac{1}{3}$ to a common denominator.

Note.—Two or more fractions are said to have a *common denominator*, when they have the *same* denominator.

QUEST.—124. How by cancelation? How does it appear that this method does not alter the value of the fraction? *Obs.* What is the advantage of this method? *Note.* What is meant by a common denominator?

Suggestion.—The object of this example is to find two other fractions, which have the same denominator, and whose values are respectively equal to the values of the given fractions, $\frac{1}{2}$ and $\frac{1}{3}$. Now, if both terms of the first fraction $\frac{1}{2}$, are multiplied by the denominator of the second, it becomes $\frac{3}{6}$, and if both terms of the second fraction $\frac{1}{3}$, are multiplied by the denominator of the first, it becomes $\frac{2}{6}$. But the fractions $\frac{3}{6}$ and $\frac{2}{6}$ have a common denominator, and are respectively equal to the given fractions, viz: $\frac{3}{6}=\frac{1}{2}$, and $\frac{2}{6}=\frac{1}{3}$. (Art. 116.) Hence,

125. To reduce fractions to a common denominator.

Multiply each numerator into all the denominators except its own for a new numerator, and all the denominators together for a common denominator.

2. Reduce $\frac{1}{2}$, $\frac{3}{4}$, and $\frac{5}{6}$ to a common denominator.

Operation.

$\left.\begin{array}{l} 1\times4\times6=24 \\ 3\times2\times6=36 \\ 5\times2\times4=40 \end{array}\right\}$ the three numerators.

$2\times4\times6=48$, the common denominator.

The fractions required are $\frac{24}{48}$, $\frac{36}{48}$, and $\frac{40}{48}$.

OBS. It is manifest that the process of reducing fractions to a common denominator, does not change their value; for, it is simply multiplying each numerator and denominator of the given fractions by the same number. (Art. 116.)

3. Reduce $\frac{3}{5}$, $\frac{1}{4}$, and $\frac{6}{7}$ to a common denominator.

Ans. $\frac{84}{140}$, $\frac{35}{140}$, $\frac{120}{140}$.

4. Reduce $\frac{1}{9}$, $\frac{2}{3}$, and $\frac{3}{4}$ to a common denominator.

Reduce the following fractions to a common denominator:

5. Reduce $\frac{2}{6}$, $\frac{1}{2}$, $\frac{4}{5}$, and $\frac{2}{3}$. 6. Reduce $\frac{3}{8}$, $\frac{4}{7}$, $\frac{6}{9}$, and $\frac{2}{5}$.

7. Reduce $\frac{7}{8}$, $\frac{5}{7}$, $\frac{6}{10}$, and $\frac{7}{12}$. 8. Reduce $\frac{9}{10}$, $\frac{6}{7}$, $\frac{12}{15}$, and $\frac{2}{5}$.

QUEST.—125. How are fractions reduced to a common denominator? *Obs.* Does the process of reducing fractions to a common denominator alter their value? Why not?

9. Reduce $\frac{12}{25}$, $\frac{63}{70}$, and $\frac{27}{40}$. 10. Reduce $\frac{8}{20}$, $\frac{70}{100}$, and $\frac{52}{25}$.
11. Reduce $\frac{7}{24}$, $\frac{35}{50}$, and $\frac{10}{25}$. 12. Reduce $\frac{3}{70}$, $\frac{65}{30}$, and $\frac{122}{250}$.

CASE VI.

13. Reduce $\frac{3}{4}$, $\frac{2}{6}$, and $\frac{5}{8}$ to the least common denominator.

Operation.

2)4 " 6 " 8
2)2 " 3 " 4
1 " 3 " 2

$2 \times 2 \times 3 \times 2 = 24$, the least com. denom.

We first find the least common multiple of all the given denominators, which is 24; (Art. 102;) and this is the least common denominator required. The next step is to reduce the given fractions to *twenty-fourths* without altering their value. This may evidently be done, by multiplying both terms of each fraction by the number of times its denominator is contained in 24. (Art. 116.) Thus 4, the denominator of the first fraction, is contained in 24, 6 times; now multiplying both terms of the fraction $\frac{3}{4}$ by 6, it becomes $\frac{18}{24}$. The denominator 6 is contained in 24, 4 times; and multiplying the second fraction $\frac{2}{6}$ by 4, it becomes $\frac{8}{24}$. The denominator 8 is contained in 24, 3 times; and multiplying the third fraction $\frac{5}{8}$ by 3, it becomes $\frac{15}{24}$. Therefore $\frac{18}{24}$, $\frac{8}{24}$, and $\frac{15}{24}$ are the fractions required. Hence,

126. To reduce fractions to their *least* common denominator.

I. *Find the least common multiple of all the denominators of the given fractions, and it will be the least common denominator.* (Art. 102.)

II. *Divide the least common denominator by the denominator of each of the given fractions, and multiply the quotient by the numerator;—the products will be the numerators required.*

QUEST. 126. How are fractions reduced to the least common denominator?

OBS. Multiplying each numerator into the number of times its denominator is contained in the least common denominator, is in effect multiplying both terms of the given fractions by the same number. For, if we multiply each denominator by the number of times it is contained in the least common denominator, the product will be equal to the least common denominator. Hence, the new fractions must be of the same value as the given fractions. (Art. 116.)

14. Reduce $\frac{2}{3}$, $\frac{3}{4}$, and $\frac{5}{6}$ to the least com. denominator.

Operation.	
2)3 " 4 " 6	2×3×2=12, the least com. denominator.
3)3 " 2 " 3	Now (12÷3)×2=8, numerator of 1st.
1 " 2 " 1	(12÷4)×3=9, " of 2d.
	(12÷6)×5=10, " of 3d.

Ans. $\frac{8}{12}$, $\frac{9}{12}$, and $\frac{10}{12}$.

15. Reduce $\frac{7}{8}$ and $\frac{9}{10}$ to the least common denominator. *Ans.* $\frac{35}{40}$ and $\frac{36}{40}$.

Reduce the following fractions to the least common denominator:

16. $\frac{3}{4}$, $\frac{5}{6}$, and $\frac{7}{9}$.
17. $\frac{3}{7}$, $\frac{1}{6}$, and $\frac{11}{14}$.
18. $\frac{8}{9}$, $\frac{7}{8}$, $\frac{5}{12}$, and $\frac{2}{24}$.
19. $\frac{1}{5}$, $\frac{3}{8}$, $\frac{7}{10}$, and $\frac{4}{20}$.
20. $\frac{2}{12}$, $\frac{1}{3}$, $\frac{1}{4}$, $\frac{1}{5}$, and $\frac{2}{6}$.
21. $\frac{5}{16}$, $\frac{7}{24}$, and $\frac{1}{36}$.
22. $\frac{2}{15}$, $\frac{9}{20}$, and $\frac{6}{120}$.
23. $\frac{11}{14}$, $\frac{6}{8}$, and $\frac{13}{18}$.
24. $\frac{5}{9}$, $\frac{3}{24}$, and $\frac{13}{45}$.
25. $\frac{6}{18}$, $\frac{9}{15}$, and $\frac{38}{60}$.

ADDITION OF FRACTIONS.

MENTAL EXERCISES.

Ex. 1. What is the sum of $\frac{1}{8}$, $\frac{2}{8}$, $\frac{3}{8}$, and $\frac{5}{8}$?

Suggestion.—Since all these fractions have the same denominator, it is plain their numerators may be added as well as so many pounds or bushels, and their sum placed over the common denominator, will be the answer required. Thus, 1 eighth and 2 eighths are 3 eighths, and 3 are 6 eighths, and 5 are 11 eighths. *Ans.* $\frac{11}{8}$, or $1\frac{3}{8}$.

QUEST.—*Obs.* Does this process alter the value of the given fractions? Why not?

2. What is the sum of $\frac{1}{4}$, $\frac{3}{4}$, $\frac{7}{4}$, and $\frac{2}{4}$?

3. What is the sum of $\frac{2}{9}$, $\frac{7}{9}$, $\frac{1}{9}$, and $\frac{8}{9}$?

4. What is the sum of $\frac{4}{13}$, $\frac{10}{13}$, $\frac{5}{13}$, and $\frac{9}{13}$?

5. What is the sum of $\frac{8}{11}$, $\frac{6}{11}$, $\frac{2}{11}$, $\frac{9}{11}$, and $\frac{10}{11}$?

6. What is the sum of $\frac{6}{25}$, $\frac{8}{25}$, $\frac{1}{25}$, $\frac{12}{25}$, and $\frac{2}{25}$?

7. What is the sum of $\frac{15}{19}$, $\frac{3}{19}$, $\frac{7}{19}$, $\frac{4}{19}$, and $\frac{10}{19}$?

8. What is the sum of $\frac{20}{64}$, $\frac{30}{64}$, $\frac{3}{64}$, $\frac{5}{64}$, and $\frac{8}{64}$?

9. What is the sum of $\frac{16}{45}$, $\frac{10}{45}$, $\frac{1}{45}$, $\frac{8}{45}$, and $\frac{5}{45}$?

10. What is the sum of $\frac{60}{100}$, $\frac{3}{100}$, $\frac{10}{100}$, and $\frac{8}{100}$?

EXERCISES FOR THE SLATE.

11. What is the sum of $\frac{2}{5}$, $\frac{1}{5}$, and $\frac{3}{5}$?

Solution.—$\frac{2}{5}+\frac{1}{5}+\frac{3}{5}=\frac{6}{5}$, or $1\frac{1}{5}$. *Ans.*

12. What is the sum of $\frac{1}{2}$, and $\frac{1}{3}$?

Suggestion.—A difficulty here presents itself; for it is manifest that 1 half added to 1 third will make neither 2 *halves* nor 2 *thirds.* (Art. 22.) This difficulty may be removed by reducing the given fractions to a common denominator. (Art. 125.) Thus,

$\left.\begin{array}{l}1\times3=3\\1\times2=2\end{array}\right\}$ the new numerators.

$2\times3=6$, the common denominator.

The fractions reduced are $\frac{3}{6}$ and $\frac{2}{6}$, and may now be added. Thus, $\frac{3}{6}+\frac{2}{6}=\frac{5}{6}$. *Ans.*

127. From these illustrations we deduce the following general

RULE FOR ADDITION OF FRACTIONS.

Reduce the fractions to a common denominator; add their numerators, and place the sum over the common denominator.

OBS. 1. *Compound fractions* must, of course, be reduced to simple ones, before attempting to reduce the given fractions to a common denominator. (Art. 123.)

QUEST.—127. How are fractions added? *Obs.* What must be done with compound fractions?

2. *Mixed numbers* may be reduced to improper fractions, then added according to the rule; or, we may add the whole numbers and fractional parts separately, and then unite their sums.

13. What is the sum of $\frac{4}{6}$, and $\frac{5}{6}$? *Ans.* $\frac{9}{6}=1\frac{3}{6}$, or $1\frac{1}{2}$.
14. What is the sum of $\frac{3}{4}$, and $\frac{5}{3}$?
15. What is the sum of $\frac{3}{8}$, $\frac{1}{2}$, and $\frac{2}{3}$?
16. What is the sum of $\frac{4}{9}$ $\frac{11}{15}$, and $\frac{1}{5}$?
17. What is the sum of $\frac{3}{12}$, $\frac{6}{7}$, and $\frac{9}{13}$?
18. What is the sum of $\frac{3}{5}$, $\frac{2}{11}$, and $\frac{6}{18}$?
19. What is the sum of $\frac{1}{10}$, $\frac{2}{7}$, and $\frac{5}{6}$?
20. What is the sum of $\frac{4}{13}$, $\frac{7}{8}$, and $\frac{12}{7}$?
21. What is the sum of $\frac{1}{3}$, $\frac{1}{6}$, $\frac{2}{8}$, and $\frac{9}{2}$?
22. What is the sum of $\frac{1}{8}$, $\frac{3}{5}$, $\frac{6}{8}$, and $\frac{8}{7}$?
23. What is the sum of $\frac{6}{8}$, $\frac{2}{3}$ of $\frac{1}{2}$, and $\frac{7}{12}$?
24. What is the sum of $\frac{3}{5}$, $\frac{2}{3}$, $\frac{7}{8}$ of $\frac{1}{4}$, and $\frac{1}{6}$?
25. What is the sum of $\frac{1}{7}$ of 3, $\frac{3}{5}$ of $\frac{1}{3}$, and $\frac{5}{8}$?
26. What is the sum of $2\frac{1}{2}$, $6\frac{1}{3}$, and $\frac{2}{3}$?
27. What is the sum of $\frac{2}{3}$ of 2, $3\frac{1}{2}$, and $5\frac{2}{7}$?
28. What is the sum of $\frac{24}{30}$, $\frac{81}{45}$, and $\frac{19}{10}$?
29. What is the sum of $35\frac{1}{3}$, $\frac{53}{12}$, and $\frac{2}{3}$ of $\frac{7}{8}$?
30. What is the sum of $\frac{25}{4}$, $6\frac{1}{2}$, $1\frac{2}{3}$, and $\frac{5}{8}$?

SUBTRACTION OF FRACTIONS.

MENTAL EXERCISES.

Ex. 1. Henry had $\frac{5}{7}$ of a watermelon, and gave away $\frac{3}{7}$ of it: how much had he left?

Solution.—3 sevenths from 5 sevenths leaves 2 sevenths. *Ans.* $\frac{2}{7}$.

2. John had $\frac{7}{9}$ of a bushel of chestnuts, and gave away $\frac{5}{9}$: how many had he left?

3. If I own $\frac{6}{7}$ of an acre of land, and sell $\frac{4}{7}$ of it, how much shall I have left?

QUEST.—*Obs.* How are mixed numbers added?

4. A man owning $\frac{5}{8}$ of a ship, sold $\frac{3}{8}$: what part of the ship had he left?

5. William had $\frac{9}{10}$ of a dollar, and spent $\frac{3}{10}$: how many tenths had he left?

6. What is the difference between $\frac{6}{15}$ and $\frac{13}{15}$?

7. What is the difference between $\frac{18}{20}$ and $\frac{13}{20}$?

8. What is the difference between $\frac{17}{28}$ and $\frac{23}{28}$?

9. What is the difference between $\frac{11}{45}$ and $\frac{20}{45}$?

10. What is the difference between $\frac{31}{100}$ and $\frac{41}{100}$?

EXERCISES FOR THE SLATE.

11. From $\frac{9}{12}$ take $\frac{4}{12}$.

Solution.—$\frac{9}{12}-\frac{4}{12}=\frac{5}{12}$. *Ans.*

12. From $\frac{5}{6}$ take $\frac{3}{4}$.

Suggestion.—A difficulty here meets the learner, similar to that which occurred in the 12th example of addition of fractions, viz: that of subtracting a fraction of one denominator from a fraction of a different denominator. He must therefore reduce the fractions to a common denominator, before the subtraction can be performed.

Thus, $5\times4=20$ }
And $3\times6=18$ } the numerators. (Art. 125.)

Also $6\times4=24$, the common denominator.

The fractions are $\frac{20}{24}$ and $\frac{18}{24}$. Now $\frac{20}{24}-\frac{18}{24}=\frac{2}{24}$. *Ans.*

128. From these illustrations we deduce the following general

RULE FOR SUBTRACTION OF FRACTIONS.

Reduce the given fractions to a common denominator; subtract the less numerator from the greater, and place the remainder over the common denominator.

OBS. *Compound fractions* must be reduced to simple ones, as in addition of fractions. (Art. 123.)

QUEST.—128. How is one fraction subtracted from another? *Obs.* What is to be done with compound fractions?

13. From $\frac{2}{3}$ take $\frac{1}{2}$. *Ans.* $\frac{1}{6}$.

14. From $\frac{8}{9}$ take $\frac{5}{9}$.

15. From $\frac{10}{12}$ take $\frac{7}{12}$.

16. From $\frac{16}{20}$ take $\frac{2}{5}$.

17. From $\frac{18}{24}$ take $\frac{5}{8}$.

18. From $\frac{37}{42}$ take $\frac{13}{125}$.

19. From $\frac{16}{21}$ take $\frac{8}{15}$.

20. From $\frac{9}{55}$ take $\frac{1}{15}$.

21. From $\frac{2}{7}$ take $\frac{2}{11}$.

22. From $\frac{27}{35}$ take $\frac{23}{48}$.

23. From $\frac{14}{75}$ take $\frac{3}{25}$.

129. *Mixed numbers* may be reduced to improper fractions, then to a common denominator and subtracted; or, the fractional part of the less number may be taken from the fractional part of the greater, and the less whole number from the greater.

24. From $8\frac{1}{3}$ take $5\frac{2}{3}$.

Operation.

$8\frac{1}{3}=\frac{25}{3}$

$5\frac{2}{3}=\frac{17}{3}$

Ans. $\frac{8}{3}=2\frac{2}{3}$.

17 thirds from 25 thirds leaves 8 thirds, which are equal to $2\frac{2}{3}$.

Or thus, $8\frac{1}{3}$

$5\frac{2}{3}$

Ans. $2\frac{2}{3}$

Note.—Since we cannot take 2 thirds from 1 third, we borrow a unit, which, reduced to *thirds* and added to 1 third, makes 4 thirds. Now 2 thirds from 4 thirds leaves 2 thirds: 1 to carry to 5 makes 6, and 6 from 8 leaves 2.

25. From $12\frac{5}{8}$ take $7\frac{1}{4}$. *Ans.* $5\frac{3}{8}$.

26. From $15\frac{3}{7}$ take $9\frac{1}{2}$.

27. From $25\frac{3}{5}$ take $17\frac{4}{5}$.

28. From $37\frac{4}{7}$ take $19\frac{3}{4}$.

29. From 2 take $\frac{3}{5}$.

Suggestion.—Since 5 fifths make a whole one, in 2 whole ones there are 10 fifths; now 3 fifths from 10 fifths leaves 7 fifths. *Ans.* $\frac{7}{5}$, or $1\frac{2}{5}$. Hence,

QUEST.—129. How are mixed numbers subtracted? 130. How is a fraction subtracted from a whole number?

130. To subtract a fraction from a whole number.

Change the whole number to a fraction having the same denominator as the fraction to be subtracted, and proceed as before. (Art. 128.)

OBS. If the fraction to be subtracted is a proper fraction, we may simply borrow a unit and take the fraction from this, remembering to diminish the whole number by 1. (Art. 36.)

30. From 6 take $\frac{2}{3}$. *Ans.* $5\frac{1}{3}$.
31. From 65 take $25\frac{3}{15}$.
32. From $\frac{2}{3}$ of $\frac{3}{4}$ take $\frac{1}{2}$ of $\frac{2}{6}$.
33. From $\frac{5}{8}$ of $\frac{3}{4}$ take $\frac{1}{6}$ of $\frac{2}{12}$.
34. From $\frac{3}{5}$ of 10 take $\frac{2}{3}$ of 6.
35. From $\frac{6}{8}$ of 24 take $\frac{6}{9}$ of 27.

MULTIPLICATION OF FRACTIONS.

MENTAL EXERCISES.

1. If a man spends $\frac{1}{8}$ of a dollar for rum in 1 day, how much will he spend in 7 days?

Suggestion.—If he spends $\frac{1}{8}$ in 1 day, in 7 days he will spend 7 times $\frac{1}{8}$; and $\frac{1}{8}\times7$ is $\frac{7}{8}$. *Ans.* $\frac{7}{8}$ of a dollar.

2. If a man spends $\frac{7}{8}$ of a dollar for rum in 1 week, how much will he spend in 4 weeks. *Ans.* $\frac{28}{8}$ or $3\frac{1}{2}$ dolls.
3. If 1 man drinks $\frac{5}{8}$ of a barrel of beer in a month, how much will 10 men drink in the same time?
4. What cost 4 yards of cloth, at $2\frac{1}{2}$ dollars per yard?

Solution.—4 yards will cost 4 times as much as 1 yard; and 4 times $\frac{1}{2}$ is 4 halves, equal to two whole ones: 4 times 2 dollars are 8 dollars, and 2 make 10 dollars.
Ans. 4 yards will cost 10 dollars.

5. What cost $5\frac{1}{3}$ bushels of peanuts, at 3 dolls. a bushel?
6. What cost $10\frac{3}{4}$ pounds of tea, at 4 shillings a pound?
7. If 1 drum of figs costs 16 shillings, what will 3 fourths of a drum cost?

Suggestion.—First find what 1 fourth will cost. Then 3 fourths will cost 3 times as much.

8. If an acre of land produces 40 bushels of corn, how many bushels will 3 eighths of an acre produce?

9. If a man can travel 50 miles in a day, how far can he travel in 2 fifths of a day? 3 fifths? 4 fifths?

10. Henry's kite line was 90 feet long, but getting entangled in a tree, he lost 3 ninths of it: how many feet did he lose?

131. We have seen that multiplying by a *whole number* is taking the multiplicand as many times as there are *units* in the multiplier. (Art. 45.) On the other hand,

If the multiplier is only a *part* of a unit, it is plain we must take only a *part* of the multiplicand. That is,

132. *Multiplying by a fraction is taking a certain* PORTION *of the multiplicand as many times as there are like portions of a unit in the multiplier.*

Multiplying by $\frac{1}{2}$, is taking 1 *half* of the multiplicand *once.* Thus, $6\times\frac{1}{2}=3$. (Art. 104. Obs.)

Multiplying by $\frac{1}{3}$, is taking 1 *third* of the multiplicand *once.* Thus, $6\times\frac{1}{3}=2$.

Multiplying by $\frac{2}{3}$, is taking 1 *third* of the multiplicand *twice.* Thus, $6\times\frac{2}{3}=4$.

OBS. If the multiplier is a *unit*, the product is *equal* to the multiplicand; if the multiplier is *greater* than a unit, the product is *greater* than the multiplicand; (Art. 45;) and if the multiplier is *less* than a unit, the product is *less* than the multiplicand.

EXERCISES FOR THE SLATE.

CASE I.

11. If a bushel of corn is worth $\frac{1}{2}$ of a dollar, how much is 5 bushels worth?

QUEST.—131. What is meant by multiplying by a whole number? 132. By a fraction? By $\frac{1}{2}$? By $\frac{1}{3}$? By $\frac{2}{3}$? By $\frac{3}{5}$? By $\frac{7}{8}$? *Obs.* If the multiplier is a unit or 1, what is the product equal to? When the multiplier is greater than 1, how is the product, compared with the multiplicand? When less, how?

Solution.—5 bushels will cost 5 times as much as 1 bushel. Now $\frac{1}{2}\times5=\frac{5}{2}$, or $2\frac{1}{2}$; that is, 5 times $\frac{1}{2}$ are 5 halves, equal to 2 and 1 half. *Ans.* $2\frac{1}{2}$ dollars.

12. Multiply $\frac{3}{4}$ by 5. *Ans* $\frac{15}{4}$, or $3\frac{3}{4}$.
13. Multiply $\frac{6}{12}$ by 8.
14. Multiply $\frac{7}{8}$ by 12.
15. Multiply $\frac{5}{15}$ by 18.
16. Multiply $\frac{12}{20}$ by 10.

17. If a pound of tea cost 6 shillings, how much will $\frac{2}{3}$ of a pound cost?

Solution.—Multiplying by $\frac{2}{3}$, is taking 1 third of the multiplicand *twice.* (Art. 132.) Now 1 third of 6 is the same as 6 thirds of 1, or $\frac{6}{3}$; and 2 thirds of 6 must be 2 times as much; that is, $\frac{6}{3}\times2=\frac{12}{3}$; and $\frac{12}{3}=4$. *Ans.*

Note.—Since the product of any two numbers will be the same, whichever is taken for the multiplier, (Art. 47,) the fraction may be taken for the multiplicand, and the whole number for the multiplier, when it is more convenient.

Thus, $\frac{2}{3}\times6=\frac{12}{3}$, or 4; and $6\times\frac{2}{3}=4$.

18. Multiply 12 by $\frac{1}{4}$. *Ans.* 3.
19. Multiply 10 by $\frac{3}{4}$.
20. Multiply 15 by $\frac{7}{8}$.
21. Multiply $\frac{5}{8}$ by 2. *Ans.* $\frac{5}{8}\times2=\frac{10}{8}$, or $1\frac{1}{4}$.

Suggestion.—Dividing the denominator of a fraction by any number, multiplies the value of the fraction by that number. (Art. 114.) Now, if we divide the denominator 8 by 2, the fraction will become $\frac{5}{4}$, which is equal to $1\frac{1}{4}$, the same as before. Hence,

133. To multiply a fraction and a whole number together.

Multiply the numerator of the fraction by the whole number, and write the product over the denominator.

Or, divide the denominator by the whole number, when this can be done without a remainder. (Art. 114.)

QUEST.—133. How multiply a fraction and a whole number together?

OBS. 1. A fraction is multiplied into a number *equal* to its *denominator* by *canceling* the denominator. (Arts. 89, 91.) Thus $\frac{4}{7}\times 7=4$.

2. On the same principle, a fraction is multiplied into *any factor* in its *denominator*, by *canceling* that factor. (Arts. 91, 114.) Thus, $\frac{3}{15}\times 3=\frac{3}{5}$.

22. Multiply $\frac{15}{25}$ by 5. *Ans.* $\frac{15}{5}$, or 3.
23. Multiply $\frac{24}{36}$ by 9.
24. Multiply $\frac{18}{25}$ by 25.
25. Multiply 36 by $\frac{12}{15}$.
26. Multiply 120 by $\frac{18}{25}$.
27. Multiply $\frac{256}{375}$ by 25.
28. Multiply $\frac{420}{650}$ by 50.
29. Multiply $9\frac{1}{2}$ by 5.

Operation.
$9\frac{1}{2}$
5
Ans. $47\frac{1}{2}$

5 times $\frac{1}{2}$ are $\frac{5}{2}$, which are equal to 2 and $\frac{1}{2}$. Set down the $\frac{1}{2}$. 5 times 9 are 45, and 2 (which arose from the fraction) make 47. Hence,

134. To multiply a mixed number by a whole one.

Multiply the fractional part and the whole number separately, and unite the products.

30. Multiply $15\frac{3}{4}$ by 7. *Ans.* $110\frac{1}{4}$.
31. Multiply $25\frac{4}{5}$ by 10.
32. Multiply $48\frac{10}{16}$ by 8.
33. Multiply 24 by $3\frac{1}{2}$.

Operation.
2)24
$3\frac{1}{2}$
72
12
Ans. 84

We first multiply 24 by 3, then by $\frac{1}{2}$, and the sum of the products is 84. Multiplying by $\frac{1}{2}$ is taking *one half* of the multiplicand *once*. (Art. 132.) But to find a half of any number, we divide the number by 2. (Art. 104. Obs.) Hence,

134. *a.* To multiply a whole by a mixed number.

Multiply first by the integer, then by the fraction, and add the products together.

34. Multiply 27 by $3\frac{1}{3}$. *Ans.* 90.
35. Multiply 63 by $10\frac{3}{7}$.
36. Multiply 75 by $12\frac{2}{9}$.

QUEST.—*Obs.* How is a fraction multiplied by a number equal to its denominator? How by any factor in its denominator? 134. How is a mixed number multiplied by a whole one? 134. *a.* How is a whole number multiplied by a mixed number?

CASE II.

37. A man owning $\frac{3}{5}$ of a ship, sold $\frac{2}{3}$ of what he owned. What part of the ship did he sell?

Analysis.—$\frac{1}{3}$ of $\frac{3}{5}$ is $\frac{3}{15}$; for, multiplying the denominator by any number, divides the value of the fraction. (Art. 113.) Now 2 thirds of $\frac{3}{5}$ is twice as much; that is, $\frac{3}{15}\times2=\frac{6}{15}$, which, reduced to its lowest terms, is $\frac{2}{5}$. *Ans.*

Or, we may reason thus: Since he owned $\frac{3}{5}$, and sold $\frac{2}{3}$ of what he owned, he must have sold $\frac{2}{3}$ of $\frac{3}{5}$ of the ship. Now $\frac{2}{3}$ of $\frac{3}{5}$ is a compound fraction, whose value is found by multiplying the numerators together for a new numerator, and the denominators for a new denominator. (Art. 123.)

Solution.—$\frac{3}{5}\times\frac{2}{3}=\frac{6}{15}$, or $\frac{2}{5}$. *Ans.* Hence,

135. To multiply a fraction by a fraction.

Multiply the numerators together for a new numerator and the denominators together for a new denominator.

OBS. It will be seen that the process of multiplying one fraction by another, is precisely the same as that of reducing compound fractions to simple ones.

38. Multiply $\frac{1}{3}$ by $\frac{3}{5}$. *Ans.* $\frac{3}{15}=\frac{1}{5}$.
39. Multiply $\frac{4}{7}$ by $\frac{2}{3}$. 40. Multiply $\frac{7}{9}$ by $\frac{5}{8}$.
41. Multiply $\frac{4}{12}$ by $\frac{5}{8}$. 42. Multiply $\frac{12}{15}$ by $\frac{6}{8}$.
43. Multiply $\frac{1}{7}$ and $\frac{2}{3}$ and $\frac{3}{4}$ and $\frac{4}{5}$ together.

Operation.

$$\frac{1}{7}\times\frac{2}{\not3}\times\frac{\not3}{\not4}\times\frac{\not4}{5}=\frac{2}{35}$$

Since the factors 3 and 4 are common both to the numerators and denominators, we may *cancel* them, and multiply the remaining factors together as in reducing compound fractions to simple ones. (Art. 124.) Hence,

---

QUEST.—135. How is a fraction multiplied by a fraction? *Obs.* To what is the process of multiplying one fraction by another similar? 136. How multiply fractions together by cancelation?

136. To multiply fractions by CANCELATION.

Cancel all the factors common both to the numerators and denominators; then multiply the remaining factors in the numerators together for a new numerator, and those remaining in the denominators for a new denominator, as in reduction of compound fractions. (Art. 1[illegible]4.)

OBS. 1. This process, in effect, is dividing the product of the numerators and that of the denominators by the same number, and therefore does not alter the value of the answer. (Art. 116.)

2. Care must be taken that the factors canceled in the numerators are *exactly equal* to those canceled in the denominators.

44. Multiply [illegible] $\frac{4}{5}$. *Ans.* $\frac{3}{5}$.
45. Multiply [illegible] by [illegible] and $\frac{3}{4}$. *Ans.* $\frac{1}{12}$.
46. Multiply [illegible] by [illegible] and $\frac{3}{7}$.
47. Multiply [illegible] by $\frac{3}{13}$ and $\frac{1}{4}$ and $\frac{13}{12}$.
48. Multiply $\frac{75}{80}$ by $\frac{16}{25}$ and $\frac{5}{3}$ and $\frac{4}{9}$ and $\frac{3}{7}$.
49. Multiply $\frac{37}{63}$ by $\frac{9}{14}$ and $\frac{7}{37}$ and $\frac{6}{21}$ and $\frac{3}{2}$.
50. Multiply $7\frac{1}{2}$ by $3\frac{1}{3}$.

Solution.—$7\frac{1}{2}$, reduced to an improper fraction, becomes $\frac{15}{2}$, and $3\frac{1}{3}$ becomes $\frac{10}{3}$. Now $\frac{15}{2}\times\frac{10}{3}=\frac{150}{6}$, or 25. *Ans.*

137. Hence, when the multiplier and multiplicand are both *mixed* numbers, they should be reduced to *improper fractions*, and then be multiplied according to the rule above.

EXAMPLES FOR PRACTICE.

1. What will 12 apples cost, at $\frac{1}{3}$ of a cent apiece?
2. If a bushel of wheat weighs $\frac{3}{5}$ of a hundred weight, how much will 10 bushels weigh?
3. If a man earns $\frac{3}{4}$ of a dollar per day, how much can he earn in 12 days?

QUEST.—*Obs.* How does it appear that this process will give the true answer? What is necessary to be observed with regard to canceling factors? 137. When the multiplier and multiplicand are mixed numbers, how proceed?

4. If a family consume $\frac{4}{7}$ of a barrel of flour in a week, how much will they consume in 15 weeks?

5. If I burn $\frac{7}{8}$ of a cord of wood in a month, how much shall I burn in 12 months?

6. If a man can reap $\frac{11}{12}$ of an acre of grain in a day, how many acres can he reap in 9 days?

7. If a pound of powder is worth 6 shillings, how much is $\frac{2}{5}$ of a pound worth?

8. If a gallon of oil is worth 7 shillings, how much is $\frac{3}{4}$ of a gallon worth?

9. When beef is 10 dollars a barrel, how much will $\frac{5}{8}$ of a barrel cost?

10. What will $\frac{4}{9}$ of a firkin of butter cost, at 15 dollars a firkin?

11. At $\frac{5}{8}$ of a dollar a cord, how much will the sawing of 20 cords of wood amount to?

12. What will 16 pounds of cheese cost, at $8\frac{1}{2}$ cents per pound?

13. What cost 9 dozen of eggs, at $12\frac{1}{2}$ cents per dozen?

14. What cost $15\frac{2}{3}$ yards of cambric, at 15 pence per yard?

15. What cost $11\frac{1}{9}$ cords of wood, at $1\frac{1}{2}$ dollar per cord?

16. At 12 cents a pound, what cost $2\frac{2}{3}$ pounds of pepper?

17. At 5 shillings a pound, what cost $12\frac{3}{8}$ pounds of tea?

18. What will 6 pounds of starch come to, at $12\frac{1}{2}$ cents per pound?

19. What will 18 ounces of nutmegs come to, at $6\frac{1}{4}$ cents an ounce?

20. At $12\frac{3}{4}$ cents a yard, what will 17 yards of cotton come to?

21. At $3\frac{11}{15}$ dollars a yard, what cost 15 yards of broadcloth?

22. What cost $15\frac{3}{4}$ yards of ribbon, at 10 cents per yard?

23. What cost 22 pocket handkerchiefs, at $\frac{12}{16}$ of a dollar apiece?

24. At $\frac{5}{10}$ of a dollar a yard, what will $\frac{3}{4}$ of a yard of lace cost?

25. At $\frac{3}{8}$ of a dollar a yard, what will $\frac{7}{9}$ of a yard of muslin come to?

26. At $\frac{3}{4}$ of a dollar a bushel, what cost $\frac{9}{10}$ of a bushel of wheat?

27. What will $\frac{6}{7}$ of a pound of tea cost, at $\frac{5}{8}$ of a dollar a pound?

28. What cost 66 bushels of apples, at $18\frac{3}{4}$ cents a bushel?

29. At $62\frac{1}{2}$ cents a yard, what cost $12\frac{1}{2}$ yards of balzorine?

30. What cost $18\frac{1}{2}$ yards of tape, at $6\frac{1}{4}$ cents per yard?

31. What cost 13 bushels of oats, at $18\frac{3}{4}$ cents per bushel?

32. What cost $31\frac{1}{2}$ yards of sheeting, at $\frac{1}{9}$ of a dollar per yard?

33. At $\frac{7}{12}$ of a dollar a quart, what cost $8\frac{1}{2}$ quarts of cherries?

34. At $3\frac{3}{4}$ shillings a yard, what cost $7\frac{1}{2}$ yards of gingham?

35. What cost $14\frac{3}{5}$ bushels of potatoes, at $18\frac{3}{4}$ cents a bushel?

36. At $7\frac{3}{8}$ shillings a yard, what cost $8\frac{3}{8}$ yards of silk?

37. At $\frac{7}{8}$ of a dollar a bushel, what cost $47\frac{4}{5}$ bushels of peaches?

38. What cost $63\frac{3}{4}$ pounds of sugar, at $9\frac{3}{4}$ cents per pound?

39. What cost $2\frac{3}{4}$ yards of velvet, at $3\frac{2}{8}$ dolls. a yard?

40. What cost $9\frac{3}{4}$ yards of calico, at $1\frac{2}{3}$ shillings a yard?

41. What cost $25\frac{1}{4}$ pounds of figs, at $15\frac{1}{2}$ cents a pound?

42. What cost $35\frac{3}{5}$ cords of wood, at $18\frac{1}{3}$ shillings per cord?

43. What cost $175\frac{1}{2}$ bushels of corn, at $\frac{3}{8}$ of a dollar a bushel?

44. What cost $8\frac{3}{4}$ tons of hay, at $15\frac{7}{8}$ dollars a ton?

45. If a man can travel $42\frac{1}{2}$ miles in one day, how far can he travel in $17\frac{1}{2}$ days?

DIVISION OF FRACTIONS.

MENTAL EXERCISES.

Ex. 1. A man divided $\frac{6}{7}$ of a pound of honey equally among his 3 children: what part of a pound did each receive?

Analysis.—1 is one third of 3; therefore 1 child must have received 1 third of 6 sevenths. 1 third of 6 sevenths is 2 sevenths. *Ans.* Each child received $\frac{2}{7}$ of a pound.

2. If 4 pounds of loaf sugar cost $\frac{8}{9}$ of a dollar, how much will 1 pound cost?

3. A father gave his 2 sons $\frac{10}{12}$ of a dollar: how many twelfths did each receive?

4. A little girl bought 5 lead pencils for $\frac{10}{12}$ of a shilling: how much did she give apiece for them?

5. A father gave $\frac{30}{32}$ parts of a vessel to his 6 sons: what part of the vessel did each receive?

6. At $\frac{1}{2}$ dollar a yard, how many yards of French muslin can you buy for 4 dollars?

Suggestion.—4 dollars will buy as many yards as 1 half is contained times in 4, or as there are halves in 4 dollars. Now since there are 2 halves in 1 dollar, in 4 dollars there are 4 times 2 halves; and 4 times 2 halves are 8 halves. *Ans.* 4 dollars will buy 8 yards.

7. At $\frac{1}{2}$ cent apiece, how many apples can I buy for 6 cents?

8. At $\frac{1}{4}$ of a dollar a pound, how many pounds of almonds can you buy for 12 dollars?

9. How many quills, at $\frac{2}{5}$ of a penny apiece, can you buy for $\frac{6}{5}$ of a penny?

Suggestion.—$\frac{6}{5}$ of a penny will buy as many quills as $\frac{2}{5}$ is contained times in $\frac{6}{5}$; and $\frac{2}{5}$ is contained in $\frac{6}{5}$, 3 times. *Ans.* 3 quills.

10. How many yards of cloth can I buy for $\frac{4}{3}$ of a cord of wood, if I give $\frac{1}{3}$ of a cord for a yard of cloth?

EXERCISES FOR THE SLATE.

CASE I.

11. If 3 bushels of oats cost $\frac{6}{8}$ of a dollar, what will 1 bushel cost?

Analysis.—1 is 1 third of 3; therefore, 1 bushel will cost 1 third part as much as 3 bushels. 1 third of $\frac{6}{8}$ is $\frac{2}{8}$.
Ans. $\frac{2}{8}$ of a dollar.

Operation.
$\frac{6}{8}\div3=\frac{2}{8}$. *Ans.*

We divide the numerator of the fraction $\frac{6}{8}$, which is the whole cost, by 3 the whole number of bushels, and place the quotient 2 over the given denominator.

12. If 4 yards of calico cost $\frac{5}{6}$ of a dollar, what will 1 yard cost?

Operation.
$\frac{5}{6}\div4=\frac{5}{6\times4}$, or $\frac{5}{24}$. *Ans.*

In this case we cannot divide the numerator of the dividend by 4 the given divisor, without a remainder. We therefore multiply the denominator by the 4, which is in effect dividing the fraction. (Art. 113.) Hence,

138. To divide a fraction by a whole number

Divide the numerator by the whole number, when it can be done without a remainder; but when this cannot be done, multiply the denominator by the whole number.

13. Divide $\frac{6}{8}$ by 3.

First Method.	*Second Method.*
$\frac{6}{8}\div3=\frac{2}{8}$, or $\frac{1}{4}$. *Ans.*	$\frac{6}{8}\div3=\frac{6}{24}$, or $\frac{1}{4}$. *Ans.*

14. Divide $\frac{12}{15}$ by 6.
15. Divide $\frac{16}{19}$ by 8.
16. Divide $\frac{15}{9}$ by 7.
17. Divide $\frac{17}{11}$ by 12.

QUEST.—138. How is a fraction divided by a whole number?

18. Divide $\frac{45}{40}$ by 9. 19. Divide $\frac{72}{85}$ by 8.

20. Divide $\frac{125}{221}$ by 25. 21. Divide $\frac{150}{238}$ by 30.

CASE II.

22. At $\frac{1}{4}$ of a dollar a pound, how many pounds of honey can be bought for $\frac{3}{4}$ of a dollar?

Suggestion.—Since $\frac{1}{4}$ of a dollar will buy 1 pound, $\frac{3}{4}$ of a dollar will buy as many pounds as $\frac{1}{4}$ is contained times in $\frac{3}{4}$. Now $\frac{1}{4}$ is contained in $\frac{3}{4}$, 3 times. *Ans.* 3 pounds.

23. At $\frac{2}{5}$ of a dollar a bushel, how much barley can be bought for $\frac{3}{4}$ of a dollar?

First Operation.

$\frac{3}{4}=\frac{15}{20}$

$\frac{2}{5}=\frac{8}{20}$

$\frac{15}{20}\div\frac{8}{20}=1\frac{7}{8}$. *Ans.*

We first reduce the given fractions to a common denominator; (Art. 125;) then divide the numerator of the dividend by the numerator of the divisor, as above.

OBS. 1. After the fractions are reduced to a common denominator, it will be perceived that no use is made of the common denominator itself. In practice, therefore, it is simply necessary to multiply the *numerator* of the dividend by the *denominator* of the divisor, and the *denominator* of the dividend by the *numerator* of the divisor, in the same manner as two fractions are reduced to a common denominator; or, what is the same in effect, invert the divisor, and proceed as in multiplication of fractions. (Art. 135.)

Note.—To *invert* a fraction is to put the numerator in the place of the denominator, and the denominator in the place of the numerator. Thus, in the example above, inverting the divisor $\frac{2}{5}$, it becomes $\frac{5}{2}$; and $\frac{3}{4}\times\frac{5}{2}=\frac{15}{8}$, or $1\frac{7}{8}$, which is the same as before.

Again, we may also illustrate the principle thus:

Second Operation.

$\frac{3}{4}\div2=\frac{3}{8}$

$\frac{3}{8}\times5=\frac{15}{8}$

And $\frac{15}{8}=1\frac{7}{8}$. *Ans.*

Dividing the dividend $\frac{3}{4}$ by 2, the quotient is $\frac{3}{8}$. (Art. 113.) But it is required to divide it by only $\frac{1}{5}$ of 2; consequently the $\frac{3}{8}$ is 5 times too small for the true quotient. Therefore $\frac{3}{8}$ multiplied by 5 will be the quotient required. Now $\frac{3}{8}\times5=\frac{15}{8}$, or $1\frac{7}{8}$, which is the same result as before,

OBS. 2. By examination the learner will perceive that this process is precisely the same in effect as the preceding; for in both cases the denominator of the dividend is multiplied by the numerator of the divisor, and the numerator of the dividend, by the denominator of the divisor. Hence,

139. To divide a fraction by a fraction.

I. If the given fractions have a common denominator;

Divide the numerator of the dividend by the numerator of the divisor.

II. When the fractions have not a common denominator;

Invert the divisor, and proceed as in multiplication of fractions. (Art. 135.)

OBS. 1. *Compound fractions* occurring in the divisor or dividend, must be reduced to simple ones, and *mixed numbers* to improper fractions.

2. The method of dividing a fraction by a fraction depends upon the obvious principle, that if two fractions have a common denominator, the numerator of the dividend, divided by the numerator of the divisor, will give the true quotient. Now multiplying the numerator of the dividend by the denominator of the divisor, and the denominator of the dividend by the numerator of the divisor, is in effect reducing the two fractions to a common denominator. The object of inverting the divisor, is simply for convenience in multiplying.

24. Divide $\frac{2}{3}$ of $\frac{4}{7}$ by $1\frac{1}{2}$.

Solution.—$\frac{2}{3}$ of $\frac{4}{7}=\frac{8}{21}$, and $1\frac{1}{2}=\frac{3}{2}$. Now $\frac{8}{21}\div\frac{3}{2}=\frac{8}{21}\times\frac{2}{3}$, or $\frac{16}{63}$. *Ans.*

25. Divide $7\frac{1}{2}$ by $2\frac{1}{4}$. *Ans.* $3\frac{1}{3}$.

26. Divide $13\frac{1}{2}$ by $\frac{3}{5}$.

27. Divide $\frac{5}{7}$ by $1\frac{2}{3}$.

28. Divide $\frac{25}{37}$ by $\frac{17}{40}$.

29. Divide $\frac{42}{57}$ by $\frac{19}{20}$.

QUEST.—139. How is one fraction divided by another when they have a common denominator? How, when they have not common denominators? *Obs.* How proceed when the divisor or dividend are compound fractions, or mixed numbers? Upon what principle does the method of dividing a fraction by a fraction, depend? Why multiply the numerator of the dividend by the denominator of the divisor, &c.? Why invert the divisor?

30. Divide $\frac{1}{3}$ of $\frac{5}{8}$ by $\frac{1}{2}$ of $\frac{2}{3}$.

Operation.

$$\begin{array}{c|l} \not 3 & 1 \\ 8 & 5 \\ 1 & \not 2 \\ \not 2 & \not 3 \\ \hline 8 & 5 = \frac{5}{8}. \textit{ Ans.} \end{array}$$

For convenience we arrange the numerators, (which answer to dividends,) on the right of a perpendicular line, and the denominators, (which answer to divisors,) on the left; then canceling the factors 3 and 2, which are common to both sides, (Art. 91. *a*,) we multiply the remaining factors in the numerators together, and those remaining in the denominators, as in the rule above. Hence,

140. To divide fractions by CANCELATION.

Having inverted the divisor, cancel all the factors common both to the numerators and denominators, and proceed as in multiplication of fractions. (Art. 136.)

OBS. Before arranging the terms of the divisor for cancelation, it is always necessary to invert them, or suppose them to be inverted.

31. Divide $4\frac{1}{2}$ by $2\frac{1}{4}$. *Ans.* 2.
32. Divide $\frac{4}{7}$ of 6 by $\frac{1}{6}$ of 4.
33. Divide $4\frac{1}{3}$ by $\frac{1}{3}$ of $\frac{33}{4}$.
34. Divide $\frac{1}{7}$ of $\frac{25}{28}$ by $\frac{1}{3}$ of $\frac{4}{7}$.
35. Divide $\frac{5}{12}$ of $\frac{3}{4}$ by $\frac{3}{7}$.
36. Divide $\frac{7}{8}$ of $15\frac{3}{4}$ by $4\frac{2}{8}$.
37. Divide $\frac{7}{9}$ by $\frac{23}{37}$ of $\frac{7}{40}$.
38. Divide $\frac{3}{5}$ by $\frac{3}{84}$ of $2\frac{5}{7}$.
39. Divide $25\frac{1}{4}$ by $\frac{1}{4}$ of 26.

CASE III.

40. A merchant sent 12 barrels of flour to supply some destitute people, allowing $\frac{2}{3}$ of a barrel to each family. How many families shared in his bounty?

Solution.—If $\frac{2}{3}$ of a barrel supplied 1 family, 12 barrels will supply as many families as $\frac{2}{3}$ is contained times in 12. Reducing the dividend 12 to the form of a fraction, it becomes $\frac{12}{1}$; now inverting the divisor, we have $\frac{12}{1}\times\frac{3}{2}=\frac{36}{2}$, or 18. *Ans.* 18 families.

QUEST.—140. How divide fractions by cancelation? How arrange the terms of the given fractions? *Obs.* What must be done to the divisor before arranging its terms?

Or, we may reason thus: $\frac{1}{3}$ is contained in 12, as many times as there are thirds in 12, viz: 36 times. Now 2 thirds are contained in 12, only half as many times as 1 third; and $36 \div 2 = 18$. *Ans.* Hence,

141. To divide a whole number by a fraction.

Reduce the whole number to the form of a fraction, (Art. 122. Obs. 1,) *and then proceed according to the rule for dividing a fraction by a fraction.* (Art. 139.)

Or, multiply the whole number by the denominator, and divide the product by the numerator.

OBS. When the divisor is a mixed number, it must be reduced to an improper fraction, then proceed as above.

41. Divide 120 by $3\frac{3}{5}$. *Ans.* $33\frac{1}{3}$.
42. Divide 35 by $\frac{2}{5}$.
43. Divide 47 by $\frac{5}{8}$.
44. Divide 165 by $\frac{7}{9}$.
45. Divide 237 by $\frac{11}{15}$.

142. From the definition of *complex* fractions, and the manner of expressing them, it will be seen that they arise from *division* of fractions. Thus the complex fraction $\frac{4\frac{1}{2}}{1\frac{1}{4}}$, is the same as $\frac{9}{2} \div \frac{5}{4}$; for, the numerator $4\frac{1}{2} = \frac{9}{2}$, and the denominator $1\frac{1}{4} = \frac{5}{4}$; but the numerator of a fraction is a dividend, and the denominator a divisor. (Art. 109.) Now $\frac{9}{2} \div \frac{5}{4} = \frac{36}{10}$, which is a simple fraction. Hence,

143. To reduce a complex fraction to a simple one.

Consider the denominator as a divisor, and proceed as in division of fractions. (Art. 139.)

46. Reduce $\frac{2\frac{1}{3}}{5\frac{3}{4}}$ to a simple fraction.

Operation.

$$\left.\begin{array}{l} 2\frac{1}{3} = \frac{7}{3}, \\ 5\frac{3}{4} = \frac{23}{4}. \end{array}\right\} \text{Now } \frac{7}{3} \div \frac{23}{4} = \frac{7}{3} \times \frac{4}{23}, \text{ or } \frac{28}{69}. \textit{ Ans.}$$

QUEST.—141. How is a whole number divided by a fraction? *Obs.* How by a mixed number? 142. From what do complex fractions arise? 143. How reduce them to simple fractions?

47. Reduce $\frac{6}{3\frac{1}{2}}$ to a simple fraction. *Ans.* $\frac{12}{7}$.

48. Reduce $\frac{5\frac{1}{2}}{6}$ to a simple fraction. *Ans.* $\frac{11}{12}$.

49. Reduce $\frac{\frac{3}{5}}{\frac{2}{3}}$ to a simple fraction. *Ans.* $\frac{9}{10}$.

50. Reduce the following complex fractions to simple ones.

$$\frac{4\frac{5}{9}}{6}.\quad \frac{8}{5\frac{1}{2}}.\quad \frac{9\frac{1}{2}}{7\frac{1}{3}}.\quad \frac{12\frac{1}{4}}{6\frac{1}{9}}.\quad \frac{18\frac{1}{3}}{12\frac{1}{2}}.\quad \frac{20\frac{3}{8}}{25\frac{3}{4}}.$$

144. To multiply complex fractions together.

First reduce the complex fractions to simple ones; (Art. 143;) *then arrange the terms, and cancel the common factors as in multiplication of simple fractions.* (Art. 136.)

OBS. 1. The terms of the complex fractions may be arranged for reducing them to simple ones, and for multiplication at the same time.

2. To divide one complex fraction by another, reduce them to simple fractions, then proceed as in Art. 139.

51. Multiply $\frac{2\frac{1}{3}}{2\frac{1}{4}}$ by $\frac{4\frac{1}{2}}{1\frac{3}{4}}$.

Operation.

3	$\not{7}$
$\not{9}$	4
$\not{2}$	$\not{9}$
$\not{7}$	$\not{4}$,2
3	8=$2\frac{2}{3}$. *Ans.*

The numerator $2\frac{1}{3}=\frac{7}{3}$. (Art. 122.) Place the 7 on the right hand and 3 on the left of the perpendicular line. The denominator $2\frac{1}{4}=\frac{9}{4}$, which must be inverted; (Art. 143;) i. e. place the 4 on the right and the 9 on the left of the line. $4\frac{1}{2}=\frac{9}{2}$, and $1\frac{3}{4}=\frac{7}{4}$, both of which must be arranged in the same manner as the terms of the multiplicand. Now, canceling the common factors, we divide the product of those remaining on the right of the line by the product of those on the left, and the quotient is $2\frac{2}{3}$. (Art. 136.)

QUEST.—144. How are complex fractions multiplied together? *Obs.* How is one complex fraction divided by another?

52. Multiply $\frac{5\frac{1}{3}}{2\frac{2}{5}}$ by $\frac{2\frac{1}{2}}{\frac{5}{12}}$.

53. Multiply $\frac{3\frac{5}{8}}{6\frac{1}{2}}$ by $\frac{5\frac{5}{6}}{2\frac{2}{3}}$.

54. Multiply $\frac{2\frac{2}{9}}{\frac{3}{5}} \times \frac{5\frac{1}{3}}{2\frac{1}{2}}$ by $\frac{\frac{3}{4}}{6\frac{2}{3}}$.

55. Multiply $\frac{3\frac{3}{4}}{2\frac{5}{6}} \times \frac{2\frac{2}{3}}{3\frac{3}{8}}$ by $\frac{\frac{3}{8}}{\frac{2}{7}}$.

EXAMPLES FOR PRACTICE.

1. At $\frac{1}{2}$ dollar per bushel, how many bushels of pears can be bought for 5 dollars?

2. At $\frac{3}{4}$ of a penny apiece, how many apples can be bought for 18 pence?

3. At $\frac{3}{8}$ of a dollar a pound, how many pounds of tea will 7 dollars buy?

4. How many bushels of pears, at $1\frac{1}{4}$ dollar a bushel, can be purchased for 15 dollars?

5. How many gallons of molasses, at $2\frac{1}{2}$ dimes per gallon, will 10 dimes buy?

6. How many yards of satinet, at $1\frac{3}{8}$ of a dollar per yard, can be purchased for 20 dollars?

7. At $4\frac{5}{8}$ dollars per yard, how many yards of cloth can be obtained for $25\frac{1}{2}$ dollars?

8. At $6\frac{1}{4}$ cents a mile, how far can you ride for $62\frac{1}{2}$ cents?

9. At $12\frac{1}{2}$ cents a pound, how many pounds of flax will $67\frac{3}{4}$ cents buy?

10. At $16\frac{1}{4}$ cents per pound, how many pounds of figs can you buy for $87\frac{1}{2}$ cents?

11. How many cords of wood, at $6\frac{1}{2}$ dollars per cord, will it take to pay a debt of $67\frac{1}{2}$ dollars?

12. How many barrels of beer, at $11\frac{3}{5}$ dollars per barrel, can be obtained for $95\frac{1}{2}$ dollars?

13. A man bought $15\frac{1}{2}$ barrels of beef for $124\frac{5}{8}$ dollars, how much did he give per barrel?

14. A man bought $13\frac{1}{2}$ pounds of sugar for $94\frac{1}{2}$ cents: how much did his sugar cost him a pound?

15. A lady bought $15\frac{3}{4}$ yards of silk for $145\frac{5}{12}$ shillings: how much did she pay per yard?

16. Bought $15\frac{1}{8}$ baskets of peaches for $24\frac{1}{4}$ dollars: how much was the cost per basket?

17. Bought $30\frac{1}{4}$ yards of broadcloth for $181\frac{1}{2}$ dollars: what was the price per yard?

18. Paid 375 dollars for $125\frac{1}{5}$ pounds of indigo: what was the cost per pound?

19. How many tons of hay, at $16\frac{1}{2}$ dollars per ton, can be bought for $196\frac{1}{3}$ dollars?

20. How many sacks of wool, at $17\frac{1}{8}$ dollars per sack, can be purchased for 1500 dollars?

21. How many bales of cotton, at $15\frac{7}{8}$ dollars per bale, can be bought for 2500 dollars?

22. Divide $145\frac{7}{12}$ by 16.
23. Divide $16\frac{75}{95}$ by 25.
24. Divide 8526 by $45\frac{9}{10}$.
25. Divide 12563 by $68\frac{11}{15}$.
26. Divide $85\frac{63}{73}$ by $18\frac{4}{9}$.
27. Divide $105\frac{2}{66}$ by $82\frac{6}{25}$.
28. Divide $\frac{3}{5}$ of $\frac{8}{15}$ by $6\frac{1}{2}$.
29. Divide $\frac{4}{7}$ of 16 by $\frac{2}{3}$ of $\frac{4}{5}$.
30. Divide $\frac{9}{10}$ of 30 by 19.
31. Divide $\frac{3}{7}$ of $\frac{2}{9}$ by 21.
32. Divide $\frac{8}{11}$ of $\frac{19}{20}$ by $\frac{2}{7}$ of 31.
33. Divide $\frac{4}{15}$ of $\frac{2}{21}$ by $\frac{1}{2}$ of $\frac{5}{8}$.

SECTION VII.

COMPOUND NUMBERS.

ART. **146.** Numbers which express things of the *same* kind or denomination, as 3 pears, 7 roses, 15 horses, are called *simple* numbers.

Numbers which express things of *different* kinds or denominations, as the divisions of *money*, *weight*, and *measure*, are called *compound* numbers. Thus 6 shillings, 7 pence; 5 pounds 2 ounces; 7 feet 3 inches, &c., are compound nnmbers.

OBS. Compound Numbers, by some late authors, are called Denominate Numbers.

QUEST.—146. What are simple numbers? What are compound numbers?

STERLING MONEY.

147. *Sterling Money* is the national currency of England.

4 farthings (*qr.* or *far.*)	make	1 penny, marked	*d.*
12 pence	"	1 shilling, "	*s.*
20 shillings	"	1 pound, or sovereign,	£.
21 shillings	"	1 guinea.	

OBS. 1. It is customary, at the present day, to express farthings in fractions of a penny. Thus, 1 qr. is written $\frac{1}{4}$ d.; 2 qrs., $\frac{1}{2}$ d.; 3 qrs. $\frac{3}{4}$ d.

2. The Pound Sterling is represented by a gold coin, called a *Sovereign*. According to *Act of Congress*, 1842, it is equal to 4 *Dollars* and 84 *cents*.

MENTAL EXERCISES.

1. In 5 pence, how many farthings?

Solution.—Since there are 4 farthings in 1 penny, in 5 pence there are 5 times as many; and 5 times 4 are 20. *Ans.* 20 farthings.

2. In 8 pence, how many farthings? In 10d.? In 12d.?
3. How many shillings are there in 3 pounds? In £5? In £8? In £10?
4. How many pence are there in 8 farthings?

Solution.—Since 4 farthings make 1 penny, 8 farthings will make as many pence, as 4 is contained times in 8; and 4 is contained in 8, 2 times. *Ans.* 2 pence.

5. How many pence in 12 farthings? In 15 qrs.? In 20 qrs.? In 25 qrs.? In 33 qrs.? In 36 qrs.?
6. How many shillings in 15 pence? In 24d.? In 30d.? In 36d.? In 60d.? In 68d.? In 75d.?
7. How many pounds in 25 shillings? In 30s.? In 40s.? In 65s.? In 80s.? In 89s.?

QUEST.—147. What is Sterling Money? Repeat the Table. *Obs.* How are farthings usually expressed? How is a pound sterling represented? What is its value in dollars and cents?

TROY WEIGHT.

Note.—Most children have very *erroneous* or *indistinct ideas* of the *weights* and *measures* in common use. It is, therefore, strongly recommended for teachers to illustrate them *practically*, by referring to some visible object of equal magnitude, or by exhibiting the ounce; the pound; the *linear* inch, foot, yard, and rod; also a *square* and *cubic* inch, foot, and yard; the pint, quart, gallon, peck, bushel, &c.

148. *Troy Weight* is used in weighing gold, silver, jewels, liquors, &c., and is generally adopted in philosophical experiments.

24 grains (*gr.*)	make	1 pennyweight,	marked	*pwt.*
20 pennyweights	"	1 ounce,	"	*oz.*
12 ounces	"	1 pound,	"	*lb.*

OBS. 1. The *standard* of Weights and Measures is different in different States of the Union. In 1834, the Government of the United States adopted a uniform standard, for the use of the several custom houses and other purposes.

2. The *standard unit* of *Weight* adopted by the Government, is the *Troy Pound* of the United States Mint, which is identical with the Imperial Troy pound of England, established by Act of Parliament, A. D. 1826.*

3. Troy Weight was formerly used in weighing articles of every kind. It was introduced into Europe about the time of the Crusades, from Cairo in Egypt, and was first adopted by the city *Troyes*, in France, from which some suppose its name was derived. Others think it was derived from *Troy-novant*, the former name of London.†

8. How many grains in 2 pennyweights? In 3 pwts? In 4 pwts?

9. How many pennyweights in 2 ounces? In 3 oz.? In 4 oz.? In 5 oz.?

10. How many ounces in 2 pounds? In 3 lbs.? In 4 lbs.? In 5 lbs.? In 6 lbs.? In 7 lbs.? In 10 lbs.?

QUEST.—148. In what is Troy Weight used? Repeat the Table. *Obs.* When was Troy Weight introduced into Europe? From what was its name derived? Do all the States have the same standard of weights and measures? What is the standard unit of weight adopted by the Government of the United States?

* Hassler on Weights and Measures, p. 10. Also, Reports of the Secretary of the Treasury, March 3, 1831; and June 20, 1832.

† Hind's Arithmetic, Art. 224. Also, North American Review, Vol. XLV.

AVOIRDUPOIS WEIGHT.

149. *Avoirdupois Weight* is used in weighing groceries and all coarse articles; as, sugar, tea, coffee, butter, cheese, flour, hay, &c., and all metals, except gold and silver.

16 drams (*dr.*)	make	1 ounce, marked	*oz.*
16 ounces	"	1 pound, "	*lb.*
25 pounds	"	1 quarter, "	*qr.*
4 quarters, or 100 lbs.	"	1 hundred weight,	*cwt.*
20 hundred weight	"	1 ton, marked	*T.*

OBS. 1. The Avoirdupois Pound of the *United States* is determined from the standard Troy Pound, and is in the ratio of 5760 to 7000;* that is,

1 pound Troy	contains	5760	grains.	
1 pound Avoirdupois	"	7000	"	Troy.
1 ounce "	"	$437\frac{1}{2}$	"	"
1 dram "	"	$27\frac{11}{32}$	"	"

2. The British Imperial Pound Avoirdupois is defined to be the weight of $27\frac{7274}{10000}$ cubic inches of distilled water, at the temperature of 61° Fahrenheit, when the barometer stands at 30°.†

3. *Gross* weight is the weight of goods with the boxes, casks, or bags which contain them.

Net weight is the weight of the goods only.

4. Formerly it was the custom to allow 112 pounds for a hundred weight, and 28 pounds for a quarter; but this practice has become nearly or quite obsolete. In buying and selling all articles of commerce estimated by weight, the laws of most of the States as well as general usage, call 100 pounds a hundred weight, and 25 pounds a quarter.

11. How many drams are there in 2 ounces? In 3 oz.? In 4 oz.? In 5 oz.?

12. How many ounces in 2 pounds? In 3 lbs.? In 4 lbs.? In 5 lbs.?

13. How many pounds in 2 quarters?

QUEST.—149. In what is Avoirdupois Weight used? Repeat the Table. Point to an object that weighs an ounce. A pound. *Obs.* How is the Avoirdupois pound of the United States determined? What is gross weight? Net weight? How many pounds were formerly allowed for a hundred weight? For a quarter?

* Reports of Secretary of Treasury, March 3, 1832: June, 20, 1832. Also, Congressional Documents of 1833.

† Hind's Arithmetic, Art. 223.

14. How many quarters in 2 hundred weight? In 3 cwt.? In 5 cwt.? In 6 cwt.?

APOTHECARIES' WEIGHT.

150. *Apothecaries' Weight* is used by apothecaries and physicians in *mixing* medicines.

20 grains (*gr.*)	make	1 scruple,	marked	*sc.*, or ℈.
3 scruples	"	1 dram,	"	*dr.*, or ʒ.
8 drams	"	1 ounce,	"	*oz.*, or ℥.
12 ounces	"	1 pound,	"	℔.

OBS. 1. The pound and ounce in this weight are the same, as the *Troy* pound and ounce; the other denominations are different.

2. Drugs and medicines are bought and sold by *avoirdupois* weight

15. In 2 scruples, how many grains? In 3 sc.?

16. In 3 drams, how many scruples? In 4 dr.? In 5 dr.? In 7 dr.?

17. In 2 pounds, how many ounces? In 3 ℔s.?

LONG MEASURE.

151. *Long Measure* is used in measuring distances or length only, without regard to breadth or depth.

12 inches (*in.*)	make	1 foot,	marked	*ft.*
3 feet	"	1 yard,	"	*yd.*
$5\frac{1}{2}$ yards, or $16\frac{1}{2}$ feet	"	1 rod, perch, or pole,	"	*r.* or *p.*
40 rods	"	1 furlong,	"	*fur.*
8 furlongs, or 320 rods	"	1 mile,	"	*m.*
3 miles	"	1 league,	"	*l*
60 geographical miles, or $69\frac{1}{2}$ statute miles	"	1 degree,	"	*deg.* or °.

360 degrees make a great circle, or the circumference of the earth.

Note.—4 inches make 1 hand; 9 inches, 1 span; 18 inches, 1 cubit; 6 feet, 1 fathom.

QUEST.—150. In what is Apothecaries' Weight used? Recite the Table. *Obs.* To what are the apothecaries' ounce and pound equal? How are drugs and medicines bought and sold? 151. In what is Long Measure used? Recite the Table.

OBS. 1. The *standard unit* of *Length* adopted by the General Government, is the *Yard* of 3 feet, or 36 inches, and is identical with the Imperial Yard of England. It is made of brass, and is determined from the scale of Troughton* at the temperature of 62° Fahrenheit.

2. Long measure is frequently called *linear*, or *lineal* measure. Formerly the inch was divided into 3 *barleycorns;* but the barleycorn is not employed as a measure at the present day. The inch is commonly divided either into *eighths* or *tenths;* sometimes, however, it is divided into *twelfths*, which are called *lines*.

19. In 3 feet, how many inches? In 3 feet and 4 in., how many inches? In 4 feet and 7 in.?

20. How many furlongs in 3 miles and 2 furlongs? How many in 4 m. and 5 fur.? In 6 m. and 7 fur.?

21. How many yards in 6 feet? In 12 ft.? In 16 ft.? In 23 ft.?

22. How many feet in 27 inches? In 36 in.? In 41 in.? In 64 in.?

23. How many yards in 12 feet? In 17 ft.? In 25 ft.? In 30 ft.?

CLOTH MEASURE.

152. *Cloth Measure* is used in measuring cloth, lace, and all kinds of goods which are bought and sold by the yard.

$2\frac{1}{4}$ inches (*in.*)	make	1 nail,	marked *na.*
4 nails, or 9 in.	"	1 quarter of a yard,	" *qr.*
4 quarters	"	1 yard,	" *yd.*
3 quarters, or $\frac{3}{4}$ of a yard	"	1 Flemish ell,	" *Fl. e.*
5 quarters, or $1\frac{1}{4}$ yard	"	1 English ell,	" *E. e.*
6 quarters, or $1\frac{1}{2}$ yards	"	1 French ell,	" *F. e.*

OBS. *Cloth* measure is a species of *long* measure. The yard is the same in both. Cloths, laces, &c., are bought and sold by the *linear* yard without regard to their width.

QUEST.—Draw a line an inch long upon the black-board. Draw one a foot, and another a yard long. How long is your desk? Your teacher's table? How wide? How long is the school room? How wide? *Obs.* What is the standard unit of Length adopted by Congress? What is Long Measure often called? 152. In what is Cloth Measure used? Repeat the Table? *Obs.* Of what is cloth measure a species? What is the kind of yard by which cloths, laces, &c. are bought and sold?

* A celebrated English artist.

24. In 3 quarters, how many nails? In 5 qrs.? In 7 qrs.?

25. How many quarters in 4 yards? In 6 yds.? In 8 yds.? In 11 yds.? In 15 yds.?

26. How many quarters in 5 Flemish ells? In 7 Fl. e.? In 10 Fl. e.?

27. How many quarters in 4 English ells? In 6 E. e.? In 9 E. e.?

28. In 10 quarters, how many yards? In 12 qrs.? In 15 qrs.? In 18 qrs.? In 24 qrs.? In 30 qrs.?

29. How many French ells in 12 quarters? In 24 qrs.? In 36 qrs.? In 45 qrs.?

SQUARE MEASURE.

153. *Square Measure* is used in measuring surfaces, or things whose length and breadth are considered without regard to heighth or depth; as, land, flooring, plastering, &c.

144 square inches (*sq. in.*)	make	1 square foot,	marked	*sq. ft.*
9 square feet	"	1 square yard,	"	*sq. yd.*
30¼ square yards, or 272¼ square feet	"	1 sq. rod, perch, or pole,	"	*sq. r.*
40 square rods	"	1 rood,	"	*R.*
4 roods, or 160 square rods	"	1 acre,	"	*A.*
640 acres	"	1 square mile,	"	*M.*

Square Inch.

OBS. 1. A *square* is a figure which has *four equal* sides, and all its angles *right angles*, as seen in the first diagram. Hence,

A *Square Inch* is a square, whose sides are each a *linear* inch in length.

A *Square Foot* is a square, whose sides are each a *linear* foot in length.

QUEST.—153. In what is Square Measure used? Recite the Table. *Obs.* What is a square? Draw a square upon the black-board. What is a square inch? A square foot?

A *Square Yard* is a square, whose sides are each a *linear* yard or three *linear* feet in length, and contains 9 *square* feet, as represented in the adjacent figure.

9 sq. ft. =1 *sq. yd.*

2. In measuring land, surveyors use a chain which is 4 rods long, and is divided into 100 links. Hence, 25 links make 1 rod, and $7\frac{92}{100}$ inches make 1 link.

This chain is commonly called *Gunter's Chain*, from the name of its inventor.

30. How many inches in 2 square feet?

31. How many square feet in 2 square yards? In 3 yds.? In 5 yds.? In 8 yds.?

32. How many square yards in 2 square rods?

33. How many roods in 80 square rods?

34. How many acres in 16 roods? In 25 R.? In 30 R.? In 48 R.?

CUBIC MEASURE.

154. *Cubic Measure* is used in measuring solid bodies, or things which have length, breadth, and thickness, such as timber, stone, boxes of goods, the capacity of rooms, ships, &c.

1728 cubic inches (*cu. in.*)	make	1 cubic foot,	marked	*cu. ft.*
27 cubic feet	"	1 cubic yard,	"	*cu. yd.*
40 feet of round, or 50 ft. of hewn timber	"	1 ton, or load,	"	*T.*
42 cubic feet	"	1 ton of shipping,	"	*T.*
16 cubic feet	"	1 foot of wood, or a cord foot,	"	*c. ft.*
8 cord feet, or 128 cubic feet	"	1 cord,	"	*C.*

OBS. 1. A pile of wood 8 feet long, 4 feet wide, and 4 feet high, contains 1 cord. For, $8 \times 4 \times 4 = 128$.

QUEST.—What is a square yard? Draw a square inch; a square foot; a square yard. 154. In what is Cubic Measure used? Recite the Table.

2. A *Cube* is a solid body bounded by *six equal squares.* It is often called a *hexaedron.* Hence,

A *Cubic Inch* is a cube, each of whose sides is a *square* inch, as represented by the adjoining figure.

Cubic Inch.

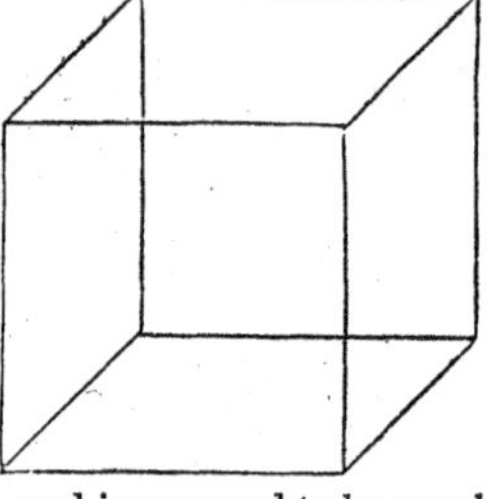

A *Cubic Foot* is a cube, each of whose sides is a *square* foot.

3. The *Cubic Ton* is chiefly used for estimating the cartage and transportation of timber. By a *ton* of *round* timber is meant, such a quantity of timber in its rough or natural state, as when hewn, will make 40 cubic feet, and is supposed to be equal in weight to 50 feet of hewn timber.

Note.—For an easy method of forming models of the cube and other regular solids, see Thomson's Legendre's Geometry, p. 222.

WINE MEASURE.

155. *Wine Measure* is used in measuring wine, alcohol, molasses, oil, and all other liquids except beer, ale, and milk.

4 gills (*gi.*)	make	1 pint,	marked	*pt.*
2 pints	"	1 quart,	"	*qt.*
4 quarts	"	1 gallon,	"	*gal.*
31½ gallons	"	1 barrel,	"	*bar.* or *bbl.*
42 gallons	"	1 tierce,	"	*tier.*
63 gallons, or 2 barrels	"	1 hogshead,	"	*hhd.*
2 hogsheads	"	1 pipe or butt,	"	*pi.*
2 pipes	"	1 tun,	"	*tun.*

OBS. The *standard unit* of *Liquid Measure* adopted by the Government, is the *English Wine Gallon* of 231 cubic inches, equal to $8\frac{339}{1000}$ pounds avoirdupois of distilled water at the maximum density, which is about 40° Fahrenheit.*

QUEST.—*Obs.* What is a cube? What is a cubic inch? A cubic foot? Draw a cubic inch upon the black-board. What is meant by a ton of round timber? 155. In what is Wine Measure used? Recite the Table. *Obs.* What is the standard unit of Liquid Measure of the United States? How many cubic inches in a wine gallon?

* Olmsted's Philosophy. Also, Hassler on Weights and Measures, p. 102.

35. In 4 pints, how many gills? In 6 pts.? In 12 pts.?

36. In 5 quarts, how many pints? In 6 qts.? In 9 qts.? In 13 qts.?

37. In 4 gallons, how many quarts? In 6 gals.? In 10 gals.? In 12 gals.?

38. In 3 hogsheads, how many barrels? In 7 hhds.?

39. How many quarts in 8 pints? In 11 pts.? In 15 pts.? In 18 pts.?

40. How many gallons in 16 quarts? In 22 qts.? In 32 qts.?

BEER MEASURE.

156. *Beer Measure* is used in measuring beer, ale, and milk.

2 pints (*pts.*)	make	1 quart,	marked	*qt.*
4 quarts	"	1 gallon,	"	*gal.*
36 gallons	"	1 barrel,	"	*bar.* or *bbl.*
$1\frac{1}{2}$ barrels or 54 gals.	"	1 hogshead,	"	*hhd.*

OBS. The beer gallon contains 282 cubic inches. In many places milk is measured by wine measure.

41. In 3 quarts, how many pints? In 7 qts.? In 15 qts.?

42. In 4 gallons, how many quarts? In 6 gallons? In 20?

DRY MEASURE.

157. *Dry Measure* is used in measuring grain, fruit, salt, &c.

2 pints (*pt.*)	make	1 quart,	marked	*qt.*
8 quarts	"	1 peck,	"	*pk.*
4 pecks or 32 qts.	"	1 bushel,	"	*bu.*
8 bushels	"	1 quarter	"	*qr.*
32 bushels	"	1 chaldron,	"	*ch.*

OBS. 1. In England, 36 bushels of coal make a chaldron.

QUEST.—156. In what is Beer Measure used? Recite the Table. 157. In what is Dry Measure used? Repeat the Table.

2. The *standard unit* of *Dry Measure* adopted by the Government, is the *Winchester Bushel* of $2150\frac{4}{10}$ cubic inches, equal to $77\frac{6274}{10000}$ pounds avoirdupois of distilled water, at the maximum density. The Winchester bushel is so called, because the standard measure was formerly kept at Winchester, England.

43. In 4 quarts, how many pints? In 6 qts.? In 10 qts.? In 14 qts.? In 18 qts.?

44. How many quarts are there in 3 pecks? In 5 pecks?

45. How many pecks in 3 bushels? In 5 bu.? In 10 bu.?

46. How many quarts in 6 pints? In 10 pts.? In 15 pts.?

47. How many bushels in 8 pecks? In 16 pks.? In 20 pks.?

TIME.

158. *Time* is naturally divided into *days* and *years;* the former are caused by the revolution of the Earth on its axis, the latter by its revolution round the sun.

60 seconds (*sec.*)	make	1 minute,	marked *min.*
60 minutes	"	1 hour,	" *hr.*
24 hours	"	1 day,	" *d.*
7 days	"	1 week,	" *wk.*
4 weeks	"	1 lunar month,	" *mo.*
12 calendar months, or 365 days and 6 hrs., (nearly)	"	1 civil year,	" *yr.*

OBS. 1. A *Solar* year is the exact time in which the earth revolves round the sun, and contains 365 days, 5 hours, 48 minutes, and 48 seconds.

2. Since the civil year contains 365 days and 6 hours, (nearly,) it is plain that in four years a whole day will be gained, and therefore every fourth year must have 366 days. This is called *Bissextile*, or *Leap Year.* The odd day is added to the month of February; in every Leap year, therefore, February has 29 days.

QUEST.—*Obs.* What is the standard unit of Dry Measure adopted by the government? 158. How is time naturally divided? Recite the Table. *Obs.* What is a solar year? How is leap year occasioned? To which month is the odd day added?

3. The following are the names of the 12 calendar months into which the civil year is divided, with the number of days in each:

January,	(Jan.)	*first*	month,	has	31	days.
February,	(Feb.)	*second*	"	"	28	"
March,	(Mar.)	*third*	"	"	31	"
April,	(Apr.)	*fourth*	"	"	30	"
May,	(May)	*fifth*	"	"	31	"
June,	(June)	*sixth*	"	"	30	"
July,	(July)	*seventh*	"	"	31	"
August,	(Aug.)	*eighth*	"	"	31	"
September,	(Sept.)	*ninth*	"	"	30	"
October,	(Oct.)	*tenth*	"	"	31	"
November,	(Nov.)	*eleventh*	"	"	30	"
December,	(Dec.)	*twelfth*	"	"	31	"

The number of days in each month may be easily remembered from the following lines:

"Thirty days hath September,
April, June, and November;
February twenty-eight alone,
All the rest have thirty-one;
Except in Leap year, then is the time,
When February has twenty-nine."

48. How many days in 3 weeks? In 4 wks.? In 5 wks.? In 7 wks.? In 9 wks.?

49. How many weeks in 14 days? In 21 days? In 32 days? In 35 days? In 40 days?

CIRCULAR MEASURE.

159. *Circular Measure* is applied to the divisions of the circle, and is used in reckoning latitude and longitude, and the motion of the heavenly bodies.

60 seconds (″)	make	1 minute,	marked	′
60 minutes	"	1 degree,	"	°
30 degrees	"	1 sign,	"	*s.*
12 signs, or 360°	"	1 circle,	"	*c.*

QUEST.—159. In what is Circular Measure used? Repeat the Table.

OBS. 1. The circumference of every circle is divided, or supposed to be divided, into 360 equal parts, called *degrees*, as in the subjoined figure.

2. Since a degree is $\frac{1}{360}$ part of the circumference of a circle, it is obvious that its length must depend on the size of the circle.

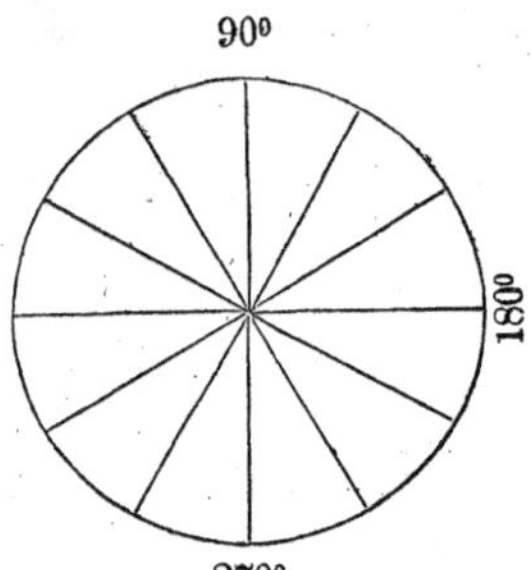

50. In 2 degrees, how many minutes? In 3 degrees?

51. In 2 signs, how many degrees? In 3 signs, how many? In 4 signs, how many?

52. How many signs in 60 degrees? In 90 degrees?

MISCELLANEOUS TABLE.

12 units	make	1 dozen, (*doz.*)
12 dozen, or 144	"	1 gross.
12 gross, or 1728	"	1 great gross.
20 units	"	1 score.
56 pounds	"	1 firkin of butter.
100 pounds	"	1 quintal of fish.
30 gallons	"	1 bar. of fish in Mass.
200 lbs. of shad or salmon	"	1 bar. in N. Y. and Conn.
196 pounds	"	1 bar. of flour.
200 pounds	"	1 bar. of pork.
14 pounds of iron, or lead	"	1 stone.
21½ stone	"	1 pig.
8 pigs	"	1 fother.
24 sheets of paper	"	1 quire.
20 quires	"	1 ream.

A sheet folded in two leaves, is called a *folio*.
A sheet folded in four leaves, is called a *quarto*, or 4*to*.
A sheet folded in eight leaves, is called an *octavo*, or 8*vo*.
A sheet folded in twelve leaves, is called a *duodecimo*, or 12*mo*.
A sheet folded in eighteen leaves, is called an 18*mo*.

OBS. Formerly 112 pounds were allowed for a quintal.

QUEST.—*Obs.* How is the circumference of every circle divided? On what does the length of a degree depend?

REDUCTION OF COMPOUND NUMBERS.

160. The process of changing *compound numbers* from one denomination into another, without altering their *value*, is called REDUCTION.

EXERCISES FOR THE SLATE.

Ex. 1. Reduce £3 to farthings.

Operation.

£3
20s. in 1 £.
———
60 shillings.
12d. in 1s.
———
720 pence.
4 far. in 1d.
———
Ans. 2880 far.

We first reduce the given pounds to shillings. This is done by multiplying them by 20, because 20s. make £1. (Art. 147.) That is, since there are 20s. in £1, in £3 there are 3 times 20s. or 60s. We now reduce the 60s. to pence, by multiplying them by 12, because 12d. make 1s. Finally, we reduce the 720d. to farthings, by multiplying them by 4, because 4 far. make 1d. The last product, 2880 far., is the answer; that is, £3=2880 far.

2. Reduce £2, 3s. 6d. and 2 far. to farthings.

Operation.

£.	*s.*	*d.*	*far.*
2	3	6	2

20s. in £1.
———
43 shillings.
12d. in 1s.
———
522 pence.
4 far. in 1d.
———
2090 far. *Ans.*

In this example there are shillings, pence, and farthings. Hence, when the pounds are reduced to shillings, the given shillings (3) must be added mentally to the product. In like manner when the shillings are reduced to pence, the given pence (6) must be added; and when the pence are reduced to farthings, the given farthings (2) must be added.

OBS. In these examples it is required to reduce higher denominations to lower; as pounds to shillings, shillings to pence, &c. This is done by *successive multiplications.*

QUEST.—160. What is Reduction? How are pounds reduced to shillings? Why multiply by 20? How are shillings reduced to pence? Why? How, pence to farthings? Why?

160. *a.* It often happens that we wish to reduce lower denominations to higher, as farthings to pence, pence to shillings, and shillings to pounds. Thus,

3. In 2880 farthings, how many pounds?

Operation.

4)2880 far.
12)720d.
20)60s.
£3 *Ans.*

First, we reduce the given farthings to pence, which is the next higher denomination. This is done by dividing them by 4. For, since 4 far. make 1d., (Art. 147,) in 2880 far. there are as many pence, as 4 is contained times in 2880; and 4 is contained in 2880, 720 times. We now reduce the 720 pence to shillings, by dividing them by 12, because 12d. make 1s. Finally, we reduce the shillings (60) to pounds, by dividing by 20, because 20s. make £1. Thus, 2880 far. =£3, which is the answer required.

4. How many pounds in 2090 farthings?

Operation.

4)2090 far.
12)522d. 2 far. over.
20)43s. 6d. over.
£2, 3s, over.
Ans. £2, 3s. 6d. 2 far.

In dividing by 4 there is a remainder of 2 far.; in dividing by 12, there is a remainder of 6d.; in dividing by 20, the quotient is £2 and 3s. over. The answer, therefore, is £2, 3s. 6d. 2 far. That is, 2090 far.=£2, 3s. 6d. 2 far.

OBS. 1. The last two examples are exactly the reverse of the first two; that is, lower denominations are required to be reduced to higher, which is done by *successive divisions.*

2. Reducing compound numbers to *lower* denominations is usually called *Reduction Descending;* reducing them to *higher* denominations, *Reduction Ascending.* The former employs multiplication; the latter division. They mutually prove each other.

QUEST.—Ex. 3. How are farthings reduced to pence? Why divide by 4? How reduce pence to shillings? Why? How shillings to pounds? Why? *Obs.* What is reducing compound numbers to lower denominations usually called? To higher denominations? Which of the fundamental rules is employed by the former? Which by the latter?

161. From the preceding illustrations we derive the following

GENERAL RULE FOR REDUCTION.

I. To reduce compound Nos. to lower denominations.

Multiply the highest denomination given, by that number which it takes of the next lower denomination to make ONE *of this higher; to the product, add the number expressed in this lower denomination in the given example. Proceed in this manner with each successive denomination, till you come to the one required.*

II. To reduce compound Nos. to higher denominations.

Divide the given denomination by that number which it takes of this denomination to make ONE *of the next higher. Proceed in this manner with each successive denomination, till you come to the one required. The last quotient, with the several remainders, will be the answer sought.*

162. PROOF.—*Reverse the operation; that is, reduce back the answer to the original denominations, and if the result correspond with the numbers given, the work is right.*

OBS. Each remainder is of the same denomination as the dividend from which it arose. (Art. 66. Obs. 2.)

STERLING MONEY. (ART. 147.)

5. In £35, 4s. 6d. how many pence?

	Operation.			*Proof.*
	£	s.	d.	12)8454 pence.
	35	4	6	20)704s. 6d.
	20			£35, 4s. 6d.
	704			
	12			
Ans.	8454d.			

QUEST.—161. How are compound numbers reduced to lower denominations? How reduced to higher denominations? 162. How is Reduction proved? *Obs.* Of what denomination is each remainder?

6. In 57600 farthings, how many pounds?

Operation.	*Proof.*
4)57600 far.	£60
12)14400 d.	20
20)1200 s.	1200 s.
£60 *Ans.*	12
	14400 d.
	4
	57600 far.

7. In £43, 12s., how many shillings?
8. In 17 shillings, how many farthings?
9. In 1176 pence, how many pounds?
10. In 12356 farthings, how many shillings?
11. In 175 pounds, how many farthings?
12. In £84, 16s. 7½d., how many farthings?
13. In 25256 pence, how many pounds?
14. In 56237 farthings, how many pounds?
15. In £25, 9s. 7½d., how many farthings?

TROY WEIGHT. (ART. 148.)

16. In 11 lbs., how many pennyweights? *Ans.* 2640 pwts.
17. In 15 ounces, how many grains?
18. In 10 lbs. 5 oz. 6 pwts., how many grains?
19. In 512 pwts., how many pounds?
20. In 2156 grains, how many ounces?
21. In 35210 grains, how many pounds?

AVOIRDUPOIS WEIGHT. (ART. 149.)

22. Reduce 25 pounds to drams. *Ans.* 6400 drams.
23. Reduce 36 cwt. 2 qrs. to pounds.
24. Reduce 5 tons, 7 cwt. 15 lbs. to ounces.
25. Reduce 3 quarters, 15 lbs. 10 oz. to drams.
26. Reduce 875 ounces to pounds.
27. Reduce 1565 pounds to hundred weight.
28. Reduce 1728 drams to pounds.

29. Reduce 5672 ounces to hundred weight.
30. Reduce 15285 pounds to tons.
31. Reduce 26720 drams to hundred weight.

APOTHECARIES' WEIGHT. (ART. 150.)

32. How many drams are there in 70 pounds?
Ans. 6720 drams.
33. How many scruples in 156 pounds?
34. How many ounces in 726 scruples?
35. How many pounds in 1260 drams?

LONG MEASURE. (ART. 151.)

36. In 96 rods, how many feet?

2)96
$5\frac{1}{2}$ yds. in 1 r.
480
48
528 yards.
3 ft. in 1 yd.
Ans. 1584 feet.

We first multiply by 5, then by $\frac{1}{2}$, and unite the two results. (Art. 134. *a.*) But to multiply by $\frac{1}{2}$, we take half of the multiplicand *once.* (Art. 132.)

37. In 45 furlongs, how many inches?
38. In 1584 feet, how many rods?

3)1584
528
2
11)1056
Ans. 96

We first reduce the feet to yards, by dividing by 3; next, reduce the yards to rods, by dividing by $5\frac{1}{2}$. But to divide by $5\frac{1}{2}$, we reduce it to halves, and also reduce the dividend (528 yds.) to halves, then divide 1056 by 11. (Art. 139, I.)

39. In 1728 inches, how many rods?
40. In 26400 feet, how many miles?
41. In 25 leagues, how many inches?
42. In 40 leagues, 6 furlongs, 2 in., how many inches?
43. In 750324 inches, how many miles?
44. How many inches in the circumference of the earth?

CLOTH MEASURE. (ART. 152.)

45. How many quarters in 45 yards?
46. How many nails in 53 Flemish ells?
47. How many nails in 81 English ells?
48. Reduce 563 quarters to yards.
49. Reduce 1824 nails to French ells.
50. Reduce 5208 nails to English ells.

SQUARE MEASURE. (ART. 153.)

51. In 1766 square rods and 19 yards, how many feet?
52. In 56 acres and 3 roods, how many square feet?
53. In 1275 square miles, how many acres?
54. How many square rods in 25640 feet?
55. How many acres in 1865 roods?
56. How many acres in 2118165½ yards?

163. The *area* of a floor, a piece of land, or any surface which has *four sides* and *four right angles*, is found by *multiplying its length and breadth together.*

Note.—The *area* of a figure is the *superficial contents* or *space* contained within the line or lines, by which the figure is bounded. It is reckoned in *square* inches, feet, yards, rods, &c.

57. How many square feet are there in a table which is 4 feet long and 3 feet wide?

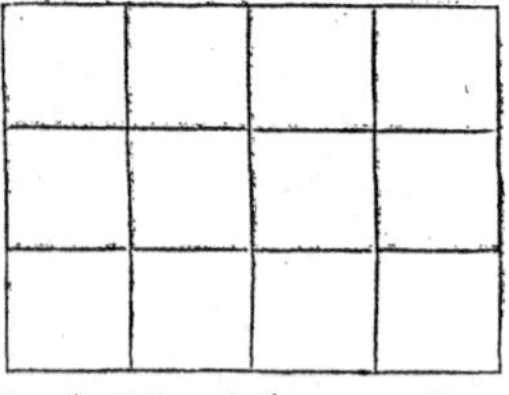

Suggestion.—Let the given table be represented by the subjoined figure, the length of which is divided into 4 equal parts, and the breadth into 3 equal parts, which we will call *linear feet.* Now it is plain that the table will contain as many sq. feet as there are squares in the given figure. But

QUEST.—163. How do you find the area or superficial contents of a surface having four sides and four right angles? *Note.* What is meant by the term area? How is it reckoned? *Obs.* What is a figure which has four sides and four right angles, called?

the number of squares in the figure is equal to the number of equal parts (linear feet) which its length contains, repeated as many times as there are equal parts (linear feet) in its breadth; that is, equal to 4×3, or 12. The table therefore contains 12 square feet.

OBS. A figure which has four sides and four right angles, like the preceding, is called a *Rectangle*, or *Parallelogram.*

58. What is the area of a garden, which is 8 rods long and 5 rods wide? *Ans.* 40 square rods.

59. How many square feet in a floor, 18 feet long and 17 feet wide?

60. How many square yards in a ceiling, 20 feet long and 18 feet wide?

61. What is the area of a field, which is 36 rods long and 25 rods wide?

62. How many acres are there in a piece of land, 80 rods long and 48 rods wide?

CUBIC MEASURE. (ART. 154.)

63. In 75 cubic feet, how many inches?

64. In 37 tons of round timber, how many inches?

65. In 28124 cubic feet, how many tons of hewn timber?

66. In 16568 cubic feet of wood, how many cords?

67. In 65 cords of wood, how many cubic feet?

164. The *solidity*, or *cubical contents* of boxes of goods, piles of wood, &c., are found by *multiplying* the *length*, *breadth*, and *thickness together*.

68. How many cubic inches are there in a box, whose length is 30 inches, its breadth 18, and its depth 15 inches? *Ans.* 8100 cu. in.

69. How many cubic inches in a block of marble, 43 inches long, 18 inches broad, and 12 inches thick?

QUEST.—164. How are the cubical contents of a box of goods, a pile of wood, &c., found?

70. How many cubic feet in a room, 16 feet long, 15 feet wide, and 9 feet high?

71. How many cubic feet in a load of wood, 8 feet long, 4 feet wide, and 3½ feet high?

72. How many cubic feet in a pile of wood, 16 feet long, 6 feet wide, and 5 feet high? How many cords?

73. How many cords of wood in a pile, 140 feet long 4½ feet wide, and 6½ feet high?

WINE MEASURE. (ART. 155.)

74. In 4624 gills, how many gallons?
75. In 24260 quarts, how many hogsheads?
76. How many pints in 15 hogsheads, and 20 gallons?
77. How many gills in 40 barrels?

BEER MEASURE. (ART. 156.)

78. How many barrels of beer in 5000 pints?
79. How many hogsheads in 7800 quarts?
80. How many quarts in 25 hogsheads, and 7 gallons?
81. How many pints in 110 gallons, 3 qts. and 1 pt.?

DRY MEASURE. (ART. 157.)

82. Reduce 536 bushels, and 3 pecks to quarts.
83. Reduce 821 chaldrons to pints.
84. Reduce 1728 pints to pecks.
85. Reduce 85600 quarts to bushels.

TIME. (ART. 158.)

86. In 15 days, 6 hours, and 9 min., how many seconds?

87. In 365 days and 6 hours, how many minutes?

88. How many seconds in a solar year?

89. Allowing 365d. 6h. to a year, how many minutes has a person lived who is 21 years old?

90. How many hours in 568240 seconds?

91. How many weeks in 8568456 minutes?

92. How many lunar months in 6925600 hours?

93. How many years in 56857200 hours?

94. How many years in 1000000000 seconds?

CIRCULAR MEASURE. (ART. 159.)

95. In 75 degrees, how many seconds?
96. In 8 signs, and 15 degrees, how many minutes?
97. In 12 signs, how many seconds?
98. In 86860 seconds, how many degrees?
99. In 567800 minutes, how many signs?
100. In 25000000 seconds, how many signs?

COMPOUND NUMBERS REDUCED TO FRACTIONS.

Ex. 1. Change 7s. 6d. to the fraction of a pound.

7s. 6d.	£1 or 20s.	
12	12	
Numerator 90d.	Denominator 240d.	

Ans. £$\frac{90}{240}$=$\frac{9}{24}$, or £$\frac{3}{8}$.

Reducing the 7s. 6d. to pence, (Art. 161, I,) we have 90d., which is the numerator of the fraction. Then reducing £1 to the same denomination as the numerator, we have 240d., which is the denominator. Consequently $\frac{90}{240}$ is the fraction required.

But $\frac{90}{240}$ may be reduced to lower terms. Thus $\frac{90}{240}$= $\frac{9}{24}$, or $\frac{3}{8}$. (Art. 120.) Hence,

165. To reduce a compound number to a common fraction of a higher denomination.

First reduce the given compound number to the lowest denomination mentioned for the numerator; then reduce a UNIT *of the denomination of the required fraction to the same denomination as the numerator, and the result will be the denominator.* (Art. 161.)

OBS. When the given number contains but one denomination, it of course requires no reduction.

2. Reduce 3s. 7d. 2 far. to the fraction of £1.

Ans. £$\frac{174}{960}$. or £$\frac{87}{480}$.

3. Reduce 9d. 3 far. to the fraction of 1s.
4. What part of a bushel is 3 pecks and 5 qts.?

QUEST.—165. How is a compound number reduced to a common fraction?

5. What part of a peck is 5 qts. and 1 pt.?
6. What part of a gallon is 3 qts. 1 pt. and 3 gills?
7. What part of 1 gallon is 1 pt and 1 gill?
8. What part of 1 hogshead is 15 gals. and 3 qts.?
9. What part of 1 ton is 5 cwt. and 2 qrs.?
10. What part of 1 hundred weight is 2 qrs. and 7 lbs.?
11. What part of 1 quarter is 1 lb. and 5 oz.?
12. What part of 1 mile is 45 rods?
13. What part of 1 mile is 10 fur. and 35 rods?
14. What part of 1 league is 1 m. 1 fur. and 1 r.?
15. What part of 1 yard is 2 qrs. and 3 nails?
16. What part of £1 is 1 penny? *Ans.* £$\frac{1}{240}$.
17. What part of £1 is $\frac{2}{3}$ of a penny?

Note.—The lowest denomination mentioned in this example, is *thirds* of a penny. Hence, £1 must be reduced to thirds of a penny for the denominator, and 2, the given number of thirds will be the numerator. *Ans.* £$\frac{2}{720}$, or £$\frac{1}{360}$.

18. What part of £1 is $5\frac{3}{4}$ shillings? *Ans.* £$\frac{23}{80}$.
19. What part of 1 day is $2\frac{1}{2}$ hours?
20. What part of 1 day is 4 h. and $8\frac{1}{2}$ min.?
21. What part of 1 hour is 3 min. and 40 sec.?
22. What part of 1 hour is $15\frac{3}{4}$ sec.?
23. What part of 1 pound is $\frac{3}{4}$ of an ounce?
24. What part of 1 ton is $\frac{5}{8}$ of a pound?
25. What part of 1 hogshead is $\frac{3}{8}$ of a gallon?
26. What part of 1 gallon is $\frac{2}{5}$ of a gill?

FRACTIONAL COMPOUND NUMBERS

REDUCED TO WHOLE NUMBERS OF LOWER DENOMINATIONS.

Ex. 1. Reduce $\frac{3}{8}$ of £1 to shillings and pence.

Operation.	
3 eighths £.	Multiply the numerator by 20, to reduce it to shillings, as in reduction. (Art 161, I.) £$\frac{3}{8}$×20s.=$\frac{60}{8}$s. or 7s. and 4 remainder. Again, multiplying the remainder 4 by 12, we have 48; and 48÷8=6d. The quotients, 7s. and 6d. are the answer required. Hence,
20	
8)60	
shil. 7, and 4 rem.	
12	
8)48	
Pence 6. *Ans.* 7s. 6d.	

166. To reduce fractional compound numbers to whole numbers.

First reduce the given numerator to the next lower denomination; (Art. 161, I;) *then divide the product by the denominator, and the quotient will be an integer of the next lower denomination. Proceed in like manner with the remainder, and the several quotients will be the whole numbers required.*

2. Reduce $\frac{3}{5}$ of £1 to shillings. *Ans.* 12s.
3. How many shillings and pence in £$\frac{1}{8}$?
4. How many shillings, &c., in £$\frac{5}{7}$?
5. In $\frac{4}{5}$ of 1 week, how many days, hours, &c. ?
6. In $\frac{7}{18}$ of 1 day, how many hours, minutes, &c. ?
7. Change $\frac{2}{5}$ of 1 league to miles, &c.
8. Change $\frac{4}{9}$ of 1 mile to furlongs, &c.
9. Reduce $\frac{7}{16}$ of 1 hundred weight to quarters, &c.
10. In $\frac{3}{7}$ of 1 ton, how many hundred weight, &c. ?
11. In $\frac{2}{3}$ of 1 bushel, how many pecks, quarts, &c. ?
12. In $\frac{4}{15}$ of 1 peck, how many quarts, &c. ?
13. Reduce $\frac{2}{67}$ of £1 to shillings.

Suggestion.—Since the numerator, when reduced to the denomination required, cannot be divided by the denominator, the division must be represented. *Ans.* $\frac{40}{67}$s.

Note.—This, in effect, is reducing $\frac{2}{67}$ of £1 to the fraction of a shilling.

14. Reduce $\frac{1}{365}$ of £1 to pence. *Ans.* $\frac{240}{365}$d.

167. From the last two examples it is manifest, that a fraction of a higher denomination may be changed to a fraction of a lower denomination, *by reducing the given numerator to the denomination of the required fraction, and placing the result over the given denominator.*

QUEST.—166. How are fractional compound numbers reduced to whole ones? 167. How is a fraction of a higher denomination changed to a fraction of a lower denomination?

15. Reduce $\frac{3}{127}$ of £1 to the fraction of a shilling. *Ans.* $\frac{60}{127}$s.

16. Reduce $\frac{7}{180}$ of 1 week to the fraction of a day.

17. Change $\frac{4}{1184}$ of 1 mile to the fraction of a rod.

18. Change $\frac{2}{100}$ of 1 rod to the fraction of a foot.

19. Change $\frac{10}{1787}$ of 1 yard to the fraction of a nail.

20. Change $\frac{1}{100000}$ of 1 ton to the fraction of a pound.

ADDITION OF COMPOUND NUMBERS.

1. What is the sum of £4, 9s. 6d. 2 far.; £3, 12s. 8d. 3 far.; and £8, 6s. 9d. 1 far.?

Operation.

£	*s.*	*d.*	*far.*	
4″	9″	6″	2	
3″	12″	8″	3	
8″	6″	9″	1	
16″	9″	0″	2	*Ans.*

Having placed the farthings under farthings, pence under pence, &c., we add the column of farthings together, as in simple addition, and find the sum is 6, which is equal to 1d. and 2 far. over. Set the 2 far. under the column of farthings, and carry the 1d. to the column of pence. The sum of the pence is 24, which is equal to 2s. and nothing over. Place a cipher under the column of pence, and carry the 2s. to the column of shillings. The sum of the shillings is 29, which is equal to £1 and 9s. over. Write the 9s. under the column of shillings, and carry the £1 to the column of pounds. The sum of the pounds is 16, the whole of which is set down in the same manner, as the left hand column in simple addition. (Art. 25.) The answer is £16, 9s. 0d. 2 far.

168. Hence, we derive the following general

RULE FOR ADDING COMPOUND NUMBERS.

1. *Write the numbers so that the same denominations shall stand under each other.*

QUEST.—168. How do you write compound numbers for addition? Which denomination do you add first? When the sum of any column is found, what is to be done with it?

II. *Beginning with the lowest denomination, find the sum of each column separately, and divide it by that number which it requires of the column added, to make* ONE *of the next higher denomination. Set the remainder under the column, and carry the quotient to the next column.*

III. *Proceed in this manner with all the other denominations except the highest, whose entire sum is set down as in simple addition.* (Art. 29.)

PROOF.—*The proof is the same as in Simple Addition.*

OBS. 1. *Fractional* compound numbers should be reduced to whole numbers of lower denominations, then added as above. (Art. 166.)

2. The process of adding numbers of different denominations, is called *Compound Addition.* It is the same as Simple Addition, except in the method of *carrying* from one denomination to another.

2. What is the sum of £$\frac{4}{5}$, $\frac{2}{3}$s. $\frac{7}{8}$d., and £$\frac{3}{8}$, $\frac{1}{6}$s.?

Ans. £1, 4s. 4d. $3\frac{1}{3}$f.

3.

£	*s.*	*d.*	*far.*
6	7	4	2
0	6	7	1
12	15	6	0

4.

£	*s.*	*d.*
10	15	8
16	11	0
25	18	9

5.

£	*s.*	*d.*
21	18	10
1	6	11
35	12	7

6.

lb.	*oz.*	*pwt.*	*gr.*
5	8	16	7
7	9	6	12
10	6	15	10
21	3	4	5

7.

oz.	*pwt.*	*gr.*
15	12	8
11	6	7
10	13	8
6	0	1

8.

lb.	*oz.*	*pwt.*
12	6	15
19	0	7
1	8	16
28	3	11

9. Add 7 lbs. 9 oz. 16 pwts. 10 grs.; 3 lbs. 10 oz. 8 pwts. 9 grs.; 8 lbs. 3 oz. 1 pwt. 4 grs.

10. A man bought a coach for £35, 12s.; a horse for £27, 8s. 10d.; a harness for £7, 16s. 11d.: what did the whole cost?

QUEST.—What is done with the last column? How prove the operation? *Obs.* How add fractional compound numbers? What is the process of adding compound numbers called? Does it differ from simple addition?

11. A merchant bought of one dairy-man 5 cwt. 11 lbs. 6 ounces of butter; of another, 3 cwt. 15 lbs. 9 oz.; of another, 7 cwt. 6 lbs. 10 oz.: how much did he buy of all?

12. A manufacturer bought of one man 73 lbs. of wool; of another, 96 lbs. 6 oz.; of another, 135 lbs. 11 oz.; of another, 320 lbs. 9 oz.; of another, 642 lbs. 3 oz.: how much wool did he buy?

13. A man sold to one customer 2 tons, 62 lbs. 10 oz. of hay; to another, 5 tons, 40 lbs. 12 oz.; to another, 3 tons, 75 lbs. 6 oz.: how much did he sell to all?

14. A man wove 7 yds. 3 qrs. 2 na. of cloth in one day; the next day, 6 yds. 1 qr. 3 na.; the next, 8 yds. 3 qrs. 1 na.; the next, 5 yds. 2 qrs. 3 na.: how much did he weave in all?

15. Bought several pieces of cotton; one contained 26 yds. 1 qr. 2 na.; another, 30 yds. 2 qrs.; another, 29½ yds. 3 na.; another, 32¼ yds. 1 na.: how many yards did they all contain?

16. A hotel keeper bought at one time, 15 bu. 2 pks. 3 qts. of oats; at another, 10 bu. 1 pk. 2 qts.; at another, 20½ bu. 6 qts.; at another, 18½ bu. 5 qts.: what was the amount of all his purchases?

17. Bought 4 loads of wheat; the first containing 23 bu. 3 pks. 5 qts.; the second, 20½ bu. 6 qts.; the third, 26¼ bu.; the fourth, 21¾ bu. 7 qts.: how many bushels did they all contain?

18. What is the sum of 16 m. 3 fur. 16 r.? 26 m. 1 fur. 33 r.; 10 m. 8 fur. 22 r.; 45 m. 7 fur. 20 r.?

19. A merchant bought 3 casks of oil; one held 2 hhds. 30 gals. 2 qts.; another, 3 hhds. 10 gals.; another, 1 hhd. 13 gals. 1 qt.: how much did they all hold?

20. Sold several lots of wine, in the following quantities; 1 pipe, 1 hhd. 21 gals. 2 qts. 1 pt.; 2 pipes, 11 gals. 3 qts. 1 pt.; 3 hhds. 15 gals. 2 qts.; 3 pipes, 10 gals. 2 qts. 1 pt.: how much was sold in all?

21. A mason plastered one room containing 45 square yards, 7 ft. 6 in.; another, 25 yds. 6 ft. 95 in.; another, 38 yds. 4 ft. 41 in.: what was the amount of plastering in all the rooms?

22. Sold 10 A. 35 r. 10 sq. ft. of land at one time; at another, 3 A. 10 r. 15 ft.; at another, 18 A. 16 r. 23 ft.: what was the amount of land sold?

23. A merchant received several boxes of goods; one contained 16 cu. ft. 61 in.; another, 25 ft. 81 in.; another 20 ft. 13 in.; another, 38 ft. 72 in.: how many cubic feet and inches did they all contain?

24. One pile of wood contains 10 C. 38 ft. 39 in.; another, 15 C. 56 ft. 73 in.; another, 30 C. 19 ft. 44 in.; another, 17 C. 84 ft. 21 in.: how much do they all contain?

SUBTRACTION OF COMPOUND NUMBERS.

Ex. 1. From £15, 7s. 6d. 3 far., subtract £6, 4s. 8d. 2 far.

Operation.

£	*s.*	*d.*	*far.*	
15 "	7 "	6 "	3	
6 "	4 "	8 "	2	
9 "	2 "	10 "	1	*Ans.*

Having placed the less number under the greater, with farthings under farthings, pence under pence, &c., we subtract 2 far. from 3 far., and set the remainder 1 far. under the column of farthings. But 8d. cannot be taken from 6d.; we therefore borrow 1 from the next higher denomination, which is shillings; and 1s. or 12d. added to the 6d. make 18d. And 8d. from 18d. leaves 10d. Since we borrowed, we must carry 1 to the next column, as in simple subtraction. 1 added to 4 makes 5; and 5 from 7, leaves 2. 6 from 15 leaves 9. *Ans.* £9, 2s. 10d. 1 far.

169. Hence, we derive the following general

RULE FOR SUBTRACTING COMPOUND NUMBERS.

I. *Write the less number under the greater, so that the same denominations may stand under each other.*

II. *Beginning with the lowest denomination, subtract the number in each denomination of the lower line from the number above it, and set the remainder below.*

QUEST.—169. How do you write compound numbers for subtraction? Where begin to subtract? When the number in the lower line is larger than that above it, what is to be done? How is the operation proved?

III. *When a number in any denomination of the lower line is larger than the number above it, borrow one of the next higher denomination and add it to the number in the upper line. Subtract as before, and carry* 1 *to the next denomination in the lower line, as in subtraction of simple numbers.* (Art. 40.)

PROOF.—*The proof is the same as in Simple Subtraction.*

OBS. 1. *Fractional* compound numbers should be reduced to whole numbers of lower denominations, then subtracted as above. (Art. 166.)
2. The process of finding the difference between numbers of different denominations is called *Compound Subtraction.* It does not differ from Simple Subtraction, except in the mode of *borrowing.*

2. From £$\frac{4}{5}$, $\frac{3}{4}$s., take £$\frac{3}{8}$, $\frac{1}{6}$s.

Solution.—£$\frac{4}{5}$, $\frac{3}{4}$s.=16s. 9d., and £$\frac{3}{8}$, $\frac{1}{6}$s.=7s. 8d.
Ans. 9s. 1d.

3. From £15, 16s. 10d. 3 far., take £7, 8s. 11d. 1 far.
4. From £56, 7s. 6d. 1 far., take £20, 3s. 10d. 3 far.

	5.	6.
From	16T. 10cwt. 3qrs. 7lbs.	125T. 7cwt. 2 qrs. 20lbs.
Take	8T. 5cwt. 1qr. 2lbs.	96T. 9cwt. 3 qrs. 12lbs.

	7.	8.
From	16gals. 3qts. 1pt. 2gi.	121hhds. 28gals. 1qt.
Take	7gals. 2qts. 0pt. 3gi.	63hhds. 21gals. 3qts.

9. Bought 2 silver pitchers, one weighing 2 lbs. 10 oz. 10 pwts. 7 grs.; the other, 2 lbs. 3 oz. 12 pwts. 5 grs.: what is the difference in their weight?

10. A merchant had 28 yds. 3 qrs. 2 na. of cloth, and sold 15 yds. 1 qr. 3 na.: how much had he left?

11. A lady bought 2 pieces of silk, one of which contained 19 yds. 2 qrs. 1 na.; the other, 15 yds. 3 qrs. 3 na.: what is the difference in the length?

12. From 25 m. 7 fur. 8 r. 12 ft. 6 in., take 16 m. 6 fur. 30 r. 4 ft. 8 in.

QUEST.—*Obs.* What is the process of subtracting compound numbers called? Does it differ from simple subtraction?

13. A man owning 95 A. 75 r. 67 sq. ft. of land, sold 40 A. 86 r. 29 ft.: how much had he left?

14. A farmer having bought 120 A. 3 R. 28 r. of land, divided it into two pastures, one of which contained 50 A. 2 R. 35 r.: how much did the other contain?

15. A tanner built two cubical vats, one containing 116 ft. 149 in., the other 245 ft. 73 in.: what is the difference between them?

16. A man having 65 C. 95 ft. 123 in. of wood in his shed, sold 16 C. 117 ft. 65 in.: how much had he left?

17. From 27 yrs. 8 mos. 3 wks. 4 ds. 13 hrs. 35 min.
Take 19 yrs. 5 mos. 6 wks. 5 ds. 21 hrs. 20 min.

18. What is the time from July 4th, 1840, to March 1st, 1845?

Operation.

Yr.	*mo.*	*d.*	
1845 "	3 "	1	
1840 "	7 "	4	
4 "	7 "	27	*Ans.*

March is the 3d month, and July the 7th. Since 4 d. cannot be taken from 1 d., we borrow 1 mo. or 30 d; then say, 4 from 31 leaves 27. 1 to carry to 7 makes 8, but 8 from 3 is impossible; we therefore borrow 1 yr. or 12 mos., and say, 8 from 15 leaves 7. 1 to carry to 0 is 1, and 1 from 5 leaves 4. Hence,

170. To find the time between two dates.

Write the earlier date under the later, placing the years on the left, the number of the month next, and the day of the month on the right, and subtract as before. (Art. 169.)

OBS. 1. The number of the month is easily determined by reckoning from January, the 1st mo., Feb. the 2d, &c. (Art. 158. Obs. 3.)

2. In finding the time between two dates, and in casting interest, 30 days are considered a month, and 12 months a year.

19. What is the time from Oct. 15th, 1835, to March 10th, 1842?

20. The Independence of the United States was de-

QUEST.—170. How do you find the time between two dates? *Obs.* In finding time between two dates, and in casting interest, how many days are considered a month? How many months a year?

clared July 4th, 1776. How much time had elapsed on the 25th of Aug. 1845?

21. A note dated Oct. 2d, 1840, was paid Dec. 25th, 1843: how long was it from its date to its payment?

22. A ship sailed on a whaling voyage, Aug. 25th, 1840, and returned April 15th, 1844: how long was her voyage?

MULTIPLICATION OF COMPOUND NUMBERS.

Ex. 1. What will 5 yards of broadcloth cost, at £2, 3s. 6d. 3 far. per yard?

Suggestion.—If 1 yard costs £2, 3s. 6d. 3 far., 5 yards will cost 5 times as much.

Operation.

£	s.	d.	far.	
2 "	3 "	6 "	3	
			5	
10 "	17 "	9 "	3	*Ans.*

Beginning with the lowest denomination, we say, 5 times 3 far. are 15 far.; now 15 far. are equal to 3d. and 3 far. over. Set the 3 far. under the denomination multiplied, and carry the 3d. to the next product. 5 times 6d. are 30d. and 3d. make 33d., equal to 2s. and 9d. Set the 9d. under the pence, and carry the 2s. to the next product. 5 times 3s. are 15s. and 2s. make 17s. As the product 17s. does not make *one* in the next denomination, we set it under the column multiplied. Finally, 5 times £2 are £10. The answer is £10, 17s. 9d. 3 far.

171. Hence, we deduce the following general

RULE FOR MULTIPLYING COMPOUND NUMBERS.

Multiply each denomination separately, beginning with the lowest, and divide each product by that number which it takes of the denomination multiplied, to make one of the

---

QUEST.—171. Where do you begin to multiply a compound number? What is done with each product? *Obs.* When the multiplier is a composite number, how proceed? What is the process of multiplying different denominations, called?

next higher; set down the remainder, and carry the quotient to the next product, as in addition of compound numbers. (Art. 168.)

OBS. 1. When the multiplier is a composite number, it is advisable to multiply first by one factor and that product by the other. (Art. 57.)
2. The process of multiplying different denominations is called *Compound Multiplication.*

2. Multiply, £5, 7s. 8d. 2 far. by 18.

Operation.

£	*s.*	*d.*	*far.*
5 "	7 "	8 "	2
			6
32 "	6 "	3 "	0
			3
96 "	18 "	9 "	0 *Ans.*

Multiply by the factors of 18, which are 6 and 3.

3. What will 5 horses cost, at £25, 10s. 6d. apiece?
4. A company of 6 persons agreed to pay £31, 5s. 8d. apiece for their passage from Hamburg to New York: what was the expense of their passage?
5. What cost 9 yards of cloth, at 18s. $9\frac{3}{4}$d. per yard?
6. What cost 6 pipes of wine, at £9, 7s. $8\frac{1}{2}$d. apiece?
7. What cost 8 cows, at £5, 10s. 6d. apiece?
8. In a solar year there are 365 days, 5 hrs. 48 min. 48 sec.: how many days, hours, &c., has a person lived who is 21 years old?
9. Bought 10 silver cups, each weighing 3 oz. 15 pwts. 10 grs.: what is the weight of the whole?
10. What is the weight of 72 silver dollars, each weighing 17 pwts. 8 grs.?
11. Bought 7 loads of hay, each weighing 1 T. 3 cwt. 3 qrs. 12 lbs.: what is the weight of the whole?
12. What is the weight of 20 hogsheads of molasses, each weighing 5 cwt. 3 qrs. 17 lbs. 10 oz.?
13. A man bought 9 oxen, weighing 1123 lbs. 15 oz. apiece: what was the weight of the whole?
14. A grocer bought 11 casks of brandy, each containing 54 gals. 3 qts. 1 pt. 2 gills: how much did they all contain?

15. If a stage-coach goes at the rate of 5 m. 2 fur. 30 r. per hour, how far will it go in 10 hours?

16. If a Railroad car goes 21 m. 2 fur. 10 r. per hour, how far will it go in 10 hours?

17. Bought 12 pieces of broadcloth, each containing 27 yds. 1 qr. 2 na.: how many yards did all contain?

18. If a man mows 3 A. 35 sq. r. per day, how many acres can he mow in 30 days?

19. How many square yards of plastering will a house which has 9 rooms require, allowing 75 yds. 18 ft. to a room?

20. A man bought 15 loads of wood, each containing 1 C. 33 ft.: how many cords did he buy?

21. A miller constructed 7 cubical bins for grain, each containing 216 feet 152 in.: what was the contents of the whole?

22. If a ship sails 2° 25′ 10″ per day, how far will she sail in 20 days?

23. Multiply 56° 42′ 11″ by 32.

24. If a brewer sells 33 gals. 2 qts. 1 pt. of beer a day, how much will he sell in 24 days?

25. If a milk-man sells 40 gals. 3 qts. 1 pt. of milk per day, how much will he sell in 60 days?

26. What cost 82 tons of iron, at £4, 15s. 6½d. per ton?

27. If 1 acre produce 33 bu. 2 pks. 5 qts. of wheat, how much will 100 acres produce?

28. If 1 suit of clothes requires 9 yds. 3 qrs. 2 na., how much will 500 suits require?

29. If 1 mile of Railroad requires 60 T. 5 cwt. 9 lbs. of iron, how much will 50 miles require?

30. How much wheat will it require to make 1000 barrels of flour, allowing 4 bu. 2 pks. 6 qts. to a barrel?

DIVISION OF COMPOUND NUMBERS.

Ex. 1. Divide £17, 6s. 9d. by 4.

Operation.

	£	s.	d.	far.
4)	17 "	6 "	9 "	0
	4 "	6 "	8 "	1

Beginning with the pounds we find 4 is contained in £17, 4 times and 1 over. Set the 4 under the pounds, and reduce the remainder £1 to shillings, which added to the 6s., make 26s. 4 in 26s., 6 times and 2s. over. Set the 6 under the shillings, and reduce the remainder 2s. to pence, which added to the 9d. make 33d. 4 in 33d., 8 times and 1d. over. Set the 8 under the pence, reduce the 1d. to farthings, and divide as before. *Ans.* £4, 6s. 8d. 1 far.

173. Hence, we deduce the following general

RULE FOR DIVIDING COMPOUND NUMBERS.

Beginning with the highest denomination, and divide each separately. Reduce the remainder, if any, to the next lower denomination, to which add the number of that denomination contained in the given example, and divide the sum as before. Proceed in this manner through all the denominations.

OBS. 1. Each partial quotient will be of the same denomination, as that part of the dividend from which it arose.

2. When the divisor exceeds 12, and is a composite number, it is advisable to divide first by one factor and that quotient by the other. (Art. 78.) If the divisor exceeds 12, but is not a composite number, long division may be employed. (Art. 77.)

3. The process of dividing different denominations, is called *Compound Division.*

QUEST.—173. Where do you begin to divide a compound number? What is done with the remainder? *Obs.* Of what denomination is each partial quotient? When the divisor is a composite number, how proceed? What is the process of dividing different denominations, called?

2. Divide £274, 4s. 6d. by 21.

Operation.

```
   £    s.   d.
3)274 " 4 " 6     Divide by the factors of 21.
 7)91 " 8 " 2
   13 " 1 " 2 Ans.
```

3. Divide £635, 17s. by 31.

Operation.

```
      £.    s.   £.   s.
31)635, 17 (20, 10 7/31.
   620
    15 rem.
    20
   317
   310
     7 rem.
```

The remainder £15, is reduced to shillings, to which we add the given shillings, making 317, and divide as before. The remainder 7s. may be reduced to pence and divided again if necessary.

4. Divide £7, 8s. 2d. by 3.
5. Divide £35, 10s. 8d. 3 far. by 6.
6. Divide £42, 17s. 3d. 2 far. by 8.
7. A man bought 5 cows for £23, 16s. 8d.: how much did they cost apiece?
8. A merchant sold 10 rolls of carpeting for £62, 12s. 9d.: how much was that per roll?
9. Paid £25, 10s. $6\frac{1}{4}$d. for 12 yards of broadcloth: what was that per yard?
10. A silver-smith melted up 2 lbs. 8 oz. 10 pwts. of silver, which he made into 6 spoons: what was the weight of each spoon?
11. The weight of 8 silver tankards is 10 lbs. 5 oz. 7 pwts. 6 grs.: what is the weight of each?
12. If 8 persons consume 85 lbs. 12 oz. of meat in a month, how much is that apiece?
13. A dairy-woman packed 95 lbs. 8 oz. of butter in 10 boxes: how much did each box contain?
16. A tailor had 76 yds. 2 qrs. 3 na. of cloth, out of which he made 8 cloaks: how much did each cloak contain?

17. A man traveled 50 m. and 32 r. in 11 hours: at what rate did he travel per hour?

18. A man had 285 bu. 3 pks. 6 qts. of grain, which he wished to carry to market in 15 equal loads: how much must he carry at a load?

19. A man had 80 A. 45 r. of land, which he laid out into 36 equal lots: how much did each lot contain?

SECTION VIII.

DECIMAL FRACTIONS.

ART. **175.** When a number or thing is divided into equal parts, those parts we have seen are called ***Fractions.*** (Art. 105.) We have also seen that these equal parts take their *name* or denomination from the *number* of parts into which the integer or thing is divided. (Art. 103.) Thus, if a unit is divided into 10 equal parts, the parts are called *tenths;* if divided into 100 equal parts, the parts are called *hundredths;* if divided into 1000 equal parts, the parts are called *thousandths*, &c.

Now it is manifest that if a *tenth* is divided into 10 equal parts, 1 of those parts will be a *hundredth;* for, $\frac{1}{10} \div 10 = \frac{1}{100}$. (Art. 138.) If a *hundredth* is divided into 10 equal parts, 1 of the parts will be a *thousandth;* for, $\frac{1}{100} \div 10 = \frac{1}{1000}$, &c. Thus a new class of fractions is obtained, which regularly decreases in value in a *tenfold ratio;* that is, a class which expresses simply *tenths, hundredths, thousandths*, &c., without the intervening parts, as in common fractions, and whose denominators are always 10, 100, 1000, &c.

176. *Fractions* which *decrease* in a *tenfold ratio*, or which express simply *tenths, hundredths, thousandths*, &c., are called DECIMAL FRACTIONS.

QUEST.—175. What are fractions? From what do the parts take their name? 176. What are decimal fractions? From what do they arise? Why called decimals?

OBS. Decimal fractions obviously arise from dividing a *unit* into *ten* equal parts, then subdividing each of those parts into *ten other* equal parts, and so on. They are called *decimals*, because they *decrease* in a *tenfold ratio*. (Art. 10. Obs. 2.)

177. Each order of integers or whole numbers, it has been shown, *increases* in value from units towards the left in a ten-fold ratio; (Art. 9;) and, conversely, each order must *decrease* from left to right in the same ratio, till we come to units' place again.

178. By extending this scale of notation below units towards the right hand, it is manifest that the *first* place on the right of units, will be *ten times* less in value than *units'* place; that the *second* will be ten times less than the *first;* the *third* ten times less than the *second*, &c.

Thus we have a series of *orders* below units, which decrease in a tenfold ratio, and exactly correspond in value with *tenths*, *hundredths*, *thousandths*, &c., in com. fractions.

179. *Decimal Fractions* are commonly expressed by writing the numerator with a point (.) before it.

OBS. If the numerator does not contain so many figures as there are ciphers in the denominator, the deficiency must be supplied by prefixing ciphers to it.

For example, $\frac{1}{10}$ is written thus .1; $\frac{2}{10}$ thus .2; $\frac{3}{10}$ thus .3; &c. $\frac{1}{100}$ is written thus .01, putting the 1 in hundredths place; $\frac{5}{100}$ thus .05; &c. That is, tenths are written in the *first* place on the right of units; hundredths in the *second* place; thousandths in the *third* place, &c.

180. *The denominator of a decimal fraction is always* 1 *with as many ciphers annexed to it as there are decimal figures in the given numerator.* (Art. 175.)

OBS. The point placed before decimals, is called the *Decimal Point*, or *Separatrix*. Its object is to distinguish the fractional parts from whole numbers.

QUEST.—177. In what manner do whole numbers increase and decrease? 178. By extending this scale below units, what would be the value of the first place on the right of units? The second? The third? With what do these orders correspond? 179. How are decimal fractions expressed. 180. What is the denominator of a decimal fraction? *Obs.* What is the point placed before decimals called?

181. The names of the different *orders of decimals* or places below units, may be easily learned from the following

DECIMAL TABLE.

Hundreds.	Tens.	*Units.*	(Decimal Point.)	Tenths.	Hundredths.	*Thousandths.*	Ten thousandths.	Hundred thousandths.	*Millionths.*	Ten millionths.	Hundred millionths.	*Billionths.*	Ten billionths.	Hundred billionths.	*Trillionths, &c.*
4	2	3	.	2	6	7	1	4	5	9	8	6	2	7	4

182. It will be seen from this table that the *value* of each figure in *decimals*, as well as in whole numbers, depends upon the *place* it occupies, reckoning from units. Thus, if a figure stands in the *first* place on the right of units, it expresses *tenths;* if in the *second*, *hundredths*, &c. each successive place or order towards the right, decreasing in value in a tenfold ratio. Hence,

183. *Each removal of a decimal figure one place from units towards the right, diminishes its value ten times.*

Prefixing a cipher, therefore, to a decimal diminishes its value *ten times;* for it removes the decimal one place farther from units' place. Thus $.4=\frac{4}{10}$; but $.04=\frac{4}{100}$, and $.004=\frac{4}{1000}$, &c.; for the denominator to a decimal fraction is 1 with as many ciphers annexed to it, as there are figures in the numerator. (Art. 180.)

Annexing ciphers to decimals does not alter their value; for, each significant figure continues to occupy the same place from units as before. Thus, $.5=\frac{5}{10}$; so $.50=\frac{50}{100}$, or $\frac{5}{10}$, by dividing the numerator and denominator by 10; (Art. 116;) and $.500=\frac{500}{1000}$, or $\frac{5}{10}$, &c.

QUEST.—181. Repeat the Decimal Table, beginning units, tenths, &c. 182. Upon what does the value of a decimal depend? 183. What is the effect of removing a decimal figure one place to the right? What then is the effect of prefixing ciphers to decimals? What, of annexing them?

OBS. 1. It should be remembered that the units' place is always the right hand place of a whole number. The effect of annexing and prefixing ciphers to decimals, it will be perceived, is the *reverse* of annexing and prefixing them to whole numbers. (Art. 58.)

2. A whole number and a decimal, written together, is called a *mixed number.* (Art. 108.)

184. To read decimal fractions.

Beginning at the left hand, read the figures as if they were whole numbers, and to the last one add the name of its order. Thus,

.5	is read	5 tenths.
.25	" "	25 hundredths.
.324	" "	324 thousandths.
.5267	" "	5267 ten thousandths.
.43725	" "	43725 hundred thousandths.
.735168	" "	735168 millionths.

OBS. In reading decimals as well as whole numbers, the *units'* place should always be made the *starting* point. It is advisable for young pupils to apply to every figure the name of its order, or the place which it occupies, before attempting to read them. Beginning at the units' place, he should proceed towards the right, thus—*units, tenths, hundredths, thousandths,* &c., pointing to each figure as he pronounces the name of its order. In this way he will very soon be able to read decimals with as much ease as he can whole numbers.

Read the following numbers:

(1.)	(2.)	(3.)	(4.)
.25	.5317	3.245	9.14712
.362	.1056	7.6071	1.06231
.451	.4308	4.3159	2.00729
.5675	.0105	3.87816	9.14051
.8432	.0007	5.91432	8.06705

(5.)	(6.)	(7.)	(8.)
25.02	56.78417	1.253456	2.000008
36.032	21.05671	0.034689	0.500072
45.7056	42.05063	7.035042	8.305001
12.07067	95.10051	9.103005	9.000001

QUEST.—*Obs.* Which is the units' place? What is a whole number and a decimal written together, called? 184. How are decimals read? *Obs.* In reading decimals, what should be made the starting point?

Note.— Sometimes we pronounce the word *decimal* when we come to the separatrix, and then read the figures as if they were whole numbers; or, simply repeat them one after another. Thus, 125.427 is read, one hundred twenty-five, *decimal* four hundred twenty-seven; or, one hundred twenty-five, *decimal* four, two, seven.

Write the fractional parts of the following numbers in decimals:

(9.)	(10.)	(11.)	12.
$2\frac{5}{10}$	$3\frac{5}{100}$	$15\frac{6435}{100000}$	$3\frac{12567}{1000000}$
$4\frac{52}{100}$	$45\frac{62}{100}$	$10\frac{534}{10000}$	$4\frac{2005}{1000000}$
$28\frac{6}{100}$	$5\frac{231}{10000}$	$2\frac{45}{10000}$	$17\frac{201}{1000000}$
$6\frac{29}{100}$	$6\frac{23}{10000}$	$300\frac{701}{100000}$	$13\frac{123567}{10000000}$

13. Write 49 hundredths; 3 tenths; 445 ten thousandths.

14. Write 36 thousandths; 25 hundred thousandths; 1 millionth.

15. Write 7 hundredths; 3 thousandths; 95 ten thousandths; 63 millionths; 26 ten millionths.

185. Decimals are *Added*, *Subtracted*, *Multiplied*, and *Divided*, in the same manner as whole numbers.

OBS. The only thing with which the learner is likely to find any difficulty, is *pointing off* the answer. To this part of the operation he should give particular attention.

ADDITION OF DECIMAL FRACTIONS.

186. Ex. 1. What is the sum of 2.5; 24.457; 123.4 and 2.369?

Operation.
2.5
24.457
123.4
2.369
152.726

Write the units under units, the tenths under tenths, hundredths under hundredths, &c.; then, beginning at the right hand or lowest order, proceed thus: 9 thousandths and 7 thousandths are 16 thousandths. Write the 6 under the column added, and carry the 1 to the next column as in addition of whole numbers. 1 to carry to 6 hundredths

---

QUEST.—*Note.* What other method of reading decimals is mentioned?

makes 7 hundredths and 5 are 12 hundredths. Set the 2 under the column and carry the 1 as before. 1 to carry to 3 tenths makes 4, and 4 are 8 tenths and 4 are 12 tenths and 5 are 17 tenths or 1 and 7 tenths. Set the 7 under the column, and carry the 1 to the next column. Finally, *place the decimal point in the amount, directly under that in the numbers added.*

187. Hence, we deduce the following general

RULE FOR ADDITION OF DECIMALS.

Write the numbers so that the same orders may stand under each other, placing tenths under tenths, hundredths under hundredths, &c. Begin at the right hand or lowest order, and proceed in all respects as in adding whole numbers. (Art. 29.)

From the right hand of the amount, point off as many figures for decimals as are equal to the greatest number of decimal places in either of the given numbers.

PROOF.—*Addition of Decimals is proved in the same manner as Simple Addition.* (Art. 28.)

Note.—The decimal point in the *answer* will always fall directly under the decimal points in the given numbers.

EXAMPLES.

(2.)	(3.)	(4.)
31.25	15.263	20.13
1.059	7.0003	117.056
126.05	213.0507	43.5
1235.6151	0.05	2185.05813
1393.9741 *Ans.*	85.306	620.30597

5. What is the sum of 2.5; 33.65 and 45.121?
6. What is the sum of 65.7; 43.09; 1.026 and 2.1765?

QUEST.—187. How are decimals added? How point off the answer? How is addition of decimals proved?

7. What is the sum of 6.15768; 1.713458 and .6573128?

8. What is the sum of .0256; 15.6941; 3.856 and .00035?

9. Add together 256.31; 29.7; 468.213; 5.6 and .75.

10. Add together 25.61; 78.003; 951.072 and 256 .3052.

11. Add together .567; 37.05; 63.501; 76.25 and .63.

12. Add together .005; 1.25; 6.456; 10.2563 and 15.434.

13. Add together 256.1; 10.15; 27.09; 35.560 and 2.067.

14. Add together 5.00257; 3.600701 and 2.10607.

15. Add together 5 tenths, 25 hundredths, 566 thousandths, and 7568 ten thousandths.

16. Add together 34 hundredths, 67 thousandths, 13 ten thousandths, and 463 millionths.

17. Add together 7 thousandths, 63 hundred thousandths, 47 millionths, and 6 tenths.

18. Add together 423 ten millionths, 63 thousandths, 25 hundredths, 4 tenths, and 56 ten thousandths.

SUBTRACTION OF DECIMAL FRACTIONS.

188. Ex. 1. From 25.367 substract 13.18.

Operation.

```
 25.367
 13.18
 ------
 12.187. Ans.
```

Having written the less number under the greater, so that units may stand under units, tenths under tenths, &c., we proceed exactly as in subtraction of whole numbers. (Art. 40.) Thus, 0 thousandths from 7 thousandths leaves 7 thousandths. Write the 7 in the thousandth's place. As the next figure in the lower line is larger than the one above it, we borrow 10. Now 8 from 16 leaves 8; set the 8 under the column, and carry 1 to the next figure. (Art. 38.) Proceed in the same manner with the other figures in the lower number. Finally, place the decimal point in the remainder directly under that in the given numbers.

189. Hence, we deduce the following general

RULE FOR SUBTRACTION OF DECIMALS.

Write the less number under the greater, with units under units, tenths under tenths, hundredths under hundredths, &c. Subtract as in whole numbers, and point off the answer as in addition of decimals. (Art. 187.)

PROOF.—*Subtraction of Decimals is proved in the same manner as Simple Subtraction.* (Art. 39.)

Note.—When there are blank places on the right hand of the upper number, they may be supplied by ciphers without altering the value of the decimal. (Art. 183.)

EXAMPLES.

2. From 15 take 1.5. *Ans.* 13.5.
3. From 256.0315 take 5.641.
4. From 15.7 take 1.156.
5. From 63.25 take 50.
6. From 201.001 take 56.04037.
7. From 1 take .125.
8. From 11.1 take .40005.
9. From .56078 take .325.
10. From 1.66 take .5589.
11. From 3.4001 take 2.000009.
12. From 1 take .000001.
13. From 256.31 take 125.4689301.
14. From 8960.320507 take 63.001.
15. From 57000.000001 take 1000.001.
16. From 75 hundredths take 75 thousandths.
17. From 6 thousandths take 6 millionths.
18. From 3252 ten thousandths take 3 thousandths.
19. From 539 take 22 thousandths.
20. From 7856 take 236 millionths.

QUEST.—189. How are decimals subtracted? How point off the answer? How is subtraction of decimals proved?

MULTIPLICATION OF DECIMAL FRACTIONS.

190. Ex. 1. Multiply .48 by .5.

Suggestion.—Multiplying by a fraction, is taking a part of the multiplicand as many times as there are like parts of a unit in the multiplier. (Art. 132.) Hence, multiplying by .5, which is equal to $\frac{5}{10}$ or $\frac{1}{2}$, is taking *half* of the multiplicand *once*. Now .48, or $\frac{48}{100} \div 2 = \frac{24}{100}$. (Art. 138.) But $\frac{24}{100} = .24$. (Art. 179.)

Operation.	
.48	
.5	
.240 *Ans.*	

We multiply as in whole numbers, and pointing off as many decimals in the product as there are decimal figures in both factors, we have .240. But since ciphers placed on the right of decimals do not affect their value, the 0 may be omitted. (Art. 183.) But $.24 = \frac{24}{100}$, which is the same result as before.

	2.	3.	4.
Multiply	8.45	96.071	456.03
By	.25	.0032	4.5
	4225	192142	228015
	1690	288213	182412
Ans.	2.1125	.3074272	2052.135

191. From the preceding illustrations we deduce the following general

RULE FOR MULTIPLICATION OF DECIMALS.

Multiply as in whole numbers, and point off as many figures from the right of the product for decimals, as there are decimal places both in the multiplier and multiplicand.

If the product does not contain so many figures as there are decimals in both factors, supply the deficiency by prefixing ciphers.

QUEST.—191. How are decimals multiplied together? How do you point off the product? When the product does not contain so many figures as there are decimals in both factors, what is to be done?

PROOF.—*Multiplication of Decimals is proved in the same manner as Simple Multiplication.* (Arts. 53, 74.)

OBS. The reason for pointing off as many decimal places in the product as there are decimals in both factors, may be illustrated thus: Suppose it is required to multiply .25 by .5. Supplying the denominators $.25=\frac{25}{100}$, and $.5=\frac{5}{10}$. (Art. 180.) Now $\frac{25}{100}\times\frac{5}{10}=\frac{125}{1000}$. (Art. 135.) But $\frac{125}{1000}=.125$; (Art. 179;) that is, the product of $.25\times.5$, contains just as many decimals as the factors themselves. In like manner it may be shown that the product of any two or more decimal numbers, must contain as many decimal figures as there are places of decimals in the given factors.

EXAMPLES.

Ex. 1. In 1 piece of cloth there are 31.7 yards: how many yards are there in 7.3 pieces?

2. In 1 barrel there are 31.5 gallons: how many gallons are there in 8.25 barrels?

3. In one rod there are 16.5 feet: how many feet are there in 35.75 rods?

4. How many cords of wood are there in 45 loads, allowing 8.25 of a cord to a load?

5. How many rods are there in a piece of land 25.35 rods long, and 20.5 rods wide?

6. If a man can travel 38.75 miles per day, how far can he travel in 12.25 days?

7. How many pounds of coffee are there in 68 sacks, allowing 961.25 pounds to a sack?

8. If a family consume .85 of a barrel of flour in a week, how much will they consume in 52.23 weeks?

9. What is the product of 10.001 into .05?

10. What is the product of 50.0065 into 1.003?

192. When the multiplier is 10, 100, 1000, &c., the multiplication may be performed by simply *removing the decimal point* as many places towards the *right*, as there are ciphers in the multiplier. (Arts. 59, 191.)

QUEST.—How is multiplication of decimals proved? 192. How proceed when the multiplier is 10, 100, 1000, &c.

11. Multiply 4.6051 by 100. *Ans.* 460.51.
12. Multiply 2.6501 by 1000.
13. Multiply .5678 by 10000.
14. Multiply .000781 by 2.40001.
15. Multiply 1.002003 by .0024.
16. Multiply .58001 by .0001003.
17. Multiply 8.001502 by .00005.
18. Multiply 85689.31 by .000001.
19. Multiply .0000045 by 69.5.
20. Multiply .0340006 by .000067.
21. Multiply .5 by 5 millionths.
22. Multiply .15 by 28 ten thousandths.
23. Multiply 25 hundredth thousandths by 7.3.
24. Multiply 225 millionths by 2.85.
25. Multiply 2367 ten millionths by 3.0002.

DIVISION OF DECIMAL FRACTIONS.

193. Ex. 1. Divide .75 by .5.

Operation.

.5).75
 1.5 *Ans.*

We divide as in whole numbers, and point off 1 decimal figure in the quotient.

Obs. We have seen in the multiplication of decimals, that the product has as many decimal figures, as the multiplier and multiplicand. (Art. 191.) Now since the dividend is equal to the product of the divisor and quotient, (Art. 65,) it follows that the dividend must have as many decimals as the divisor and quotient together; consequently, as the dividend has two decimals, and the divisor but one, we must point off one in the quotient; that is, we must point off as many decimals in the quotient, as the decimal places in the dividend exceed those in the divisor.

2. Divide .289 by 2.4.

Operation.

2.4).289(.12+*Ans.*
 24
 49
 48
 1 rem.

Since the divisor contains two figures, we substitute long division for short, and point off the quotient as before.

Note.—When there is a remainder, the sign + should be annexed to the quotient, to show that it is not complete.

3. Divide 1.345 by .5. *Ans.* 2.69.
4. Divide .063 by 9.

Operation.
9).063
.007 *Ans.*

In this example the dividend has three more places of decimals than the divisor; hence the quotient must have three places of decimals. We must, therefore, prefix two ciphers to the quotient.

194. From these illustrations we deduce the following general

RULE FOR DIVISION OF DECIMALS.

Divide as in whole numbers, and point off as many figures for decimals in the quotient, as the decimal places in the dividend exceed those in the divisor. If the quotient does not contain figures enough, supply the deficiency by prefixing ciphers.

PROOF.—*Division of Decimals is proved in the same manner as Simple Division.* (Art. 73.)

OBS. 1. When the number of decimals in the divisor is the *same* as that in the dividend, the quotient will be a *whole* number.

2. When there are *more* decimals in the divisor than in the dividend, annex as many ciphers to the dividend as are necessary to make its decimal places *equal* to those in the divisor. The quotient thence arising will be a whole number. (Obs. 1.)

3. After all the figures of the dividend are divided, if there is a remainder, ciphers may be annexed to it and the division continued at pleasure. The ciphers annexed must be regarded as decimal places belonging to the dividend.

Note.—For ordinary purposes, it will be sufficiently exact to carry the quotient to three or four places of decimals; but when great accuracy is required, it must be carried farther.

QUEST.—194. How are decimals divided? How point off the quotient? How is division of decimals proved? *Obs.* When the number of decimal places in the divisor is equal to that in the dividend, what is the quotient? When there are more decimals in the divisor than in the dividend, how proceed? When there is a remainder, what may be done?

EXAMPLES.

1. If 1.7 of a yard of cloth will make a coat, how many coats will 10.2 yards make?
2. In 6.75 cords of wood, how many loads are there, allowing .75 of a cord to a load?
3. If a man mows 3.2 acres of grass per day, how long will it take him to mow 39.36 acres?
4. If 23.25 bushels of barley grow on an acre, how many acres will 556 bushels require?
5. In 74.25 feet, how many rods?
6. In 99.225 gallons of wine, how many barrels?
7. If a man chops 3.75 cords of wood per day, how many days will it take him to chop 91.476 cords?
8. If a man can travel 35.4 miles per day, how long will it take him to travel 244.26 miles?
9. A dairy-man has 187.5 pounds of butter, which he wishes to pack in boxes containing 12.5 pounds apiece: how many boxes will it require?
10. In 3.575, how many times .25?

195. When the divisor is 10, 100, 1000, &c., the division may be performed by simply *removing the decimal point* in the dividend as many places towards the *left*, as there are ciphers in the divisor, and it will be the quotient required. (Arts. 80, 194.)

11. Divide 756.4 by 100. *Ans.* 7.564.
12. Divide 1268.2 by 1000. *Ans.* 1.2682.
13. Divide 1 by 1.25.
14. Divide 1 by 562.5.
15. Divide .012 by .005.
16. Divide 2 by .0002.
17. Divide 5 by .000001.
18. Divide 13.2 by .75.
19. Divide .0248 by .04.
20. Divide 2071.31 by 65.3.

QUEST.—195. When the divisor is 10, 100, 1000, &c., how may the division be performed?

REDUCTION OF DECIMALS.

CASE I.

Ex. 1. Change the decimal .25 to a common fraction.

Suggestion.—Supplying the denominator, $.25=\frac{25}{100}$. (Art. 180.) Now $\frac{25}{100}$ is expressed in the form of a common fraction, and as such may be reduced to lower terms, and be treated in the same manner as any other common fraction. Thus $\frac{25}{100}=\frac{5}{20}$, or $\frac{1}{4}$. Hence,

196. To reduce a Decimal to a Common Fraction.

Erase the decimal point; then write the decimal denominator under the numerator, and it will form a common fraction, which may be treated in the same manner as other common fractions.

2. Change .125 to a common fraction, and reduce it to the lowest terms. *Ans.* $\frac{1}{8}$.

3. Reduce .66 to a common fraction, &c.
4. Reduce .75 to a common fraction, &c.
5. Reduce .375 to a common fraction, &c.
6. Reduce .525 to a common fraction, &c.
7. Reduce .025 to a common fraction, &c.
8. Reduce .875 to a common fraction, &c.
9. Reduce .0625 to a common fraction, &c.
10. Reduce .000005 to a common fraction, &c.

CASE II.

Ex. 1. Change $\frac{3}{5}$ to a decimal.

Suggestion.—Multiplying both terms by 10 the fraction becomes $\frac{30}{50}$. Again dividing both terms by 5, it becomes $\frac{6}{10}$. (Art. 116.) But $\frac{6}{10}=.6$, (Art. 179,) which is the decimal required.

QUEST.—196. How are Decimals reduced to Common Fractions?

Now since we make no use of the denominator 10 after it is obtained, we may omit the process of getting it; for if we annex a cipher to the numerator and divide it by 5, we shall obtain the same result.

Operation.

5)3.0
  .6

A decimal point is prefixed to the quotient, to distinguish it from a whole number.

PROOF.—.6 reduced to a common fraction is $\frac{6}{10}$; (Art. 196;) and $\frac{6}{10}=\frac{3}{5}$. (Art. 120.)

2. Reduce $\frac{1}{8}$ to a decimal.

Operation.

8)1.000
  .125

Annex ciphers to the numerator and proceed as before. Hence,

197. To reduce a Common Fraction to a Decimal.

Annex ciphers to the numerator and divide it by the denominator. Point off as many decimal figures in the quotient, as you have annexed ciphers to the numerator.

OBS. 1. If there are not as many figures in the quotient as you have annexed ciphers to the numerator, supply the deficiency by prefixing ciphers to the quotient.

2. The reason of this process may be illustrated thus. Annexing a cipher to the numerator multiplies the fraction by 10. (Arts. 59, 133.) If, therefore, the numerator with a cipher annexed to it, is divided by the denominator, the quotient will obviously be ten times too large. Hence, in order to obtain the true quotient, or a decimal equal to the given fraction, the quotient thus obtained must be divided by 10, which is done by pointing off one figure. (Art. 80.) Annexing 2 ciphers to the numerator multiplies the fraction by 100; annexing 3 ciphers by 1000, &c., consequently, when 2 ciphers are annexed, the quotient will be 100 times too large, and must therefore be divided by 100; when three ciphers are annexed, the quotient will be 1000 times too large, and must be divided by 1000; &c. (Art. 80.)

QUEST.—197. How are Common Fractions reduced to Decimals? *Obs.* When there are not so many figures in the quotient as you have annexed ciphers, what is to be done?

3. Reduce $\frac{3}{2}$ to decimals. *Ans.* 1.5.
4. Reduce $\frac{3}{4}$, and $\frac{4}{5}$ to decimals.
5. Reduce $\frac{3}{20}$, and $\frac{7}{25}$ to decimals.
6. Reduce $\frac{3}{8}$, $\frac{1}{5}$, and $\frac{3}{5}$ to decimals.
7. Reduce $\frac{4}{5}$, $\frac{5}{6}$, and $\frac{2}{20}$ to decimals.
8. Reduce $\frac{4}{25}$, $\frac{8}{20}$, and $\frac{3}{75}$ to decimals.
9. Reduce $\frac{5}{8}$, $\frac{2}{5}$, and $\frac{1}{20}$ to decimals.
10. Reduce $\frac{12}{480}$, and $\frac{5}{1785}$ to decimals.
11. Reduce $\frac{6}{240}$, and $\frac{3}{1000}$ to decimals.
12. Reduce $\frac{1}{3}$ to a decimal. *Ans.* .333333+.
13. Reduce $\frac{128}{999}$ to a decimal. *Ans.* .128128128+.

198. It will be seen that the last two examples cannot be exactly reduced to decimals; for there will continue to be a remainder after each division, as long as we continue the operation.

In the 12th, the remainder is always 1; in the 13th, after obtaining three figures in the quotient, the remainder is the same as the given numerator, and the next three figures in the quotient are the same as the first three, when the same remainder will recur again.

The same remainders, and consequently the same figures in the quotient, will thus continue to recur, as long as the operation is continued.

199. Decimals which consist of the same figure or set of figures continually repeated, as in the last two examples, are called ***Periodical*** or ***Circulating Decimals;*** also, ***Repeating Decimals***, or ***Repetends.***

CASE III.

Ex. 1. Reduce 7s. 6d. to the decimal of a pound.

Suggestion.—First, reduce 7s. 6d. to pence for the numerator, and £1 to pence for the denominator of a com-

---

QUEST.—199. What are Periodical or Repeating Decimals?

mon fraction, and we have £$\frac{90}{240}$. (Art. 165.) Now $\frac{90}{240}$ reduced to a decimal is £.375. *Ans.* Hence,

200. To reduce a compound number to the decimal of a higher denomination.

First reduce the given compound number to a common fraction; (Art. 165;) *then reduce the common fraction to a decimal.* (Art. 197.)

2. Reduce 5s. 4d. to the decimal of £1. *Ans.* £.2666+.
3. Reduce 15s. 6d. to the decimal of £1.
4. Reduce 12s. 6d. to the decimal of £1.
5. Reduce 9d. to the decimal of 1 shilling.
6. Reduce 7d. 2 far. to the decimal of a shilling.
7. Reduce 1 pt. to the decimal of a quart.
8. Reduce 18 hours to the decimal of a day.
9. Reduce 9 in. to the decimal of a yard.
10. Reduce 2 ft. 6 in. to the decimal of a yard.
11. Reduce 6 furlongs to the decimal of a mile.
12. Reduce 13 oz. 8 dr. to the decimal of a pound.

CASE IV.

Ex. 1. Reduce £.123 to shillings, pence, and farthings.

Operation.

	£.123
	20
shil.	2.460
	12
pence	5.520
	4
far.	2.080

Ans. 2s. 5d. 2 f.

Multiply the given decimal by 20, as if it were a whole pound, because 20s. make £1, and point off as many figures for decimals, as there are decimal places in the multiplier and multiplicand. (Art. 191.) The product is in shillings and a decimal of a shilling. Then multiply the decimal of a shilling by 12, and point off as before, &c. The numbers on the left of the decimals, viz: 2s. 5d. 2 far. form the answer. Hence,

---

QUEST.—200. How is a compound number reduced to the decimal of a higher denomination?

201. To reduce a decimal compound number to whole numbers of lower denominations.

Multiply the given decimal by that number which it takes of the next lower denomination to make ONE *of this higher, as in reduction,* (Art. 161, I,) *and point off the product, as in multiplication of decimal fractions.* (Art. 191.) *Proceed in this manner with the decimal figures of each succeeding product, and the numbers on the left of the decimal point in the several products, will constitute the whole number required.*

2. Reduce £.125 to shillings and pence. *Ans.* 2s. 6d.
3. Reduce .625s. to pence and farthings.
4. Reduce £.4625 to shillings and pence.
5. Reduce .756 gallons to quarts and pints.
6. Reduce .6254 days to hours, minutes, and seconds.
7. Reduce .856 cwt. to quarters, &c.
8. Reduce .6945 of a ton to hundreds, &c.
9. Reduce .7582 of a bushel to pecks, &c.
10. Reduce .8237 of a mile to furlongs, &c.
11. Reduce .45683 of an acre to roods and rods.
12. Reduce .75631 of a yard to quarters and nails.

FEDERAL MONEY.

202. FEDERAL MONEY is the currency of the United States. The denominations are, *Eagles, Dollars, Dimes, Cents, and Mills.*

TABLE.

10 mills (*m.*)	make	1 cent,	marked *ct.*
10 cents	"	1 dime,	" *d.*
10 dimes	"	1 dollar,	" *doll.* or $.
10 dollars	"	1 eagle,	" *E.*

OBS. Federal Money was established by Congress, Aug. 8th, 1786. Previous to this, English or sterling money was the principal currency of the country.

QUEST.—201. How are decimal compound numbers reduced to whole ones? 202. What is Federal Money? Recite the Table. *Obs.* When and by whom was it established?

Note.— Many foreign coins are still in circulation. Indeed some of the rates of postage established by the government, were, until recently, adapted to foreign coins. To the 28th Congress belongs the honor of abolishing these *anti-national rates*, and of establishing others in Federal Money.

203. The national coins of the United States are of three kinds, viz: gold, silver, and copper.

1. The gold coins are the *eagle, the double eagle,* half eagle, quarter eagle, and gold dollar.**

The eagle contains 258 grains of *standard* gold; the double eagle, half eagle, and quarter eagle, like proportions.

2. The silver coins are the *dollar, half dollar, quarter dollar,* the *dime,* and *half dime.*

The dollar contains 412½ grains of *standard* silver; the others, like proportions.

3. The copper coins are the *cent,* and *half cent.*

The cent contains 168 grains of *pure* copper; the half cent, a like proportion.

Mills are not coined.

OBS. 1. The fineness of gold used for coin, jewelry, and other purposes, also the gold of commerce, is estimated by the number of parts of gold which it contains. Pure gold is commonly supposed to be divided into 24 equal parts, called *carats.* Hence, if it contains 10 parts of *alloy,* or some *baser* metal, it is said to be 14 carats fine; if 5 parts of alloy, 19 carats fine; and when absolutely pure, it is 24 carats fine.†

2. The present *standard* for both *gold* and *silver coins* of the United States, by Act of Congress, 1837, is 900 parts of pure metal by weight to 100 parts of alloy. The alloy of gold coin is composed of silver and copper, the silver not to exceed the copper in weight. The alloy of silver coin is pure copper.

204. All accounts in the United States are kept in

QUEST.—203. Of how many kinds are the coins of the United States? What are they? What are the gold coins? The silver coins? The copper? *Obs.* How is the fineness of gold estimated? Into how many carats is pure gold supposed to be divided? When it contains 10 parts of alloy, how fine is it said to be? 5 parts of alloy? 2 parts? 4 parts? What is the standard for the gold and silver coins of the United States? What is the alloy of gold coins? What of silver coins? 204. In what are accounts kept? How would you express 5 eagles? 7 E. and 5 dolls.? 10 E.? How express 6 dimes? 8 dimes? 10 dimes?

* By Act of Congress, Feb. 20th, 1849. † Silliman's Chemistry.

dollars, *cents*, and *mills*. Eagles are expressed in dollars, and dimes in cents. Thus, instead of 5 eagles, we say, 50 dollars; instead of 7 eagles and 5 dollars, we say, 75 dollars, &c. So, instead of 6 dimes, we say, 60 cents; instead of 8 dimes and 7 cents, we say, 87 cents, &c.

205. It will be seen from the Table that Federal Money is based upon the *Decimal* system of Notation; that its denominations increase and decrease from right to left and left to right in a *tenfold ratio*, like whole numbers and decimals.

206. The Dollar is regarded as the *unit; cents* and *mills* are fractional parts of the dollar, and are distinguished from it by a *decimal point* or *separatrix* (.) in the same manner as common decimals. (Art. 179.) *Dollars* therefore occupy *units'* place of simple numbers; *eagles*, or tens of dollars, *tens'* place, &c. *Dimes*, or tenths of a dollar, occupy the place of *tenths* in decimals; *cents* or hundredths of a dollar, the place of *hundredths; mills*, or thousandths of a dollar, the place of *thousandths; tenths* of a mill, or ten thousandths of a dollar, the place of *ten thousandths*, &c.

OBS. 1. Since *dimes* in business transactions are expressed in *cents*, *two places* of decimals are assigned to cents. If therefore the number of cents is *less* than 10, a *cipher must always be placed on the left hand of them;* for cents are *hundredths* of a dollar, and hundredths occupy the second decimal place. (Art. 181.) For example, 4 cents are written thus .04; 7 cents thus .07; 9 cents thus .09, &c.

2. Mills occupy the *third* place of decimals; for they are *thousandths* of a dollar. Consequently, when there are no cents in the given sum, *two ciphers* must be placed before the mills. Hence,

207. To read any sum of Federal Money.

Call all the figures on the left of the decimal point dollars; the first two figures after the point, are cents; the

---

QUEST.—205. How do the denominations of Federal Money increase and decrease? Upon what is it based? 206. What is regarded as the unit in Federal Money? What are cents and mills? How are they distinguished from dollars? 207. How do you read Federal Money? *Obs.* What other mode of reading Federal Money is mentioned?

third figure denotes mills; the other places on the right are decimals of a mill. Thus, $3.25232 is read, 3 dollars, 25 cents, 2 mills, and 32 hundredths of a mill.

OBS. Sometimes all the figures after the point are read as decimals of a dollar. Thus, $5.356 is read, "5 and 356 thousandths dollars."

Read the following sums of Federal Money:

1.	2.	3.
$250.56	$44.081	$3.7542
105.863	60.05	0.6054
200.057	75.003	4.0151
506.507	20.501	6.0057
850.071	30.065	8.0106

Write the following sums in Federal Money:

4. 63 dollars, and 85 cents. *Ans.* $63.85.
5. 150 dollars, and 73 cents.
6. 201 dollars, and 9 cents.
7. 300 dollars, 5 cents, and 3 mills.
8. 4 dollars, 6 cents, and 8 mills.
9. 100 dollars, 7 cents, 5 mills, and 3 tenths of a mill.
10. 1000 dollars, 6 mills, and 36 hundredths of a mill.

Note.—In business transactions, when dollars and cents are expressed together, the cents are frequently written in the form of a common fraction. Thus, $76.45 are written $76\frac{45}{100}$ dollars.

REDUCTION OF FEDERAL MONEY.

CASE I.

Ex. 1. How many cents are there in 75 dollars?

Suggestion.—Since in 1 dollar there are 100 cents, in 75 dollars there are 75 times as many. And $75\times100=7500$. *Ans.* 7500 cents.

2. In 9 cents, how many mills? *Ans.* 90 mills.
3. In 25 dollars, how many mills? *Ans.* 25000 mills.

Note.—To multiply by 10, 100, &c., is simply annexing as many ciphers to the multiplicand, as there are ciphers in the multiplier. (Art. 59.) Hence,

208. *To reduce dollars to cents, annex two ciphers.*
To reduce dollars to mills, annex three ciphers.
To reduce cents to mills, annex one cipher.

OBS. To reduce dollars and cents to cents, *erase the sign of dollars and the separatrix.* Thus, $25.36 reduced to cents, becomes 2536 cents.

4. In $5, how many cents?
5. How many mills in $364?
6. How many mills in $621?
7. How many cents in $6245?
8. Reduce $75.26 to cents.
9. Reduce $625.48 to cents.

CASE II.

10. In 4500 cents, how many dollars?

Suggestion.—Since 100 cents make 1 dollar, 4500 cents will make as many dollars as 100 is contained times in 4500. And 4500÷100=45. *Ans.* $45.

11. In 150 mills, how many cents? *Ans.* 15 cents.
12. In 25000 mills, how many dollars? *Ans.* $25.

Note.—To divide by 10, 100, &c., is simply cutting off as many figures from the right of the dividend as there are ciphers in the divisor. (Art. 80.) Hence,

209. *To reduce cents to dollars, cut off two figures on the right.*

To reduce mills to dollars, cut off three figures on the right.

To reduce mills to cents, cut off one figure on the right.

OBS. The figures cut off are cents and mills.

QUEST.—208. How are dollars reduced to cents? Dollars to mills? Cents to mills? *Obs.* Dollars and cents to cents? 209. How are cents reduced to dollars? Mills to dollars? Mills to cents? *Obs.* What are the figures cut off?

13. In 325 cents, how many dollars? *Ans.* $3.25.
14. In 423 mills, how many cents? *Ans.* 42c. 3m.
15. In 4320 mills, how many dollars?
16. How many dollars in 63500 cents?
17. How many cents in 4890 mills?

210. Since Federal Money is expressed according to the decimal system of notation, it is evident that it may be subjected to the same operations and treated in the same manner as decimal fractions.

ADDITION OF FEDERAL MONEY.

Ex. 1. A man bought a cow for $15.75, a calf for $2.375, a sheep for $3.875, and a load of hay for $8.68: how much did he pay for all?

Operation.

$15.75
2.375
3.875
8.68
$30.680 *Ans.*

We write the dollars under dollars, cents under cents, &c. Then add each column separately, and point off as many figures for cents and mills, in the amount, as there are places of cents and mills in either of the given numbers.

211. Hence, we derive the following general

RULE FOR ADDING FEDERAL MONEY.

Write dollars under dollars, cents under cents, &c., so that the same orders or denominations may stand under each other. Add each column separately, and point off the amount as in addition of decimal fractions. (Art. 187.)

OBS. If either of the given numbers have no cents expressed, it is customary to supply their place by ciphers.

2. A farmer sold a firkin of butter for $9.28, a cheese for $1.17, a quarter of veal for 56 cents, and a bushel of wheat for $1.12: how much did he receive for the whole?

QUEST.—211. How is Federal Money added? How point off the amount? *Obs.* When any of the given numbers have no cents expressed, how is their place supplied?

3. A man bought a hat for $5.375, a cloak for $35.68, and a pair of boots for $4.75: how much did he pay for all?

4. What is the sum of $37.565, $85.20, $90.03, and $150.638?

5. What is the sum of $10.385, $46.238, $190.62 and $23.036?

6. What is the sum of $23.005, $16.03, $110.738, and $131.26?

7. What is the sum of 63 dolls. and 4 cts., 86 dolls. and 10 cts., and 47 dolls. and 37 cts.?

8. What is the sum of $608.05, $365.205, $2.268, and $47.006?

9. What is the amount of 11 dolls. 3 cts. and 5 mills, 16 dolls. and 8 mills, 49 dolls. 7 cts. and 8 mills?

10. What is the amount of 100 dolls. and 61 cts., 51 dolls. and 3 cts., 65 dolls. 8 cts. and 3 mills?

11. What is the amount of 95 dolls. 67 cts. and 8 mills, 120 dolls. 45 cts., 101 dolls. 7 cts. and 9 mills?

12. A lady bought a bonnet for $6.67, a pair of gloves for $0.625, a pair of shell combs for $0.75, and a cap for $2.50: what was the amount of her bill?

SUBTRACTION OF FEDERAL MONEY.

Ex. 1. A man bought a horse for $56.50, and a cow for $23.38: how much more did he pay for his horse than his cow?

Operation.

$56.50
23.38
$33.12 *Ans.*

We write the less number under the greater, placing dollars under dollars, &c., then subtract, and point off the answer as in subtraction of decimals.

212. Hence, we derive the following general

RULE FOR SUBTRACTING FEDERAL MONEY.

Write the less number under the greater, with dollars under dollars, cents under cents, &c., then subtract, and point off the remainder as in subtraction of decimal fractions. (Art. 189.)

OBS. If either of the given numbers have no cents expressed, it is customary to supply their place by ciphers.

2. A man owing $57.35, paid $17.93: how much does he still owe? *Ans.* $39.42.

3. A grocer bought two hogsheads of molasses for $68.90, and sold it for $79.26: how much did he gain by the bargain?

4. A man owed a debt of $105, and paid all but $23.67: how many dollars did he pay?

5. A merchant bought a quantity of silks for $237.63, and sold it for $196.03: how much did he lose?

6. A drover bought a flock of sheep for $357, and sold them for $17.33 less than he paid for them: how much did he sell them for?

7. What is the difference between 365 dolls. 7 cts. and 208 dolls. 20 cts.?

8. From 1 cent subtract 6 mills.

9. From 1 dollar, 6 cts. and 7 mills, take 89 cts. and 3 mills.

10. From 96 dollars, 6 cents, take 41 dolls., 63 cents, and 8 mills.

11. From 100 dollars, 10 cents, and 3 mills, take 1 cent and 5 mills.

12. From 1000 dollars, 6 cents, take 100 dolls. and 5 mills.

MULTIPLICATION OF FEDERAL MONEY.

213. In Multiplication of Federal Money, as well as in simple numbers, the multiplier must always be considered an *abstract number*. (Art. 45. Obs. 2.)

Ex. 1. How much will 5 yards of cloth cost, at $1.75 per yard?

QUEST.---212. How is Federal Money subtracted? How point off the remainder? *Obs.* When either of the given numbers have no cents expressed, how is their place supplied? 213. In Multiplication of Federal Money, what must one of the given factors be considered?

Operation.
$1.75
5
$8.75 *Ans.*

If 1 yard cost $1.75, 5 yards will obviously cost 5 times as much. Hence, we multiply the price of 1 yard by the number of yards, and point off two figures for decimals in the product. (Art. 191.)

2. How much will 15.8 yards of fringe cost, at 12 cents per yard?

Operation.
15.8
.1 2
$1.89 6

Reasoning as before, 15.8 yards will cost 15.8 times 12 cents. But in performing the multiplication, it is more convenient to take the 12 for the multiplier, and the result will be the same as if it was placed for the multiplicand. (Art. 47.) Point off the product as before.

214. Hence, when the *price* of *one* article, one pound, *one* yard, &c., is given to find the *cost* of any *number* of articles, pounds, yards, &c.

Multiply the price of one article and the number of articles together, and point off the product as in multiplication of decimals. (Art. 191.)

3. Multiply $45.035 by 6.2. *Ans.* $279.217.

215. From the preceding illustrations we derive the following general

RULE FOR MULTIPLYING FEDERAL MONEY.

Multiply as in simple numbers, and point off the product as in multiplication of decimal fractions. (Art. 191.)

OBS. 1. When the price or the quantity contains a common fraction, the fraction should be changed to a common decimal. (Art. 197.)

2. In business operations, when the mills in the answer are 5, or over, it is customary to call them a cent; when under 5, they are disregarded.

QUEST.—214. When the price of 1 article, 1 pound, &c., is given, how is the cost of any number of articles found? 215. What is the rule for Multiplication of Federal Money? *Obs.* When the price or quantity contains a common fraction, what should be done with it?

4. What will 10 lbs. of beef cost, at $6\frac{1}{2}$ cents a pound?

Solution.—$6\frac{1}{2}$ cts.=.065, and .065×10=.65.

Ans. 65 cents.

5. What cost 14 lbs. of starch, at $10\frac{1}{2}$ cts. per pound?
6. What cost $15\frac{1}{2}$ pounds of sugar, at $9\frac{1}{2}$ cts. a pound?
7. What cost 25 gals. of molasses, at $18\frac{3}{4}$ cts. a gallon?
8. What cost $23\frac{1}{4}$ lbs. of raisins, at $8\frac{1}{2}$ cts. per pound?
9. What cost $33\frac{1}{2}$ lbs. of candles, at $12\frac{1}{2}$ cts. per pound?
10. What cost $16\frac{1}{4}$ lbs. of hyson tea, at $56\frac{1}{4}$ cts. a pound?
11. What will 83 lbs. of beef cost, at $4.62½ per hund.?

Analysis.—83 pounds are $\frac{83}{100}$ of 100 pounds; therefore 83 pounds will cost $\frac{83}{100}$ of $4.625; and $\frac{83}{100}$ of $4.625=$\frac{4.625\times83}{100}$.

Operation.

```
 $4.625
     83
 ------
 13 875
3 70 00
-------
$3.83 875 Ans.
```

We multiply the price of 100 ($4.625) by 83, the given number of pounds, and the product $383.875, is the cost of 83 lbs. at $4.625 per *pound.* But the price is $4.625 per *hundred;* consequently, the product $383.875 is 100 times too large, and must therefore be divided by 100, to give the true answer. But to divide by 100, we simply remove the decimal point two places toward the left. (Art. 195.)

12. What will 825 feet of boards cost, at $6.75 per 1000?

Operation.

```
   6.75
   8 25
  -----
  33 75
 135 0
5400
-------
$5.568 75
```

Reasoning as before, 825 feet will cost $\frac{825}{1000}$ of $6.75. We multiply the price of 1000 feet by the given number of feet, and divide the product by 1000. To divide by 1000, we remove the decimal point three places towards the left. (Art. 195.) Hence,

216. To find the cost of articles bought and sold by the 100, or 1000.

Multiply the given price by the given number of articles; then if the price is for 100, *divide the product by* 100; *but if the price is for* 1000, *divide it by* 1000. (Art. 195.)

13. At $4.50 per 1000, what will 1250 bricks cost?

14. A farmer sold a quarter of beef, weighing 256.5 pounds, at $5.37½ per 100: how much did he receive for it?

15. At $4.62½ per hundred, what will 1675 pounds of pork cost?

16. What cost 2129 feet of spruce boards, at $18.25 per 1000?

17. How much will 456¾ yards of shirting cost, at 12½ cts. per yard?

18. What cost 156 lbs. of chocolate, at 15½ cents a pound?

19. What cost 235 pounds of cheese, at 6¼ cents a pound?

20. What cost 175 dozens of eggs, at 10½ cents per dozen?

21. At 47½ cents per bushel, what will be the cost of 300 bushels of corn?

22. What will 153 lbs. of sugar cost, at 8½ cents per pound?

23. What will 1500 pounds of butter cost, at $8.50 per hundred?

24. What cost 28500 feet of timber, at $3.76 per 100?

25. What cost 8230 feet of mahogany, at $70.20 per 1000?

26. What cost 7630 hemlock shingles, at $3.50 per 1000?

27. What cost 15024 pine shingles, at $8.37 per 1000?

28. At 16½ cts. a pound, what will 219½ pounds of honey cost?

QUEST.—216. How do you find the cost of articles bought and sold by the 100, or 1000?

29. At \$2.67$\frac{3}{4}$ per yard, what will 400 yards of cloth cost?

30. At \$5$\frac{3}{4}$ per barrel, what will 1560 barrels of flour cost?

DIVISION OF FEDERAL MONEY.

Ex. 1. A man bought 6 hats for \$25.68: how much did they cost apiece?

Operation.

6)25.68
\$4.28 *Ans.*

If 6 hats cost \$25.68, 1 hat will cost *one sixth* of \$25.68. Divide as in simple numbers, and point off two decimal figures in the quotient. (Art. 194.)

Proof.

\$4.28
6
\$25.68

If 1 hat costs \$4.28, 6 hats will cost 6 *times* as much; and \$4.28×6=\$25.68, which is the given cost. Hence,

217. When the number of articles, pounds, yards, &c., and the *cost* of the *whole* are given, to find the price of *one* article, *one* pound, &c.

Divide the whole cost by the whole number of articles, and point off the quotient as in division of decimal fractions. (Art. 194.)

2. How many yards of cloth, at \$3.13 per yard, can be bought for \$20.345?

Operation.

3.13)20.345(6.5 *Ans.*
18 78
1 565
1 565

Since \$3.13 will buy 1 yard \$20.345 will buy as many yards as \$3.13 is contained times in \$20.345. Divide as in simple numbers, and point off one decimal figure in the quotient. (Art. 194.)

Proof.—\$3.13×6.5=\$20.345. Hence,

QUEST.—217. When the number of articles, pounds, &c., and the cost of the whole are given, how is the cost of one article found?

218. When the *price* of *one* article, pound, yard, &c., and the *cost* of the *whole* are given, to find the number of articles, &c.

Divide the whole cost by the price of one, and point off the quotient as in Art. 217.

3. Divide $149.625 by $2.375. *Ans.* 63.

4. If $75 are divided equally among 18 men, how much will each receive?

Operation.

```
18)75($4.16 6 Ans.
   72
   ----
   3000
   18
   ----
   120
   108
   ----
    120
    108
    ----
     12 rem.
```

After dividing the $75 by 18, there is a remainder of 3 dollars, which must be reduced to cents and mills, (Art. 208,) and then be divided as before. The ciphers thus annexed must be regarded as decimals; consequently there will be three decimal figures in the quotient.

219. From the preceding illustrations we derive the following general

RULE FOR DIVIDING FEDERAL MONEY.

Divide as in simple numbers, and point off the quotient as in division of decimal fractions. (Art. 194.)

OBS. After all the figures of the dividend are divided, if there is a remainder, ciphers may be annexed to it, and the operation may be continued as in division of decimals. (Art. 194. Obs. 3.) The ciphers thus annexed must be regarded as decimal places of the dividend.

5. How many pounds of cheese, at 7 cts. a pound, can you buy for $1.47?

QUEST.—218. When the price of 1 article, 1 pound, &c., and the cost of the whole are given, how is the number of articles found? 219. What is the rule for Division of Federal Money? *Obs.* When there is a remainder after all the figures of the dividend are divided, how proceed?

6. A man paid $0.75 for the use of a horse and buggy to go 8 miles: how much was that per mile?

7. How many quarts of cherries, at 7 cents a quart, can you buy for $1.12?

8. How many pounds of figs, at 14 cents a pound, can you buy for $3.57?

9. How many watermelons, at 12½ cts. apiece, can be bought for $3?

10. How many pen-knives, at 20 cts. apiece, can be bought for $7.20?

11. At 17½ cts. a quart, how many quarts of molasses can be bought for $4.40?

12. A man bought 50 pair of thick boots for $175: how much did he give a pair?

13. A man paid $485.50 for 260 sheep: how much did he give per head?

14. At $2.50 a cord, how many cords of wood can I buy for $165?

15. At $4.75 per barrel, how many barrels of flour can I buy for $8.50?

16. If a man's income is $1.68 per day, how much is it per hour?

17. If a man pays $3.62½ per week for board, how long can he board for $188.50?

18. Suppose a man's income is $500 a year, how much is that per day?

19. Suppose a man's interest money is $28.80 per day how much is it per minute?

20. A mason received $94.375 for doing a job, which took him 75½ days: how much did he receive per day?

21. At $1.12½ per bushel, how many bushels of wheat can be bought for $523.75?

22. If $1285.20 were divided equally among 125 men, what would each receive?

23. If $1637.10 were divided equally among 150 men, what would each receive?

24. The salary of the President of the United States is $25000 a year: how much does he receive per day?

APPLICATIONS OF FEDERAL MONEY.

BILLS, ACCOUNTS, &C.

220. A *Bill*, in mercantile operations, is a paper, containing a written statement of the items, and the price or amount of goods sold.

Ex. 1. What is the cost of the several articles, and what the amount, of the following bill?

BOSTON, May 25th, 1845.

James Brown, Esq.

Bought of Fairfield & Lincoln,

5 yds. Broadcloth,	at	$3.25	- -
3 yds. Cambric,	"	.12½	- -
3 doz. Buttons,	"	.15	- -
6 skeins Sewing Silk,	"	.06¼	- -
4 yds. Wadding,	"	.08	- -

Amount, $17.77.

Received Pay't,

Fairfield & Lincoln.

(2.)

NEW HAVEN, Sept. 2d, 1845.

Hon. R. S. Baldwin.

Bought of Durrie & Peck,

4 Lovell's Young Speaker,	at	$ .62½
5 Olmsted's Rudiments,	"	.58
6 Morse's Geography,	"	.50
8 Webster's Spelling Book,	"	.10
3 Day's Algebra,	"	1.25

What was the cost of the several articles, and what the amount of his bill?

(3.)

NEW YORK, Aug. 18th, 1845.

John Jacob Astor, Esq.

Bought of G. W. Lewis & Co.

25 lbs. Sugar,	at $ .09	- - - - - -	
50 lbs. Coffee,	" .11	- - - - - -	
12 lbs. Tea,	" .75	- - - - - -	
14 lbs. Raisins,	" .14	- - - - - -	
9 doz. Eggs,	" .10	- - - - - -	
15 lbs. Butter,	" .12½	- - - - - -	

What was the cost of the several articles, and what the amount of his bill?

(4.)

PHILADELPHIA, June 3d, 1845.

W. A. Sanford, Esq.

To James Conrad, Dr.

For 28 yds. Silk,	at $1.25	- -
" 22 yds. Muslin,	" .56	- -
" 16 pair Cotton Hose,	" .37½	- -
" 35 " Silk "	" 1.10	- -
" 25 " Shoes,	" 1.25	- -

What was the cost of the several articles, and how much is due on his account?

(5.)

CINCINNATI, July 1st, 1845.

Messrs. Holmes & Homer

To H. W. Morgan & Co., Dr.

For 100 bbls. Flour,	at $4.50	-
" 50 " Pork,	" 8.25	- - -
" 25 " Beef,	" 9.75	- - -
" 112 kegs Lard,	" 3.25	- - -
" 25 bush. Corn,	" .34	- - -

What was the cost of the several articles, and how much is due on his account?

(6.)

NEW ORLEANS, Aug. 12th, 1845.

F. C. Emerson, Esq.

To W. H. Arnold & Co., Dr.

For	35 hhds. Molasses,	at	$12.60	- -
"	2100 lbs. Sugar,	"	.05½	- -
"	14000 lbs. Cotton,	"	.07½	- -
"	1350 lbs. Coffee,	"	.06¼	- -
"	31200 lbs. Rice,	"	.08	- -
"	150 boxes Oranges,	"	4.12½	- -

CREDIT.

By 500 Clocks,	- at	$5.00	- -
" Note to balance account,	- -	- -	-

What was the amount of charges, and what the amount of the note?

SECTION IX.

PERCENTAGE.

ART. **222.** The terms *Percentage* and *Per Cent.* signify a certain *allowance on a hundred;* that is, a certain *part* of a hundred, or simply *hundredths.* Thus the expressions 2 per cent., 4 per cent., 6 per cent., &c., of any number or sum of money, signify 2 hundredths ($\frac{2}{100}$,) 4 hundredths ($\frac{4}{100}$,) 6 hundredths ($\frac{6}{100}$,) &c. of that number or sum. For example,

1 per cent. of $100, is $\frac{1}{100}$ of that sum, which is 1 dollar;
2 per cent. of $100, is $\frac{2}{100}$ of that sum, which is 2 dollars;
4 per cent. of $100, is $\frac{4}{100}$ of that sum, which is 4 dollars;
6 per cent. of $100, is $\frac{6}{100}$ of that sum, which is 6 dollars, &c.

Hence, universally,

QUEST.—222. What do the terms percentage and per cent. signify? What is meant by 2 per cent., 4 per cent., &c., of any sum? What then does any given percentage of any number or sum of money imply? *Obs.* From what are the terms percentage and per cent. derived?

222. *a. Any given percentage of any number, or sum of money, implies so many units for every* 100 *units; so many dollars for every* 100 *dollars; so many cents for every* 100 *cents; so many pounds for every* 100 *pounds, &c.*

Note.—The terms *Percentage* and *Per Cent.* are derived from the Latin *per* and *centum*, signifying *by the hundred.*

MENTAL EXERCISES.

Ex. 1. A boy found a purse containing 8 dollars, and on returning it, the owner gave him 6 per cent. of the money: how much did the boy receive?

Suggestion.—6 per cent. is 6 cents for every 100 cents. (Art. 222.) If, then, he received 6 cents for 1 dollar, (100 cents,) for 8 dollars, he must have received 8 times 6 cents, or 48 cents. *Ans.* 48 cents.

2. What is 6 per cent. of 10 dollars? *Ans.* 60 cents.

3. The printer's boy collected 12 dollars on some bills, for which his employer gave him 4 per cent. for his services: how much did he receive?

4. What is 4 per cent. of 8 dollars? 6 dollars? 10 dollars? 7 dollars? 12 dollars? 15 dollars?

5. A man borrowed 12 dollars for a year, and agreed to give 6 per cent. for the use of it: how much did he pay?

6. What is 6 per cent. of 10 dollars? 4 dollars? 6 dollars? 8 dollars? 11 dollars? 15 dollars?

7. A market-man sold 20 dollars' worth of butter for one of his neighbors, who paid him 5 per cent. for his trouble: how much did he receive?

8. What is 5 per cent. of 12 dollars? 10 dollars? 7 dollars? 15 dollars? 20 dollars?

9. A farmer bought a cow for 12 dollars, and sold it again so that he gained 10 per cent. by his bargain: how much did he gain?

10. What is 10 per cent. of 12 dollars? 20 dollars? 14 dollars? 18 dollars? 13 dollars? 15 dollars?

11. What is 11 per cent. of 1 dollar? 4 dollars? 10 dollars? 7 dollars? 6 dollars? 9 dollars? 12 dollars?

12. A boy having 100 canary birds, lost 7 per cent. of them by disease: how many of them did he lose?

13. What is 7 per cent. of 200 dollars? 300 dollars? 500? 900? 700?

Suggestion.—Since 7 per cent. is 7 dollars on 100 dollars, for 200 dollars it is twice as much, or 14 dollars, &c.

14. A merchant invested 500 dollars in a certain adventure, and gained 6 per cent. on the sum invested: how much did he gain?

15. What is 3 per cent. of 100 dollars? 300? 500? 600? 900?

16. What is 5 per cent. of 200 dollars? 400? 300? 700? 1000? 800? 1100?

17. What is 8 per cent. of 200 dollars? 500? 600? 300? 700? 400? 1200?

18. What is 6 per cent. of 300 dollars? 500? 400? 700? 1100? 900? 1200?

19. What is 7 per cent. of 400 dollars? 300? 500? 900? 600? 1000?

20. What is 10 per cent. of 500 dollars? 300? 700? 1200? 1500? 2000?

21. What is 8 per cent. of 200 dollars? 400? 800? 900? 1000?

22. What is 9 per cent. of 300 dollars? 600? 500? 800? 700?

23. What is 12 per cent. of 8 dollars? 7 dollars? 6 dollars?

24. What is 20 per cent. of 4 dollars? 5 dollars? 6 dollars?

25. What is 15 per cent. of 3 dollars? 4 dollars? 6 dollars?

26. What is 18 per cent. of 3 dollars? 5 dollars? 4 dollars?

27. What is 25 per cent of 4 dollars? 5 dollars? 8 dollars?

28. What is 30 per cent. of 3 dollars? 5 dollars? 9 dollars?

EXERCISES FOR THE SLATE.

223. We have seen that *hundredths* are *decimal* expressions, occupying the first two places of figures on the right of the decimal point. (Arts. 181, 182.) Now, since *percentage* and *per cent.* signify *hundredths*, it is manifest they may properly be expressed by decimals.

PERCENTAGE TABLE.

1 per cent.	is written thus:	.01
2 per cent.	" " "	.02
3 per cent.	" " "	.03
4 per cent.	" " "	.04
6 per cent.	" " "	.06
7 per cent.	" " "	.07
10 per cent.	" " "	.10
12 per cent.	" " "	.12
25 per cent.	" " "	.25
50 per cent.	" " "	.50
75 per cent.	" " "	.75
99 per cent.	" " "	.99
100 per cent.	" " "	1.00
103 per cent.	" " "	1.03
125 per cent., &c.	" " "	1.25
$\frac{1}{2}$ per cent., that is, $\frac{1}{2}$ of 1 per cent.	" " "	.005
$\frac{1}{4}$ per cent., that is, $\frac{1}{4}$ of 1 per cent.	" " "	.0025
$\frac{3}{4}$ per cent., that is, $\frac{3}{4}$ of 1 per cent.	" " "	.0075
$13\frac{1}{8}$ per cent.	" " "	.13125
$25\frac{3}{8}$ per cent.	" " "	.25375

OBS. 1. It will be seen from the preceding Table, that when the given per cent. is *less* than 10, a cipher must be prefixed to the figure expressing it, in the same manner as when the number of cents is less than 10. (Art. 206. Obs. 1.)

When the given per cent. is *more* than 100, it must plainly require a mixed number to express it. (Art. 183. Obs. 2.)

2. Parts of 1 per cent. may be expressed either by a *common* fraction, or by *decimals*. Thus, the expression $17\frac{5}{8}$ per cent., is equivalent to .17625 per cent.

3. The *first two* decimal figures properly denote the *per cent.*, for they are *hundredths;* the other decimals being *parts of hundredths*, express *parts* of 1 per cent.

QUEST.—223. How may percentage or per cent. be expressed? *Obs.* When the given per cent. is less than 10, how is it written? When more than 100, how?

1. Write 1 per cent., 2 per cent., 3 per cent., 5 per cent., 8 per cent., and 9 per cent., in decimals.

2. Write 13 per cent.; 15; 30; 50; 75; 49; 73; 85.

3. Write $\frac{1}{2}$ per cent.; $\frac{1}{4}$; $\frac{1}{5}$; $\frac{3}{4}$; $\frac{4}{5}$; $\frac{3}{10}$; $\frac{1}{8}$; $\frac{1}{3}$; $\frac{1}{6}$; $\frac{1}{7}$.

4. Write $3\frac{1}{2}$ per cent.; $5\frac{2}{5}$; $16\frac{1}{8}$; 125; $331\frac{1}{4}$; $462\frac{1}{2}$.

5. A merchant handed some bills amounting to $400 to a constable, and gave him 3 per cent. for collecting them: how much did the constable receive for his services?

Analysis.—Since 3 per cent. is $\frac{3}{100}$, the constable must have received $\frac{3}{100}$ of $400, or 3 dollars for every 100 dollars he collected. Now $\frac{1}{100}$ of $400 is $\frac{400}{100}$, which is equal to $4; and 3 hundredths is 3 times $4, or $12.

Operation.
$400
.03
$12.00. *Ans.*

Since $\frac{3}{100}$=.03, we have only to multiply the given number of dollars by .03 and it will give the answer in cents, which must be reduced to dollars by pointing off 2 decimals. (Art. 209.)

Note.—It is important for the learner to observe, that the *amount* of money *collected*, is made the *basis* upon which the percentage is computed. That is, the constable is entitled to 3 dollars, as often as he *collects* 100 dollars, and *not* as often as he *pays over* 100 dollars, as is frequently supposed. For in the latter case he would receive only $\frac{3}{103}$, instead of $\frac{3}{100}$ of the sum in question. This distinction is important, especially in calculating percentage on large sums. Hence,

225. To calculate percentage on any number or sum of money.

Multiply the given number or sum by the given per cent. expressed decimally; and point off the product as in multiplication of decimal fractions. (Art. 191.)

OBS. If the per cent. contains a common fraction which cannot be expressed decimally, first multiply by the decimal, then by the common fraction of the given per cent., and point off the sum of their products as above.

QUEST.—*Note.* When it is said that a man receives a certain per cent. for collecting money, upon what is the per cent. calculated? 225. How do you calculate percentage? *Obs.* If the per cent. contains a common fraction which cannot be expressed decimally, how proceed?

6. What is 2 per cent. of \$350? *Ans.* \$7.
7. What is 4 per cent. of \$63? *Ans.* \$2.52.
8. What is 3 per cent. of \$145.25? *Ans.* \$4.3575.
9. What is $\frac{1}{2}$ per cent. of \$180.42? (Art. 223. Obs. 2.)
10. What is $\frac{1}{4}$ per cent. of \$827.63?
11. What is $\frac{3}{4}$ per cent. of \$128.632?
12. What is $\frac{1}{5}$ per cent.f o \$90.45?
13. What is 10 per cent. of \$600.451?
14. What is 12 per cent. of \$2500.63?
15. What is 20 per cent. of \$2250.84?
16. What is $3\frac{1}{5}$ per cent. of \$436?

Suggestion.—$3\frac{1}{5}$ per ct. is equal to .032. (Art. 223. Obs. 2.)

17. What is $2\frac{1}{3}$ per cent. of \$144?

Operation.

\$144
$.02\frac{1}{3}$
\$2.88=2 per ct.
48=$\frac{1}{3}$ per cent.
\$3.36 *Ans.*

Since $\frac{1}{3}$ per cent. cannot be exactly expressed by decimals, we first multiply by .02, and then by $\frac{1}{3}$; (Art. 134. *a*;) and point off two decimal places. (Art. 225. Obs.)

18. What is $4\frac{1}{2}$ per cent. of \$257?
19. What is $8\frac{3}{4}$ per cent. of \$673?
20. A merchant having deposited \$200 in a bank, afterwards drew out $10\frac{1}{2}$ per cent. of it: how many dollars did he draw out?
21. A merchant makes a deposit of \$1864 and draws out 25 per cent. of it: how much has he left in the bank?
22. A merchant shipped 865 boxes of lemons; on the passage home, 15 per cent. of them were thrown overboard: how many boxes did he lose; and how many had he left?
23. How much is $6\frac{1}{3}$ per cent. of \$1000?
24. How much is 7 per cent. of \$1526.33?
25. How much is $8\frac{1}{6}$ per cent of \$16.325?
26. A young man worth \$1,500, lost $31\frac{1}{4}$ per cent. of it in gambling: how much did he lose; and how much had he left?

27. A merchant bought a cargo of flour for $1230, and paid $4\frac{1}{2}$ per cent. for bringing it home: what was the whole cost of his flour?

28. What is $37\frac{1}{2}$ per cent. of $100? Of $2537.50?

29. What is 112 per cent. of $150?

30. What is 125 per cent. of $635?

31. What is 250 per cent. of $17.35?

32. Which is the most, 7 per cent. of $1000, or 6 per cent. of $1100?

33. What is the difference between 6 per cent. and 7 per cent. of $12000?

34. What is the difference between 9 per cent. of $2000, and 6 per cent. of $3000?

35. What is $17\frac{1}{5}$ per cent. of $10000?

36. What is $20\frac{1}{2}$ per cent. of $10500?

37. A man gave his two sons $10000 apiece; the elder added $15\frac{1}{2}$ per cent. to his the first year, and the younger spent $15\frac{1}{2}$ per cent. of his: what was the difference of their property at the end of the first year?

38. A labouring man earning $225 a year, laid up $23\frac{1}{2}$ per cent. of it: how much did he spend?

39. A man having deposited $856.25 in a savings bank, drew out $31\frac{1}{4}$ per cent. of it: how much had he left in the bank?

40. A farmer owning 3560 sheep, lost 50 per cent. of them by disease: how many had he left?

APPLICATIONS OF PERCENTAGE.

226. PERCENTAGE, or the method of reckoning by *hundredths*, is applied to various calculations in the practical concerns of life. Among the most important of these are Commission, Brokerage, the Rise and Fall of Stocks, Interest, Discount, Insurance, Profit and Loss, Duties, and Taxes. Its principles, therefore, should be thoroughly understood by every scholar.

QUEST.—226. To what are the principles of percentage applied? What are some of the most important of these calculations?

COMMISSION, BROKERAGE, AND STOCKS.

227. *Commission* is the *per cent.* or *sum charged* by agents for their services in buying and selling goods, or transacting other business.

OBS. An *Agent* who buys and sells goods for another, is called a *Commission Merchant, a Factor, or Correspondent.*

228. *Brokerage* is the *per cent.* or *sum charged* by money dealers, called *Brokers*, for negotiating *Bills of Exchange*, and other monetary operations, and is of the same nature as Commission.

229. By the term *Stocks*, is meant the *Capital* of moneyed institutions, as incorporated Banks, Manufactories, Railroad and Insurance Companies; also, the funds of Government, State Bonds, &c.

OBS. Stocks are usually divided into portions of $100 each, called *shares;* and the owners of these shares are called *Stockholders.*

230. The original *cost* or *valuation* of a share is called its *nominal*, or *par value;* the *sum* for which it can be sold, is its *real value*, which varies at different times.

OBS. 1. The *rise* or *fall* of Stocks is reckoned at a certain per cent. of its *par value.* The term *par* is a Latin word, which signifies *equal*, or a *state of equality.*

2. When stocks sell for their original cost or valuation, they are said to be *at par;* when they sell for more than cost, they are said to be *above par*, or *at an advance;* when they do not sell at cost, they are said to be *below par*, or *at a discount.*

3. Persons who deal in Stocks are usually called *Stock Brokers*, or *Stock Jobbers.*

231. The *commission* or *allowance* made to factors and brokers, and the *rise* and *fall* of stocks, are usually reckoned at a *certain percentage* on the *amount* of money

---

QUEST.—227. What is commission? *Obs.* What is an agent who buys and sells goods for another usually called? 228. What is brokerage? 229. What is meant by stocks? How are stocks usually divided? What are the owners of the shares called? 230. What is the par value of stocks? What the real value? *Obs.* What is the meaning of the term par? When are stocks at par? When above par? When below? 231. How are commission, brokerage, and the rise or fall of stocks reckoned?

employed in the transaction, or on the *par value* of the given shares. Hence,

232. To compute commission, brokerage, and the advance or discount on stocks.

Multiply the given sum by the given per cent. expressed in decimals, and point off the product as in Percentage.

When the commission is to be deducted in advance from a specified sum and the balance invested, divide the given amount by $1 increased by the per cent. commission, and the quotient will be the part to be invested. Subtract the part invested from the given amount, and the remainder will be the commission required.

EXAMPLES.

Ex. 1. A commission merchant sold a quantity of corn for $236, and charged 2 per cent. commission: how much did he receive for his services? *Ans.* $4.72.

2. A tax-gatherer collected $533.56, for which he was to have 3 per cent. commission: what did he receive?

3. An auctioneer sold $860.45 worth of goods, at $2\frac{1}{2}$ per cent. commission: how much did he receive?

4. An agent sold a quantity of oil for $265.35, and charged $2\frac{1}{4}$ per cent. commission: how much did the agent receive; and how much the owner?

5. Sold goods to the amount of $356, at $4\frac{1}{3}$ per cent. commission: how much did I receive for my services?

6. Bought goods amounting to $480, at $3\frac{1}{4}$ per cent. commission: what is the amount of my commission?

7. What is the commission on $163.625, at $6\frac{1}{4}$ per ct.?

8. What is the commission, at $5\frac{1}{5}$ per cent. for purchasing flour to the amount of $1365.25?

9. I send my agent $1000 to be laid out in cotton, and pay him $5\frac{3}{4}$ per cent. commission: what is his commission; and how many dollars worth of cotton shall I receive?

10. A man sent a broker $10478.13 to lay out in stocks after deducting his brokerage, at $\frac{1}{2}$ per cent.: what was the brokerage; and how much stock did he receive?

QUEST.—232. How compute commission, brokerage, &c. on any given sum? How, when the commission is to be deducted in advance?

11. A merchant negotiated a bill of exchange of $5000 with a broker, and agreed to give him 7 per cent.: how much did the broker receive?

12. What is the brokerage on $8265, at 5½ per ct.?

13. What is the brokerage on $6524.13, at 8 per cent?

14. What is the commission on $146.356, at 20 per cent.?

15. What is the commission on $1625.75, at 25 per cent.?

16. What is the commission on $25026.10, at 15 per cent.?

17. What is the brokerage on $50265.95, at 3½ per cent.?

18. What is the brokerage on $38212.085, at 1¼ per cent.?

19. What is the brokerage on $752600, at 1 per cent.?

20. What is the brokerage on $1000000, at ½ per cent.?

21. Bought 10 shares of bank stock, for which I agreed to pay 4 per cent. advance: how much did the stock cost me?

Suggestion.—The stock manifestly cost me its par value, viz: $1000, together with 4 per cent. of it. (Art. 229. Obs.) Now $1000×.04=$40; and $1000+$40=$1040. *Ans.*

22. A man bought 5 shares of the Boston and Providence Railroad stock, at 5½ per cent. advance: what did his stock cost him?

23. A stock broker bought 15 shares of the New York and Erie Railroad stock at par, and sold them at 15 per cent. discount: what did they come to?

24. Sold 29 shares in the American Manufacturing Co., at 16 per cent. advance: what did they come to?

25. A stock jobber bought 45 shares of the Auburn and Rochester Railroad stock at 3 per cent. discount, which he sold at 7 per cent. advance: how much did he make by the transaction?

26. A widow invested $9000 in the Commonwealth Bank stock at par, and finally sold it at 75 per cent. discount: how much did she lose?

27. A man owned 53 shares of the Long Island Railroad stock, which he sold at auction, at 13 per cent. advance: how much did they come to?

28. Bought 38 shares in the Union Gas Co. at 7 per cent. advance, and sold them at 5 per cent. discount: how much was my loss?

INTEREST.

233. INTEREST is the sum paid for the *use of money* by the borrower to the lender. It is reckoned at a given *per cent. per annum;* that is, so many dollars are paid for the use of $100 for *one year;* so many cents for 100 cents: so many pounds for £100; &c.

OBS. The learner should be careful to notice the distinction between *Commission* and *Interest.* The *former* is reckoned at a certain *per cent.* without regard to time; (Art. 231;) the latter is reckoned at a certain *per cent.* for *one year;* consequently, for *longer* or *shorter* periods than one year, *like proportions* of the percentage for one year are taken.

The term *per annum,* signifies *for a year.*

234. The *money lent,* or that for which interest is paid, is called the *principal.*

The *per cent.* paid per annum, is called the *rate.*

The *sum* of the principal and interest, is called the *amount.* Thus, if I borrow $100 for 1 year, and agree to pay 5 per cent. for the use of it, at the end of the year I must pay the lender the $100 the sum which I borrowed and $5 interest, making $105. The principal in this case, is $100; the interest $5; the rate 5 per cent.; and the amount $105.

QUEST.—232. What is Interest? How is it reckoned? *Obs.* What is meant by the term per annum? 234. What is meant by the principal? The rate? The amount? 235. How is the rate usually determined? Is it the same everywhere?

235. The *rate* of interest is usually established by law. It varies in different countries and in different parts of our own country.

OBS. 1. The *legal rate* of interest in New England, New Jersey, Pennsylvania, Delaware, Maryland, Virginia, North Carolina, Tennessee, Kentucky, Ohio, Indiana, Illinois, Missouri, and Arkansas, is 6 per cent.

In New York, South Carolina, Michigan, Wisconsin, and Iowa, it is 7 per cent.

In Georgia, Alabama, Mississippi, and Florida, it is 8 per cent.; and in Louisiana but 5 per cent.

On debts and judgments in favour of the *United States*, interest is computed at 6 per cent.

2. In *England* and *France* the legal rate is 5 per cent.; in *Ireland*, 6 per cent. In *Italy* about the commencement of the 13th century, it varied from 20 to 30 per cent.

236. Any rate of interest *higher* than the legal rate, is called *usury*, and the person exacting it is liable to a heavy penalty.

Any rate *less* than the legal rate may be taken, if the parties concerned so agree.

OBS. 1. When no rate is mentioned, the rate established by the laws of the State in which the transaction takes place, is always understood to be the one intended by the parties.

2. The term *per annum*, is seldom expressed in connexion with the *rate per cent.*, but it is always understood; for the *rate* is the *per cent.* paid *per annum*. (Art. 234.)

Ex. 1. What is the interest of $15 for 1 year, at 4 per cent.?

Suggestion.—4 per cent. is $\frac{4}{100}$; that is, $4 for $100, 4 cents for 100 cents, &c. (Art. 222. *a.*) Now as the interest of $1 (100 cents) for a year, is 4 cents, the interest of $15 for the same time, is 15 times as much. And 15 times 4 cents are 60 cents. *Ans.* 60 cents.

QUEST.—*Obs.* What is the legal rate of interest in New England, New Jersey, &c.? What is the legal rate of interest in New York, South Carolina, &c.? In Georgia, Alabama, &c.? On debts due the United States? What is the legal rate of interest in England and France? Ireland? 236. What is any rate higher than the legal rate called? What is the consequence of exacting usury? Is it safe to take less than legal interest? *Obs.* When no rate is mentioned, what rate is understood?

Operation.	
$15 Prin.	We multiply the principal by the given rate per cent. expressed in decimals, as in per centage; (Art. 225;) and point off as many decimals in the product as there are decimal places in both factors.
.04 Rate.	
$.60 Int.	

2. What is the interest of $45 for 1 year, at 3 per cent.? $1.35 *Ans.*

3. What is the interest of $32.125 for 1 year, at 4½ per cent.?

Operation.	
$32.125 Prin.	4½ per cent. expressed in decimals is .045. (Art. 223.) Multiply, &c. as above, and point off 6 decimals in the product. (Art. 191.) The fractions of a mill may be omitted in the answer. Hence,
.045 Rate.	
160625	
128500	
$1.445625 *Ans.*	

237. To find the interest of any sum, at any given rate for 1 year.

Multiply the principal by the given rate per cent. expressed in decimals, and point off the product as in multiplication of decimal fractions. (Art. 191.)

The amount is found by adding the principal and interest together. (Art. 234.)

OBS. 1. In adding the principal and interest, care must be taken to add dollars to dollars, cents to cents, &c. (Art. 211.)

2. When the rate per cent, is *less* than 10, a cipher must always be prefixed to the figure denoting it. (Art. 223. Obs. 1.) It is highly important that the principal and the rate should both be *written* correctly, in order to prevent mistakes in pointing off the product.

4. What is the interest of $75.21 for 1 year, at 6 per cent.? $4.5126. *Ans.*

5. What is the interest of $100 for 1 year, at 5 per cent.? at 6 per cent.? at 4 per cent.? at 7 per cent.?

QUEST.—237. How do you compute interest for 1 year? How find the amount? *Obs.* What precaution is necessary in adding the principal and interest together? When the rate is less than 10 per cent., how is it written?

6. What is the interest of $35.31 for 1 year, at 6 per cent.?

7. What is the interest of $50.10 for 1 year, at 7 per cent.?

8. What is the interest of $63 for 1 year, at $5\frac{1}{2}$ per cent.?

9. What is the interest of $136.75 for 1 year, at $4\frac{1}{2}$ per cent.?

10. What is the interest of $260.61 for 1 year, at 6 per cent.? What is the amount?

Ans. $15.636 int. $276.246 amount.

11. What is the interest of $140.25 for 1 year, at 7 per cent.? What is the amount?

12. What is the interest of $163.40 for 1 year, at 8 per cent.? What is the amount?

13. What is the interest of $400 for 1 year, at 6 per cent.? What is the amount?

14. What is the amount of $500 for 1 year, at 7 per cent.?

15. What is the amount of $1000 for 1 year, at 8 per cent.?

16. What is the interest of $100 for 3 years, at 6 per cent. per annum?

Operation.

$100 Prin.
.06 Rate.
$6.00 Int. 1 y.
3 No. of y.
$18.00 Int. for 3 y.

The interest for 3 years is plainly 3 times as much as for 1 year. We therefore first find the interest for 1 year as above, which is $6; then multiplying this by 3, gives the interest for 3 years. Hence,

238. To compute the interest of any sum for a given number of years.

First find the interest of the given sum for 1 *year, at the given rate;* (Art. 237;) *then multiply the interest of* 1 *year by the given number of years.*

QUEST.—238. How is interest computed for any number of years?

17. At 5 per cent. per annum, what is the interest of $45 for 4 years? *Ans.* $9.

18. At 6 per cent., what is the interest of $200 for 5 years? What is the amount?

19. At 7 per cent., what is the interest of $250 for 10 years? What is the amount?

20. At 8 per cent., what is the interest of $340.50 for 3 years? What is the amount?

21. At 6 per cent. per annum, what is the interest of $100 for 1 month?

Operation.

```
     $100 Prin.
      .06 Rate.
12)6.00 Int. for 1 y.
    $.50 Int. for 1 m.
```

1 month is $\frac{1}{12}$ of 12 months or a year, therefore the interest for 1 month will be $\frac{1}{12}$ as much as the interest for 1 year. Now the interest of $100 for 1 year is $6, and $\frac{1}{12}$ of $6, is 50 cts. In like manner any number of months may be considered a fractional part of a year, and the interest for them may be computed in the same way. Hence,

239. To compute the interest of any sum for a given number of months.

First find the interest for 1 *year as above; then take such a fractional part of* 1 *year's interest, as is denoted by the given number of months.*

Thus, for 1 month take $\frac{1}{12}$ of 1 year's interest; for 2 months, $\frac{2}{12}$ or $\frac{1}{6}$; for 3 months, $\frac{3}{12}$ or $\frac{1}{4}$; for 4 months, $\frac{4}{12}$ or $\frac{1}{3}$; for 6 months, $\frac{6}{12}$ or $\frac{1}{2}$; &c.

22. At 5 per cent., what is the interest of $600 for 6 months. *Ans.* $15.

23. At 7 per cent., what is the interest of $250 for 4 months?

QUEST.—239. How is interest computed for months? For 2 months, what part would you take? For 3 months? 4 months? 5 months? 6 months? 7 months? 8 months? 9 months? 10 months? 11 months?

24. What is the interest of \$375.31 for 3 months, at 6 per cent.?

25. What is the interest of \$60 for 7 months, at 8 per cent.? What is the amount?

26. What is the interest of \$96 for 10 months, at 6 per cent.? What is the amount?

27. At 6 per cent., what is the interest of \$600 for 1 day?

Operation.

	\$600 Prin.
	.06 Rate.
12)	\$36.00 In. for 1 y.
30)	3.00 In. for 1 m.
Ans.	\$0.10 In. for 1 d.

1 day is $\frac{1}{30}$ of 30 days, or a month; hence the interest for 1 day will be $\frac{1}{30}$ of the interest for 1 month. If, therefore, we find the interest for 1 month, and take $\frac{1}{30}$ of this, it will evidently be the interest for 1 day. In like manner, any number of days may be considered a fractional part of a month, and the interest for them may be found in the same way. Hence,

240. To compute the interest of any sum for a given number of days.

First find the interest for 1 month as above, then take such a fractional part of 1 month's interest as is denoted by the given number of days. Thus for 1 day take $\frac{1}{30}$ of 1 month's interest; for 2 days, $\frac{2}{30}$, or $\frac{1}{15}$; for 3 days. $\frac{3}{30}$, or $\frac{1}{10}$; for 10 days, $\frac{1}{3}$; for 20 days, $\frac{2}{3}$; &c.

28. At 4 per cent., what is the interest of \$470 for 10 days? *Ans.* \$0.522.

29. What is the interest of \$1000 for 1 y. 1 m. and 1 d., at 6 per cent.?

30. What is the interest of \$42.50 for 2 years and 6 months, at 7 per cent.?

31. What is the interest of \$69.46 for 1 year and 8 months, at 8 per cent.?

Quest.—240. How is interest computed for days? For 2 days, what part would you take? For 5 days? 7 days? 12 days? 25 days?

241. From the foregoing principles we may deduce the following general

RULE FOR COMPUTING INTEREST.

I. FOR ONE YEAR. *Multiply the principal by the given rate, and from the product point off as many figures for decimals, as there are decimal places in both factors.* (Art. 237.)

II. FOR TWO OR MORE YEARS. *Multiply the interest of* 1 *year by the given number of years.* (Art. 238.)

III. FOR MONTHS. *Take such a fractional part of* 1 *year's interest, as is denoted by the given number of months.* (Art. 239.)

IV. FOR DAYS. *Take such a fractional part of* 1 *month's interest, as is denoted by the given number of days.*

OBS. 1. In calculating interest, a month, whether it contains 30 or 31 days, or even but 28 or 29, as in the case of February, is usually assumed to be *one twelfth* of a year.

2. In calculating interest 30 days are considered a month; consequently the interest for 1 day, or any number of days under 30, is so many *thirtieths* of a month's interest. (Art. 170. Obs. 2.)

This practice seems to have been originally adopted on account of its convenience. Though not strictly accurate, it is sanctioned by custom, and is everywhere allowed by law.

32. What is the interest of $45.23 for 1 year and 2 months, at 5 per cent.?

33. What is the interest of $43.01 for $2\frac{1}{2}$ years, at 7 per cent.?

34. What is the interest of $215.135 for 2 years and 3 months, at 6 per cent.?

35. At 8 per cent., what is the interest of $75.98 for 3 years?

36. At $5\frac{1}{2}$ per cent., what is the interest of $939 for 4 years?

37. At 6 per cent., what is the interest of $137.50 for 6 months?

QUEST.—241. What is the general rule for computing interest? *Obs.* In reckoning interest, what part of a year is a month considered? How many days are considered a month? Is this practice strictly accurate?

38. At 7 per cent., what is the interest of $1500 for 10 days?

39. At 20 per cent., what is the interest of $3000 for 3 days?

40. At 12½ per cent., what is the interest of $1736.25 for 6 months?

SECOND METHOD OF COMPUTING INTEREST.

242. There is another method of computing interest, which is very simple and convenient in its application, particularly when the interest is required for *months* and *days*, at 6 *per cent.*

INTEREST OF $1 FOR MONTHS.

243. We have seen, (Art. 237,) that,

For 1 year,	the interest of $1 is	6 cents,	which is	$.06;
" 1 month,	" "	is $\frac{1}{12}$ of 6 cents,	" "	.005;
" 2 months,	" "	is $\frac{2}{12}$, or $\frac{1}{6}$ of 6 cents,	" "	.01;
" 3 months,	" "	is $\frac{3}{12}$, or $\frac{1}{4}$ of 6 cents,	" "	.015;
" 4 months,	" "	is $\frac{4}{12}$, or $\frac{1}{3}$ of 6 cents,	" "	.02;
" 5 months,	" "	is $\frac{5}{12}$ of 6 cents,	" "	.025;
" 6 months,	" "	is $\frac{6}{12}$, or $\frac{1}{2}$ of 6 cents,	" "	.03;

That is, *the interest of* $1 *for* 1 *month, at* 6 *per cent., is* 5 *mills; and for every* 2 *months, it is* 1 *cent, &c.* Hence,

244. To find the interest of $1 for any number of months, at 6 per cent.

Multiply the interest of $1 *for* 1 *month,* ($.005,) *by the given number of months, and point off* 3 *decimal figures in the product.* (Art. 191.)

1. At 6 per cent., what is the interest of $1 for 7 months?

QUEST.—244. How may the interest of $1 be found for any number of months, at 6 per cent.?

2. At 6 per cent., what is the interest of $1 for 8 months?

3. At 6 per cent., what is the interest of $1 for 9 months? For 10 months? For 11 months?

4. At 6 per cent., what is the interest of $1 for 14 months? For 15 months? For 18 months?

INTEREST OF 1$ FOR DAYS.

245. Since the interest of $1 for 1 mo. (30 d.) is 5 mills, or $.005, (Art. 243,)

For 6 days ($\frac{1}{5}$ of 30 d.) the interest of $1 is $\frac{1}{5}$ of 5 mills, or $.001;
" 12 days ($\frac{2}{5}$ of 30 d.) " " is $\frac{2}{5}$ of 5 mills, or .002;
" 18 days ($\frac{3}{5}$ of 30 d.) " " is $\frac{3}{5}$ of 5 mills, or .003;
" 24 days ($\frac{4}{5}$ of 30 d.) " " is $\frac{4}{5}$ of 5 mills, or .004;
" 3 days ($\frac{1}{10}$ of 30 d.) " " is $\frac{1}{10}$ of 5 mills, or .0005:

That is, the interest of $1 for *every* 6 *days*, is 1 *mill*, or $.001; and for any number of days, it is as many *mills, or thousandths of a dollar, as* 6 *is contained times* in the given number of days. Hence,

246. To find the interest of $1 for any number of days, at 6 per cent.

Divide the given number of days by 6, *and set the first quotient figure in thousandths' place, when the days are* 6, *or more than* 6; *but in ten thousandths' place, when they are less than* 6.

OBS. For 60 days (2 mo.) the interest of $1 is 1 cent; (Art. 243;) in this case, therefore, the first quotient figure must occupy *hundredths'* place.

5. What is the interest of $1 for 1 day, at 6 per cent. expressed decimally? *Ans.* $.000166+

6. What is the interest of $1 for 9 days, at 6 per cent.? 22 days? 4 days? 14 days?

QUEST.—246. How may the interest of $1 be found for any number of days, at 6 per cent.?

7. What is the interest of $1 for 10 days, at 6 per cent.? 16 days? 20 days? 24 days? 27 days? 28 days?

8. What is the interest of $1 for 1 year, 5 months, and 3 days, at 6 per cent.?

Solution.—For 1 year,	the interest of $1 is	$.06
" 5 months,	" " "	.025
" 3 days,	" " "	.0005
	Ans.	$.0855

9. What is the interest of $1 for 2 years, 7 months, and 20 days, at 6 per cent.?

10. What is the interest of $1 for 3 years, 1 month, and 15 days, at 6 per cent.?

11. What is the interest of $145 for 6 months, and 24 days, at 6 per cent.?

Operation.
$145 Prin.
$3\frac{2}{5}=\frac{1}{2}$ the mo.
435 Int. for 6 mo.
58 " 24 d.
$4.93 *Ans.*

Since the int. of $1 for 1 mo. is 5 mills, or 1 cent for 2 mo., it is manifest the answer may be found by multiplying the prin. by *half* the number of months, regarding the *days* as a *fractional part* of a mo.; for, the int. of $1 is equal to *half as many cents* as there are months in the given time.

247. From these illustrations we may derive a

SECOND RULE FOR COMPUTING INTEREST.

Multiply the principal by the interest of $1 *for the given time, and point off the product as before.* (Art. 241.)

Or, multiply the principal by half the number of months, and point off two more decimals in the product than there are decimal figures in the multiplicand.

OBS. 1. In the latter method, the years must be reduced to months, and the days to the fraction of a month, then take *half* of them.

The interest at any other rate, *greater*, or *less* than 6 per cent. may be found by adding to, or subtracting from the interest at 6 per cent., such a *fractional part of itself*, as the required rate exceeds or falls short of 6 per cent. Thus, if the required rate is 7 per cent., first find the interest at 6 per cent., then add $\frac{1}{6}$ of it to itself; if 5 per cent., subtract $\frac{1}{6}$ of it from itself, &c.

QUEST.—247. What is the second method of computing interest? *Obs.* When the rate is greater or less than 6 per cent., how proceed?

2. When it is required to compute the interest on a *note*, we must first find the *time* for which the note has been on interest, by subtracting the *earlier* from the *later* date; (Art. 170;) then cast the interest on the face of the note for the time, by either of the preceding methods. (Arts. 241, 247.)

13. What is the interest of $300 for 4 months, and 18 days, at 7 per cent.?

Operation.

$300 Prin.
.023 Int. of $1 for the time.
900
600
6)$6.900=Int. at 6 per ct.
1 150=$\frac{1}{6}$ of 6 per cent.
Ans. $8.050 Int. at 7 per ct.

The required rate is 1 per cent. more than 6 per cent.; we therefore find the interest at 6 per cent., and add $\frac{1}{6}$ of it to itself.

14. At 5 per cent., what is the interest of $256.25 for 9 months and 15 days?

15. What is the interest of $450 from Jan. 1st, 1844, to March 13th, 1845, at 6 per cent.?

Operation.

	Yr.	*mo.*	*d.*
	1845	3	13
	1844	1	1
Time	1	2	12

$450 Principal.
.072 Int. of $1 for the time.
900
31 50
$32.400 *Ans.*

EXAMPLES FOR PRACTICE.

1. What is the interest of $45.25 for 8 months, at 6 per cent.?

2. What is the interest of $167.375 for 6 months, at 6 per cent.?

3. What is the interest of $93.86 for 3 months and 15 days, at 6 per cent.?

4. What is the interest of $110 for 1 month and 20 days, at 6 per cent.?

5. At 7 per cent., what is the interest of $158.91 for 1 year and 3 months?

QUEST.—247. How compute the interest on a note?

6. At 7 per cent., what is the amount of $217 for 1 year and 8 months?

7. At 6 per cent., what is the amount of $348.10 for 2 years and 1 month?

8. At 7 per cent., what is the interest of $400 for 1 year and 6 months?

9. At 7 per cent., what is the amount of $213.01 for 9 months?

10. At 5 per cent., what is the amount of $603 for 2 years and 5 months?

11. What is the amount of $861 for 8 months and 24 days, at 6 per cent.?

12. What is the amount of $1236 for 3 months and 14 days, at 7 per cent.?

13. What is the interest of $1400 for 1 year, 1 month and 9 days, at 7 per cent.?

14. What is the interest of $469.20 for 27 days, at 8 per cent.?

15. What is the amount of $705 for 5 years, at 9 per cent.?

16. What is the amount of $1000 for 10 years, at 5 per cent.?

17. What is the amount of $1650.06 for 20 years, at 7 per cent.?

18. What is the amount of $2500 for 7 years, at 15 per cent.?

19. At $4\frac{1}{2}$ per cent., what is the interest of $17000 for $1\frac{1}{2}$ years?

20. At $7\frac{1}{4}$ per cent., what is the interest of $1625.81 for 45 days?

21. At $12\frac{1}{2}$ per cent., what is the amount of $165.13 for 33 days?

22. At 7 per cent., what is the amount of $8531 for 63 days?

23. At 6 per cent., what is the amount of $16021 for 93 days?

24. What is the interest on a note of $65, dated Jan. 10th, 1844, to May 16th, 1845, at 6 per cent.?

25. What is the interest of $170 from June 19th, 1840, to July 1st, 1841, at 7 per cent.?

26. What is the interest of $105.63 from Feb. 22d, 1839, to Aug. 10th, 1840, at 5 per cent.?

27. What is the interest of $234 from April 10th, 1834, to Oct. 1st, 1835, at 6 per cent.?

28. What is the interest of $195.22 from June 25th, 1838, to March 31st, 1840, at 6 per cent.?

29. What is the interest of $391 from Sept. 1st, 1840, to Nov. 30th, 1841, at 8 per cent.?

30. What is the interest of $510.83 from March 21st, 1842, to Dec. 30th, 1842, at 7 per cent.?

31. At 6 per cent., what is the interest of $469.65 from August 10th, 1843, to Feb. 6th, 1844?

32. At 7 per cent., what is the amount due on a note of $285, dated March 15th, 1844, and payable Sept. 18th, 1845?

33. At 6 per cent., what is the amount due on a note of $391, dated Oct. 9th, 1844, and payable March 1st, 1845?

34. At 5 per cent., what is the amount of $623 from Feb. 19th, 1844, to Aug. 10th, 1844?

35. At 4 per cent., what is the amount of $589.20 from January 10th, 1844, to January 13th, 1845?

36. At 4 per cent., what is the amount of $731.27 from July 1st, 1844, to April 4th, 1845?

37. What is the interest of $849 from July 4th, 1841, to July 7th, 1845, at 6 per cent.?

38. What is the interest of $966 from Jan. 1st, 1842, to March 20th, 1844, at 7 per cent.?

39. What is the interest of $1539 from May 21st, 1842, to Aug. 19th, 1843, at 6 per cent.?

40. What is the amount of $1100 from June 15th, 1840, to Aug. 3d, 1845, at 5 per cent.?

41. What is the amount of $1 for 50 years, at 6 per ct.? At 7 per cent.?

42. What is the amount of one cent for 500 years, at 7 per cent.?

PARTIAL PAYMENTS.

248. When *partial payments* are made and endorsed upon Notes and Bonds, the rule for computing the interest adopted by the *Supreme Court of the United States*, is the following.

I. "*The rule for casting interest, when partial payments have been made, is to apply the payment, in the first place, to the discharge of the interest then due.*

II. "*If the payment exceeds the interest, the surplus goes towards discharging the principal, and the subsequent interest is to be computed on the balance of principal remaining due.*

III. "*If the payment be less than the interest, the surplus of interest must not be taken to augment the principal; but interest continues on the former principal until the period when the payments, taken together, exceed the interest due, and then the surplus is to be applied towards discharging the principal; and interest is to be computed on the balance as aforesaid.*"

Note.—The above rule is adopted by *Massachusetts*, *New York*, and the other States of the Union, with but few exceptions. It is given in the language of the distinguished Chancellor Kent.—*Johnson's Chancery Reports*, Vol. I. p. 17.

$850. NEW HAVEN, Jan. 1st, 1841.

43. For value received, I promise to pay George Howland, or order, eight hundred and fifty dollars, on demand, with interest at 6 per cent.

JOHN HAMILTON.

The following payments were endorsed on this note:

July 1st, 1841, received $100.62.
Dec. 1st, 1841, received $15.28.
Aug. 13th, 1842, received $175.75.

What was due on taking up the note, Jan. 1st, 1843?

QUEST.—248. What is the general method of casting interest on Notes and Bonds, when partial payments have been made?

Operation.

Principal, - - - - - - -		$850.00
Interest to first payment, July 1st, (6 months,)		25.50
Amount due on note July 1st, - - -		$875.50
1st payment, (to be deducted from amount,)		100.62
Balance due July 1st, - - - - -		$774.88
Int. on Bal. to 2d pay't Dec. 1st, (5 mo.,)	$19.37	
2d pay't (which is less than the interest then due,)	15.28	
Surplus interest unpaid Dec. 1st,	$4.09	
Int. continued on Bal. from Dec. 1st, 1842, to Aug. 13th, (8 mo., 12 d.,)	32.54	36.63
Amount due Aug. 13th, 1842.		$811.51
3d payment (being greater than the interest now due) is to be deducted from the am't.		175.75
Balance due Aug. 13th, - - - -		$635.76
Int. on Bal. to Jan. 1st, (4 mo., 18d.,) -		14.62
Bal. due on taking up the note, Jan. 1st, 1843,		$650.38

$500. NEW YORK, May 10th, 1842.

44. For value received, I promise to pay James Monroe, or order, five hundred dollars on demand, with interest at 7 per cent.

HENRY SMITH.

The following sums were endorsed upon it:

Received, Nov. 10th, 1842, $75.
Received, March 22d, 1843, $100.

What was due on taking up the note, Sept. 28th, 1843?

$692.35. BOSTON, Aug. 15th, 1843.

45. Three months after date, I promise to pay John Warren, or order, six hundred and ninety-two dollars and thirty-five cents, with interest at 6 per cent., value received.

SAMUEL JOHNSON.

Endorsed, Nov. 15th, 1843, $250.375.
" March 1st, 1844, $65.625.

How much was due July 4th, 1845?

$1000. PHILADELPHIA, June 20th, 1841.

46. Six months after date, I promise to pay Messrs. Carey, Hart & Co., or order, one thousand dollars, with interest at 5 per cent., value received.

HORACE PRESTON.

Endorsed, Jan. 10th, 1844, $125.
" June 16th, 1844, $93.
" Feb. 20th, 1845, $200.

What was the balance due Aug. 1st, 1845?

CONNECTICUT RULE.

249. "Compute the interest on the principal to the time of the first payment; if that be one year or more from the time the interest commenced, add it to the principal, and deduct the payment from the sum total. If there be after payments made, compute the interest on the balance due to the next payment, and then deduct the payment as above; and in like manner, from one payment to another, till all the payments are absorbed; provided the time between one payment and another be one year or more. But if any payments be made before one year's interest hath accrued, then compute the interest on the principal sum due on the obligation, for one year, add it to the principal, and compute the interest on the sum paid, from the time it was paid up to the end of the year; add it to the sum paid, and deduct that sum from the principal and interest added as above.*

"If any payments be made of a less sum than the interest arisen at the time of such payment, no interest is to be computed, but only on the principal sum for any period."—*Kirby's Reports.*

THIRD RULE.

249. *a.* First find the amount of the given principal for the whole time; then find the amount of each of the several payments from the time it was endorsed to the time of settlement. Finally, subtract the amount of the several payments from the amount of the principal, and the remainder will be the sum due.

* If a year does not extend beyond the time of payment; but if it does, then find the amount of the principal remaining unpaid, up to the time of settlement, likewise the amount of the endorsements from the time they were paid to the time of settlement, and deduct the sum of these several amounts from the amount of the principal.

Note.—It will be an excellent exercise for the pupil to cast the interest on each of the preceding notes by each of the above rules.

47. What is the interest of £175, 10s. 6d. for 1 year, at 5 per cent.?

Operation.

£175.525 Prin.
.05 Rate.
£8.77625 Int. for 1 yr.
20
s. 15.52500
12
d. 6.30000
4
far. 1.20000
Ans. £8, 15s. $6\frac{1}{4}$d.

We first reduce the 10s. 6d. to the decimal of a pound, (Art. 200,) then multiply the principal by the rate and point off the product as in Art. 241. The figure 8 on the left of the decimal point is pounds, and those on the right are decimals of a pound, and must be reduced to shillings, pence, and farthings. (Art. 201.) Hence,

250. To compute the interest on pounds, shillings, &c.

Reduce the given shillings, pence, and farthings to the decimal of a pound; (Art. 200;) *then find the interest as on dollars and cents; finally, reduce the decimal figures in the answer to shillings, pence, and farthings.* (Art. 201.)

48. What is the interest of £56, 15s. for one year and 6 months, at 6 per cent.? *Ans.* £5, 2s. $1\frac{3}{4}$d.

49. What is the interest of £75, 12s. 6d. for 1 year and 3 months, at 7 per cent.?

50. What is the interest of £96, 18s. for 2 years and 6 months, at $4\frac{1}{2}$ per cent.?

51. What is the amount of £100 for 2 years and 4 months, at 5 per cent.?

52. What is the amount of £430, 16s. 10d. for 1 year and 5 months, at 6 per cent.?

QUEST.—250. How is interest computed on pounds, shillings, &c.?

PROBLEMS IN INTEREST.

251. It will be observed that there are *four parts* or *terms* connected with each of the preceding operations, viz: *the principal, the rate per cent., the time, and the interest, or the amount.* These parts or terms have such a relation to each other, that if any *three* of them are given, the *other* may be found. The questions, therefore, which may arise in interest, are numerous; but they may be reduced to a few *general principles*, or ***Problems.***

OBS. 1. The term ***Problem***, in its common acceptation, means a question proposed, which requires a solution.

2. A number or quantity is said to be *given*, when its value is stated, or may be easily inferred from the conditions of the question under consideration. Thus, when the principal and interest are known, the *amount* may be said to be *given*, because it is merely the *sum* of the principal and interest. So, if the principal and the amount are known, the *interest* may be said to be *given*, because it is the *difference* between the amount and the principal.

252. To find the *interest* on any given sum, as in the foregoing examples, the *principal*, the *rate per cent.*, and the *time* are always given. This is the ***First*** and *most important* ***Problem*** in interest. The other ***Problems*** will now be illustrated.

PROBLEM II.*

To find the RATE PER CENT., *the principal, the interest, and the time being given.*

1. A man loaned $75 to one of his neighbors for 4 years, and received $24 interest: what was the rate per cent.?

QUEST.—251. How many terms are connected with each of the preceding examples? What are they? Are they all given? When three are given, can the fourth be found? *Obs.* What is a problem? When is a number or quantity said to be given? 252. What terms are given when it is required to find the interest?

* Should this and the following Problems be deemed too difficult for beginners, they can be omitted till review.

Analysis.—The interest of $75 at 1 per cent. for 1 year, is $.75, and for 4 years it is $.75×4=$3. (Art. 238.) Now since $3 is 1 per cent. interest on the principal for the given time, $24 must be $\frac{24}{3}$ of 1 per cent., which is equal to 8 per cent. (Art. 121.)

Or, we may reason thus: If $3 is 1 per cent. on the principal for the given time, $24 must be as many per cent. as $3 is contained times in $24; and $24÷$3=8.

Ans. 8 per cent.

PROOF.—$75×.08=$6.00, the interest for 1 year at 8 per cent., and $6×4=$24, the interest of $75 for 4 years at 8 per cent. Hence,

253. To find the *rate per cent.* when the principal, interest, and time are given.

First find the interest of the principal at 1 *per cent. for the given time; then make the interest thus found the denominator and the given interest the numerator of a common fraction, which being reduced to a whole or mixed number, will give the required per cent.* (Art. 121.)

Or, simply divide the given interest by the interest of the principal at 1 *per cent. for the given time, and the quotient will be the per cent.*

2. If I borrow $300 for 2 years, and pay $42 interest, what rate per cent. do I pay?

Operation.
$6)$42
7 *Ans.* 7 per ct.

The interest of $300 for 2 yrs. at 1 per cent., is $6. (Art. 238.)

PROOF.—$300×.07×2=$42.

3. If I borrow $460 for 3 years, and pay $82.80 interest, what is the rate per cent.?

4. A man loaned $500 for 8 months, and received $40 interest: what was the rate per cent.?

5. At what rate per cent. must $450 be loaned, to gain $56.50 interest in 1 year and 6 months?

QUEST.—253. When the principal, interest, and time are given, how is the rate per cent. found?

6. At what per cent. must $750 be loaned, to gain $225 in 4 years?

7. A man has $8000 which he wishes to loan for $600 per annum for his support: at what per cent. must he loan it?

8. A gentleman deposited $1250 in a savings bank, for which he received $31.25 every 6 months; what per cent. interest did he receive on his money?

9. A capitalist invested $9260 in Railroad stock, and drew a semi-annual dividend of $416.70: what rate per cent. interest did he receive on his money?

10. A man built a hotel at an expense of $175000, and rented it for $8750 per annum: what per cent. interest did his money yield him?

PROBLEM III.

To find the PRINCIPAL, *the interest, the rate per cent., and the time being given.*

11. What sum must be put at interest, at 6 per cent., to gain $30 in two years?

Analysis.—The interest of $1 for 2 years at 6 per cent., (the given time and rate,) is 12 cents. Now 12 cents interest is $\frac{12}{100}$ of its principal $1; consequently, $30 the given interest, must be $\frac{12}{100}$ of the principal required. The question therefore resolves itself into this: $30 is $\frac{12}{100}$ of what number of dollars? If $30 is $\frac{12}{100}$, $\frac{1}{100}$ is $\frac{1}{12}$ of $30, which is 2\frac{1}{2}$; and $\frac{100}{100}$=2\frac{1}{2}$×100, which is $250, the principal required.

Or, we may reason thus: Since 12 cents is the interest of 1 dollar for the given time and rate, 30 dollars must be the interest of as many dollars for the same time and rate, as 12 cents is contained times in 30 dollars. And $30÷.12=250. *Ans.* $250.

PROOF.—$250×.06=$15.00, the interest for 1 year at the given per cent., and $15×2=$30, the given interest. Hence,

254. To find the *principal*, when the interest, rate per cent., and time are given.

Make the interest of $1 for the given time and rate, the numerator, and 100 *the denominator of a common fraction ; then divide the given interest by this fraction ; and the quotient will be the principal required.* (Art. 141.)

Or, simply divide the given interest by the interest of $1 for the given time and rate expressed in decimals ; and the quotient will be the principal.

12. What sum put at interest will produce $13.30 in 6 months, at 7 per cent. ?

Operation.
$.035)$13.300
380. *Ans.* $380.

The int. of $1 for 6 mo. at 7 per cent. is $.035. (Art. 239.)

13. A father bequeaths his son $500 a year: what sum must be invested, at 5 per cent. interest, to produce it?

14. What sum must be put at 6 per cent. interest, to gain $350 interest semi-annually ?

15. A gentleman retiring from business, loaned his money at 7 per cent., and received $1200 interest a year: how much was he worth ?

PROBLEM IV.

To find the TIME, *the principal, the interest, and the rate per cent. being given.*

16. A man loaned $80 at 5 per cent., and received $10 interest: how long was it loaned ?

Analysis.—The interest of $80 at 5 per cent. for 1 year is $4. (Art. 237.) Now, since $4 interest requires the principal 1 year at the given per cent., $10 interest will require the same principal $\frac{10}{4}$ of 1 year, which is equal to $2\frac{1}{2}$ years. (Art. 121.)

QUEST.---254. When the interest, rate per cent., and time are given, how is the principal found ?

Or, we may reason thus: If \$4 interest requires the use of the given principal 1 year, \$10 interest will require the same principal as many years as \$4 is contained times in \$10. And \$10÷\$4=2.5. *Ans.* 2.5 years. Hence,

255. To find the *time* when the principal, interest, and rate per cent. are given.

Make the given interest the numerator, and the interest of the principal for 1 *year at the given rate the denominator of a common fraction, which being reduced to a whole or mixed number, will give the time required.*

Or, simply divide the given interest by the interest of the principal at the given rate for 1 *year, and the quotient will be the time.*

OBS. If the quotient contains a decimal of a year, it should be reduced to months and days. (Art. 201.)

17. How long will it take \$100, at 5 per cent., to double itself; that is, to gain \$100 interest?

Operation. The interest of \$100 for 1 year, at 5 per cent., is \$5. (Art. 237.)

\$5)\$100

20 *Ans.* 20 years.

PROOF.—\$100×.05×20=\$100. (Art. 238.)

TABLE.

Showing in what time any given principal will double itself at any rate, from 1 to 20 per cent. Simple Interest.

Per cent.	Years.	Per cent.	Years.	Per cent.	Years.	Per cent.	Years.
1	100	6	$16\frac{2}{3}$	11	$9\frac{1}{11}$	16	$6\frac{1}{4}$
2	50	7	$14\frac{2}{7}$	12	$8\frac{1}{3}$	17	$5\frac{15}{17}$
3	$33\frac{1}{3}$	8	$12\frac{1}{2}$	13	$7\frac{9}{13}$	18	$5\frac{5}{9}$
4	25	9	$11\frac{1}{9}$	14	$7\frac{1}{7}$	19	$5\frac{5}{19}$
5	20	10	10	15	$6\frac{2}{3}$	20	5

QUEST.—255. When the principal, interest, and rate per cent. are given, how is the time found? *Obs.* When the quotient contains a decimal of a year, what should be done with it?

18. In what time will $500, at 6 per cent., produce $100 interest?

19. How long will it take $100, at 6 per cent., to double itself?

20. How long will it take $100, at 7 per cent., to double itself?

21. How long will it take $7250, at 10 per cent., to double itself?

COMPOUND INTEREST.

256. *Compound Interest* is the interest arising not only from the principal, but also from the *interest itself*, after it becomes due.

OBS. 1. Compound Interest is often called *interest upon interest.*

2. When the interest is paid on the *principal only*, it is called *Simple Interest.*

Ex. 1. What is the compound interest of $500 for 3 years, at 6 per cent.?

Operation.

	$500	principal.
$500×.06=	$ 30	Int. for 1st year.
	530	Amt. for 1 year.
$530×.06=	31.80	Int. for 2d year.
	561.80	Amt. for 2 years.
$561.80×.06=	33.70	Int. for 3d year.
	$595.50	Amt. for 3 years.
	500.00	Prin. deducted.
Ans.	$95.50	compound Int. for 3 years.

QUEST.—256. From what does compound interest arise? *Obs* What is compound interest often called? What is Simple Interest?

257. Hence, to calculate compound interest.

Cast the interest on the given principal for 1 year, or the specified time, and add it to the principal; then cast the interest on this amount for the next year, or specified time, and add it to the principal as before. Proceed in this manner with each successive year of the proposed time. Finally, subtract the given principal from the last amount, and the remainder will be the compound interest.

2. What is the compound interest of $350 for 4 years, at 6 per cent.?

3. What is the compound interest of $865 for 5 years, at 7 per cent.?

4. What is the amount of $250 for 6 years, at 5 per cent. compound interest?

5. What is the amount of $1000 for 3 years, at 4 per cent. compound interest, payable semi-annually?

6. What is the amount of $1200 for 2 years, at 6 per cent. compound interest, payable quarterly?

7. What is the amount of $800 for 3 years, at 5 per cent. compound interest, payable semi-annually?

8. What is the amount of $1500 for 5 years, at 7 per cent. compound interest?

9. What is the amount of $2000 for 2 years, at 3 per cent. compound interest, payable quarterly?

10. What is the amount of $3500 for 6 years, at 6 per cent. compound interest?

Note.—This and the next two examples may be solved either by the rule, or by the Table below.

11. What is the amount of $1860 for 8 years, at 7 per cent. compound interest?

12. What is the amount of $20000 for 10 years, at 3 per cent. compound interest?

QUEST.—257. How is compound interest calculated?

TABLE,

Showing the amount of $1, or £1, at 3, 4, 5, 6, and 7 per cent., com-
mn pound interest, for any number of years, from 1 to 35. nm

Yrs.	3 per cent.	4 per cent.	5 per cent.	6 per cent.	7 per cent.
1.	1.030,000	1.040,000	1.050,000	1.060,000	1.07,000
2.	1.060,900	1.081,600	1.102,500	1.123,600	1.14,490
3.	1.092,727	1.124,864	1.157,625	1.191,016	1.22,504
4.	1.125,509	1.169,859	1.215,506	1.262,477	1.31,079
5.	1.159,274	1.216,653	1.276,282	1.338,226	1.40,255
6.	1.194,052	1.265,319	1.340,096	1.418,519	1.50,073
7.	1.229,874	1.315,932	1.407,100	1.503,630	1.60,578
8.	1.266,770	1.368,569	1.477,455	1.593,848	1.71,818
9.	1.304,773	1.423,312	1.551,328	1.689,479	1.83,845
10.	1.343,916	1.480,244	1.628,895	1.790,848	1.96,715
11.	1.384,234	1.539,454	1.710,339	1.898,299	2.10,485
12.	1.425,761	1.601,032	1.795,856	2.012,196	2.25,219
13.	1.468,534	1.665,074	1.885,649	2.132,928	2.40,984
14.	1.512,590	1.731,676	1.979,932	2.260,904	2.57,853
15.	1.557,967	1.800,944	2.078,928	2.396,558	2.75,903
16.	1.604,706	1.872,981	2.182,875	2.540,352	2.95,216
17.	1.652,848	1.947,900	2.292,018	2.692,773	3.15,881
18.	1.702,433	2.025,817	2.406,619	2.854,339	3.37,293
19.	1.753,506	2.106,849	2.526,950	3.025,600	3.61,652
20.	1.806,111	2.191,123	2.653,298	3.207,135	3.86,968
21.	1.860,295	2.278,768	2.785,963	3.399,564	4.14,056
22.	1.916,103	2.369,919	2.925,261	3.603,537	4.43,040
23.	1.973,587	2.464,716	3.071,524	3.819,750	4.74,052
24.	2.032,794	2.563,304	3.225,100	4.048,935	5.07,236
25.	2.093,778	2.665,836	3.386,355	4.291,871	5.42,743
26.	2.156,592	2.772,470	3.555,673	4.549,383	5.80,735
27.	2.221,289	2.883,369	3.733,456	4.822,346	6.21,386
28.	2.287,928	2.998,703	3.920,129	5.111,687	6.64,883
29.	2.356,566	3.118,651	4.116,136	5.418,388	7.11,425
30.	2.427,262	3.243,398	4.321,942	5.743,491	7.61,225
31.	2.500,080	3.373,133	4.538,039	6.088,101	8.14,571
32.	2.575,083	3.508,059	4.764,941	6.453,386	8.71,527
33.	2.652,335	3.648,381	5.003,189	6.840,590	9.32,533
34.	2.731,905	3.794,316	5.253,348	7.251,025	9.97,811
35.	2.813,862	3.946,089	5.516,015	7.686,087	10.6,765

258. To calculate compound interest by the preceding Table.

Find the amount of $1, or £1 for the given number of years by the table, multiply it by the given principal, and the product will be the amount required.

Subtract the principal from the amount thus found, and the remainder will be the compound interest.

13. What is the compound interest of $200 for 10 years, at 6 per cent? What is the amount?

Operation.

$1.790848 Amt. of $1 for 10 years by table.
200 the given principal.

$358.169600 amount required.
$200 principal to be subtracted.

Ans. $158.1696 interest required.

14. What is the amount of $350 for 12 years, at 4 per cent.?

15. What is the amount of $469 for 15 years, at 3 per cent.? What the interest?

16. What is the interest of $500 for 24 years, at 6 per cent.?

17. What is the interest of $650 for 30 years, at 7 per cent.?

DISCOUNT.

259. DISCOUNT is the *abatement* or *deduction* made for the payment of money before it is *due.* For example, if I owe a man $100, payable in one year without interest, the *present worth* of the note is less than $100; for, if $100 were put at interest for 1 year, at 6 per cent., it would amount to $106; at 7 per cent., to $107; &c. In consideration, therefore, of the *present payment* of the note, justice requires that he should make some *abatement* from it. This abatement is called *Discount.*

QUEST.—258. How is compound interest computed by the Table? 259. What is discount? What is the present worth of a debt, payable at some future time, without interest?

The *present worth* of a debt payable at some future time without interest, is that sum which, being put at legal interest, *will amount to the debt*, at the time it becomes due.

Ex. 1. What is the present worth of $545, payable in 1 year and 6 months without interest, when money is worth 6 per cent. per annum?

Analysis.—The *amount*, we have seen, is the sum of the principal and interest. (Art. 234.) Now the amount of $1 for 1 year and 6 months, at 6 per cent., is $1.09; (Art. 237;) that is, the amount is $\frac{109}{100}$ of the principal $1. The question then resolves itself into this: $545 is $\frac{109}{100}$ of what principal? If $545 is $\frac{109}{100}$, $\frac{1}{100}$ is 545÷109, or $5; and $\frac{100}{100}$=$5×100, which is $500.

Or, we may reason thus: Since $1.09 (amount) requires $1 principal for the given time, $545 (amount) will require as many dollars as $1.09 is contained times in $545; and $545÷$1.09=$500. That is, the *present worth* of $545, payable in 1 year and 6 months, is $500, which is the answer required.

PROOF.—$500×.09=$45, the interest for 1 year and 6 months; and $500+$45=$545 the given amount. (Art. 247.) Hence,

260. To find the *present worth* of any sum, payable at a future time without interest.

First find the amount of $1 *for the time, at the given rate, as in simple interest;* (Art. 247;) *then divide the given sum by this amount, and the quotient will be the present worth.*

The present worth subtracted from the debt, will give the true discount.

OBS. This process is often classed among the Problems of Interest, in which the amount, (which answers to the given sum or debt,) the rate per cent., and the time are given, to find the *principal*, which answers to the *present worth.*

QUEST.—260. How do you find the present worth of a debt? How find the discount?

2. What is the present worth of $250.38, payable in 8 months, when money is worth 6 per cent. per annum? What is the discount?

Operation.

```
1.04)250.38(240.75
       208
       ---
        423
        416
        ---
         780
         728
         ---
          520
          520
          ---
```

The amount of $1 for the given time and rate, is $1.04. (Art. 247.) Dividing the given sum by this amount, the quotient $240.75, is the *present worth*. And $250.38—240.75=$9.63, the *discount*.

Ans. { $240.75 the present worth; $9.63 the discount.

3. What is the present worth of $475, payable in 1 year, when money is worth 7 per cent. per annum?

4. What is the present worth of $175, payable in 2 years, when money is worth 7 per cent. per annum?

5. What is the present worth of $1000, payable in 4 months, when the rate of interest is 6 per cent.?

6. What is the discount on $750, due 6 months hence, when interest is 5 per cent. per annum?

7. A man sold a farm for $1800, payable in 15 months: what is the present worth of the debt, allowing the rate to be 6 per cent.?

8. I have a note of $1150.33, payable in 9 months: what is its present worth at 7 per cent. interest per annum?

9. A merchant sold goods amounting to $840.75, payable in 6 months: how much discount should he make for cash down, when money is worth 7 per cent.?

10. What is the discount on a draft of $2500, payable in 3 months, at $4\frac{1}{2}$ per cent. per annum?

11. What is the present worth of $5000, payable in 2 months, at 6 per cent. per annum?

12. What is the difference between the discount on $500 for 1 year, and the interest of $500 for 1 year, at 6 per cent.?

BANK DISCOUNT.

261. It is customary for *Banks* in discounting a note or draft, to deduct in advance the *legal interest* on the given sum from the time it is discounted to the time when it becomes due.

Bank discount, therefore, is the same as simple interest paid *in advance*. Thus, the *bank discount* on a note of $106, payable in 1 year at 6 per cent., is $6.36, while the *true discount* is but $6. (Art. 260.)

OBS. 1. The difference between *bank discount* and *true discount*, is the interest of the true discount for the given time. On small sums for a short period this difference is trifling, but when the sum is large, and the time for which it is discounted is long, the difference is considerable.

2. Taking *legal interest in advance*, according to the general rule of law, is *usury*. An exception is generally allowed, however, in favor of notes, drafts, &c., which are payable in *less* than a *year*.

The Safety Fund Banks of the State of New York, though the legal rate of interest is 7 per cent., are not allowed by their charters to take over 6 per cent. discount in advance on notes and drafts which mature within 63 days from the time they are discounted.*

262. According to custom, a note or draft is not presented for collection until *three* days after the time specified for its payment. These three days are called *days of grace*. It is customary to charge interest for them. Banks, therefore, always calculate the interest for *three* days more than the time stated in the note.

13. What is the bank discount on a note of $500, payable in 1 year, at 6 per cent.? What is the present worth?

QUEST.—261. How do banks usually reckon discount? What then is bank discount? *Obs.* What is the difference between bank discount and true discount? Is this difference worth noticing? How is taking interest in advance generally regarded in law? What exception to this rule is allowed? 262. When is it customary to present notes and drafts for collection? What are these 3 days called? Is it customary to charge interest for the days of grace?

* Revised Statutes of New York, Vol. I. p. 741.

Operation.

The interest of \$500 for 1 year is \$30.
The " " " 3 days' grace, is 0.25
Therefore the discount is \$30.25
And the present worth is \$500—\$30.25=\$469.75.

Note.—Interest should be reckoned on the *three days grace* in each of the following examples, except the last two.

14. What is the bank discount on a draft of \$250, payable in 4 months, at 7 per cent.?

15. What is the bank discount on a draft of \$375, payable in 30 days, at 6 per cent.?

16. What is the bank discount on a note of \$1000, payable in 60 days, at 5 per cent.?

17. What is the present worth of \$1160, payable in 90 days, discounted at a bank at 6 per cent.?

18. What is the present worth of \$750.36, payable in 5 months, at 4½ per cent.?

19. What is the bank discount of \$1825.60, payable in 4 months and 15 days, at 6 per cent.?

20. What is the present worth of a draft of \$1292, payable in 60 days, at 7 per cent. discount?

21. What is the present worth of a draft of \$5000, payable in 15 days, at 6 per cent. discount?

22. What is the present worth of a draft of \$15000, payable in 3 days, at 6 per cent. discount?

23. What is the present worth of \$1326, payable in 10 months, at 5½ per cent. discount?

24. What is the bank discount, at 7 per cent., on a note of \$836.81, payable in 90 days?

25. What is the bank discount, at 8 per cent., on a draft of \$1261.38, payable in 60 days?

26. What is the bank discount, at 6½ per cent., on a draft of \$10000, payable in 30 days?

27. What is the difference between the true discount and bank discount on \$1000, payable in 5 years, at 6 per cent?

28. What is the difference between the true discount and bank discount on \$100000, payable in 1 year, at 7 per cent.?

INSURANCE.

263. INSURANCE is *security* against *loss* or *damage* of property by fire, storms at sea, and other casualties. This security is usually effected by contract with Insurance Companies, who, for a stipulated sum, agree to restore to the owners the amount insured on their houses, ships, and other property, if destroyed or injured during the specified time of insurance.

264. The *written instrument* or *contract* is called the *Policy.*

The *sum* paid for insurance is called the ***Premium.***

The premium paid is a *certain per cent.* on the amount of property insured for 1 year, or during a voyage at sea, or other specified time of risk. Hence,

265. To compute Insurance for 1 year, or the specified time.

Multiply the sum insured by the given rate per cent., as in interest. (Art. 237.)

OBS. 1. Insurance on ships and other property at sea is sometimes effected by contract with individuals. It is then called *out-door insurance.*

2. The insurers, whether an incorporated company or individuals, are often termed *Underwriters.*

Ex. 1. How much premium must a mechanic pay annually for the insurance of his shop and tools worth $350, at $1\frac{1}{2}$ per cent.?

Solution.—$\$350 \times .015 = \$5.25$. *Ans.*

2. What amount of premium must be paid annually for insuring a house worth $875, at $\frac{3}{4}$ per cent.?

3. Shipped a box of books valued at $1000, from New

QUEST.—263. What is Insurance? 264. What is meant by the policy? The premium? 265. How is insurance computed? *Obs.* When insurance is effected with individuals, what is it called? What are the insurers sometimes called?

York to New Orleans, and paid $1\frac{1}{4}$ per cent. insurance: what was the amount of premium?

4. A powder mill worth \$925, was insured at $15\frac{1}{2}$ per cent.: what was the annual amount of premium?

5. A merchant shipped a lot of goods worth \$1560, from Boston to Natchez, and paid $1\frac{3}{4}$ per cent. insurance: what amount of premium did he pay?

6. A gentleman obtained a policy of insurance on his house and furniture to the amount of \$2500, at $3\frac{1}{4}$ per cent. per annum: what premium did he pay a year?

7. A man owning a sixteenth of a whale ship, which cost him \$2750, got it insured, at $7\frac{1}{2}$ per cent. for the voyage: how much did he pay?

8. A man owning a schooner worth \$3800, obtained insurance upon it, at $5\frac{1}{3}$ per cent. for the season: what amount of premium did he pay?

9. A crockery merchant having a stock of goods valued at \$7500, paid 2 per cent. for insurance: how much premium did he pay a year?

10. A merchant shipped \$3765 worth of flour, from Cincinnati to New York, and paid $1\frac{1}{5}$ per cent. insurance: how much premium did he pay?

11. What is the annual premium for insuring a store worth \$7350, at $\frac{4}{5}$ per cent.?

12. An importer effected insurance on a cargo of tea worth \$65000, from Canton to Philadelphia, at 3 per cent.: how much did his insurance cost him?

13. A manufacturer obtained insurance to the amount of \$76500 on his stock and buildings, at $\frac{3}{4}$ per cent.: how much premium did he pay annually?

14. A policy was obtained on a cargo of goods valued at \$95600, shipped from Liverpool to New York, at $2\frac{1}{2}$ per cent.: what was the amount of premium?

15. The owners of the whale ship George Washington obtained a policy of \$58000 on the ship and cargo, at $7\frac{1}{4}$ per cent. for the voyage: what was the amount of premium?

16. A gentleman paid \$60 annually for insurance on his house and furniture, which was 2 per cent. on its value: what amount of property was covered by the policy?

Note.—This example is similar to those of Problem III, in interest. (Art. 254.)

Solution.—Since the rate of insurance is 2 per cent. or .02, it is plain that $60 is $\frac{2}{100}$ of the amount insured. Now if $60 is $\frac{2}{100}$, $\frac{1}{100}$ is half as much, or $30; and $\frac{100}{100}$ is $30×100, or $3000. Or thus: 60÷.02=3000 *Ans.* $3000.

PROOF.—$3000×.02=$60, which was the annual premium paid.

17. If I pay $250 premium on silks, from Havre to New York, at 1½ per cent., what amount of property does my policy cover?

18. A merchant paid $1200 premium, at $2\frac{1}{5}$ per cent. on a ship and cargo from London to Baltimore, which was lost on the voyage: what amount should he recover from the Insurance Company?

19. If a man pays $60 premium annually for the insurance of his house, which is worth $3000, what rate per cent. does he pay?

Note.—This example is similar to those of Problem II, in interest. (Art. 253.)

Solution.—$60÷$3000=.02. *Ans.* 2 per cent.

PROOF.—$3000×.02=$60, which is the premium paid.

20. A merchant paid $40 premium for insuring $5000 on his stock: what rate per cent. did he pay?

21. If a man pays $75 for insuring $15000, what rate per cent. does he pay?

22. If the owner pays $2800 for insuring a ship worth $40000, what rate per cent. does he pay?

23. A blacksmith owns a shop worth $720: what amount must he get insured annually, at 10 per cent., so that in case of loss, both the value of the shop and the premium may be repaid?

Analysis.—Since the rate of insurance is 10 per cent., on a policy of $100, the owner would actually receive but $90; for he pays $10 for insurance. The question then resolves itself into this: $720 is $\frac{90}{100}$ of what sum?

If 720 is $\frac{90}{100}$, $\frac{1}{100}$ is $720 \div 90 = 8$, and $\frac{100}{100}$ is $8 \times 100 = 800$. *Ans.* $800.

PROOF.—$800×.10=$80, the premium he would pay, and $800—$80=$720, which is the value of his shop.

24. If I send an adventure to China worth $6250, what amount of insurance, at 8 per cent., must I obtain, that in case of a total wreck I may sustain no loss by the operation?

25. What amount of insurance must be effected on $11250, at 5 per cent., in order to cover both the premium and property insured?

PROFIT AND LOSS.

266. PROFIT and Loss in commerce, signify the sum *gained* or *lost* in ordinary business transactions. They are reckoned at a certain per cent. on the *purchase price*, or *sum paid* for the articles under consideration.

MENTAL EXERCISES.

1. A merchant bought a barrel of flour for $6, and sold it at a profit of 10 per cent.: how much did he sell it for?

Suggestion.—Since he made 10 per cent. profit, if we add 10 per cent. to the *purchase* price, it will give the *selling* price. Now 10 per cent. of $6 is 60 cents; (Art. 225,) which added to $6, make $6.60.
Ans. He sold it for $6.60.

2. A grocer bought a box of oranges for $5, and sold it, at 12 per cent. profit: how much did he receive for his oranges?

3. A farmer bought a ton of hay for $9, and sold it

QUEST—266. What is meant by profit and loss? How are they reckoned?

for 10 per cent. more than he gave: how much did he sell it for?

4. Bought a sleigh for $12, and sold it at a loss of 8 per cent.: how much did I receive for the sleigh?

Solution.—8 per cent. of $12, is 96 cents; and $12—96 cents leaves $11.04. *Ans.*

5. Bought a box of honey for $5, and having lost a portion of it, sold the remainder, at 11 per cent. loss: how much did I receive for it?

6. A shop-keeper bought a piece of calico for $7, and sold it, at 12 per cent. profit: how much did he sell it for?

7. A lad bought a sheep for $3, and on his way home was offered 15 per cent. for his bargain: how much was he offered for his sheep?

8. A farmer bought a colt for $20, and offered to sell it for 5 per cent. less than he gave: how much did he ask for it?

9. A gentleman bought a horse for $100; after using it awhile, he sold it, at 7 per cent. loss: how much did he get for his horse?

10. A man bought a building lot for $150, and in consequence of the rise of property, sold it for 10 per cent. advance: how much did he get for it?

11. A hack-man bought a carriage for $200, and after using it for one season, sold it for 15 per cent. less than he gave for it: how much did he sell it for?

12. A man bought a house for $800, and sold it the next day for 10 per cent. advance: how much did he sell it for?

EXERCISES FOR THE SLATE.

CASE I.

1. A merchant bought a quantity of grain for $75, and sold it for 8 per cent. profit: how much did he gain by the bargain?

Solution.—$75×.08=$6.00. (Art. 225.) Hence,

267. To find the *amount of profit* or *loss*, when the purchase price and rate per cent. are given.

Multiply the purchase price by the given per cent. as in percentage; and the product will be the amount gained or lost by the transaction. (Art. 225.)

2. A man bought a sleigh for $60, and afterwards sold it for 10 per cent. less than cost: how much did he lose?

3. A grocer bought a cask of oil for $96.50, and retailed it, at a profit of 6 per cent.: how much did he make on his oil?

4. A pedlar bought a lot of goods for $215, and retailed them, at 20 per cent. advance: how much was his profit?

5. A merchant bought a cargo of coal for $450, which he afterwards sold for 12½ per cent. less than cost: what was the amount of his loss?

6. A manufacturer purchased $1000 worth of wool, and after making it up, sold the cloth for 25 per cent. more than the cost of the materials: how much did he receive for his labor?

CASE II.

7. A man bought a span of horses for $350, and wished to dispose of them for 12 per cent. profit: how much must he sell them for?

Operation.
$350 purchase price.
.12 per cent. profit.
$42.00 gained.
Ans. $392 selling price.

Reasoning as before, he must sell them for the *purchase price*, together with 12 per cent. of that price. Having found 12 per cent. of $350, (Art. 225,) add it to the cost, and the sum $392, is manifestly the *selling price.*

8. A stage proprietor bought a coach for $480; find-

QUEST.—267. How is the amount of profit or loss found, when the cost and rate per cent, are given?

ing it damaged, he was willing to sell it, at 5 per cent. loss: at what price would he sell it?

Operation.
$480 purchase price.
.05 per cent. loss.
$24.00 sum lost.
Ans. $456 selling price.

Having found the sum lost, (Art. 225,) subtract it from the cost, and the remainder is obviously the selling price. Hence,

268. To find how any article must be sold, in order to *gain* or *lose* a given rate per cent.

First find the amount of profit or loss on the purchase price at the given rate, as in the last Case; then the amount thus found added to, or subtracted from the purchase price, as the case may be, will give the selling price required.

9. A merchant bought a firkin of butter for $22.75: how much must he sell it for in order to gain 15 per cent. by his bargain?

10. Bought a chest of tea for $37.50: for how much must I sell it, in order to make 18 per cent. by the operation?

11. Bought a quantity of produce for $89.33, which I propose to sell, at 20 per cent. loss: how much must I receive for it?

12. A drover bought a flock of sheep for $275, and taking them to market, sold them, at 25 per cent. advance: how much did he sell them for?

13. A merchant had a quantity of groceries on hand, which cost him $367.13; for the sake of closing up his business he sold them, at 15 per cent. less than cost: how much did he get for them?

14. A man bought a farm for $875, and was offered 33 per cent. advance for his bargain: how much was he offered?

15. A merchant bought a cargo of cotton for $30000;

QUEST.—268. What is the method of finding how an article must be sold, in order to gain or lose a given per cent.?

the price declining, he sold it at $2\frac{1}{2}$ per cent. less than cost: for how much did he sell it?

CASE III.

16. A man bought a cow for $25, which he afterwards sold for $29: what per cent. profit did he make?

Analysis.—Subtracting the cost from the selling price, shows that he gained $4. Now 4 dollars are $\frac{4}{25}$ of 25 dollars; hence, he gained $\frac{4}{25}$ of his *outlay*, or the purchase price of the cow. And $\frac{4}{25}$ reduced to a decimal, is 16 *hundredths*, which is the same as 16 per cent. (Arts. 197, 223. Obs. 3.)

Or, we may reason thus: If 25 dollars (outlay) gain 4 dollars, 1 dollar (outlay) will gain $\frac{1}{25}$ of 4 dollars. Now $4÷25 is equal to 16 hundredths of a dollar. But 16 hundredths is the same as 16 per cent. Hence,

269. To find the rate per cent. of profit or loss, when the *cost* and *selling prices* are given.

First find the amount gained or lost, as the case may be, by subtraction; then make the gain or loss the numerator and the purchase price the denominator of a common fraction, which being reduced to a decimal, will give the per cent. required. (Art. 197.)

Or, simply annex ciphers to the profit or loss, and divide it by the cost; the quotient will be the per cent.

OBS. 1. As *per cent.* signifies *hundredths*, we have seen that the *first two* decimal figures which occupy the place of hundredths, are properly the per cent.; the other decimals are *parts* of 1 per cent. After obtaining two decimal figures, there is sometimes an advantage in placing the remainder over the divisor, and annexing it to the decimals thus obtained. (Art. 223. Obs. 3.)

2. It should be remembered that the percentage which is *gained* or *lost*, is always calculated on the *purchase* price, or the *sum paid* for

---

QUEST.—269. How is the rate per cent. of profit or loss found, when the cost and selling price are given? *Obs.* What figures properly signify the per cent.? Why? What do the other decimal figures on the right of hundredths denote? On what is the per cent. gained or lost calculated?

the article, and not on the *selling* price, or *sum received*, as it is often supposed.

17. A merchant bought a piece of cloth for $2.75 per yard, and sold it for $3.25: what per cent. did he gain?

Solution.—Since he gained 50 cents on a yard, his gain was $\frac{50}{275}$ of the cost. And $\frac{50}{275}=.18\frac{2}{11}$.

Ans. $18\frac{2}{11}$ per cent.

18. A boy purchased a book for 20 cents, and sold it for 30 cents: what per cent. did he make?

19. A merchant bought a box of sugar, at 6 cents a pound, and sold it for $7\frac{1}{2}$ cents a pound: what per cent. was his profit?

20. A grocer bought eggs at 9 cents, and sold them for 12 cents per dozen: what per cent. was his profit?

21. A man bought a hat for $4.50, and sold it for $6: what per cent. did he gain?

22. A jockey bought a horse for $73, and sold him for $68: what per cent. did he lose?

23. A merchant bought a quantity of goods for $155.63, and sold them for $148.28: what per cent. did he lose?

24. A gentleman bought a house for $3500, and sold it for $150 more than he gave: what per cent. was his profit?

25. A speculator laid out $7500 in land, and afterwards sold it for $10000: what per cent. did he make?

26. A drover bought a herd of cattle for $1175, and sold them for $1365: what per cent. did he gain; and how much did he make by the operation?

27. A merchant bought $10000 worth of wool, and sold it for $12362: what per cent.; and how much was his profit?

CASE IV.

28. A jockey sold a horse for $250, which was 25 per cent. more than it cost him: how much did he pay for the horse?

Analysis.—It will be observed that the selling price ($250) is equal to the *cost* and the *amount gained* added together. Now considering the cost a unit or 1, the gain which is a certain per cent. of the cost, (Art. 266,) is $\frac{25}{100}$, consequently $1+\frac{25}{100}=\frac{125}{100}$, (Art. 127,) will denote the sum of the cost and the gain. The question therefore resolves itself into this: 250 is $\frac{125}{100}$ of what number? If 250 is $\frac{125}{100}$, $\frac{1}{100}$ is 2; and $\frac{100}{100}$ is 100 times 2, or 200.

Or, we may simply divide 250 by the fraction $\frac{125}{100}$. (Art. 141.) The quotient 200 is the cost required.

PROOF.—$200×.25=$50; and $200+$50=$250, the selling price?

29. A merchant sold a quantity of goods for $180, which was 10 per cent. less than cost: how much did the goods cost him?

Analysis.—It will be observed that the selling price ($180) is equal to the cost diminished by the sum lost. Now reasoning as in the last example, $1-\frac{10}{100}=\frac{90}{100}$ will denote the cost diminished by the loss. The question now is this: 180 is $\frac{90}{100}$ of what number? If 180 is $\frac{90}{100}$, $\frac{1}{100}$ is 2, and $\frac{100}{100}$ is 200. Or thus: $\$180 \div \frac{90}{100}=\$200$. *Ans.*

PROOF.—$200×.10=$20, and $200—$20=$180, the selling price. Hence,

270. To find the *cost* when the *selling price* and the per cent. *gained* or *lost* are given.

Make the given per cent. added to or subtracted from 100, *as the case may be, the numerator, and* 100 *the denominator of a common fraction; then divide the selling price by this fraction; and the quotient will be the cost required.*

OBS. 1. It is not unfrequently supposed that if we find the percentage on the selling price at the given rate, and add the percentage thus found to, or subtract it from the selling price, as the case may be, the sum or remainder will be the cost. This is a mistake, and leads

---

QUEST.—270. How is the cost found, when the selling price and the rate per cent. gained or lost, are given? *Obs.* What mistake is sometimes made in finding the cost? How may it be avoided?

to serious errors in the result. It will easily be avoided by remembering, that the basis on which *profit* and *loss* are calculated, is always the *purchase price*, or *sum paid* for the articles under consideration. (Art. 269. Obs. 2.)

30. A grocer sold a hogshead of molasses for $24, and gained 20 per cent. on the cost: what was the cost of the molasses?

31. A merchant sold a piece of broadcloth for $85, which was 10 per cent. less than the cost: what was the cost of it?

32. A butcher sold a yoke of oxen for $125, and thereby made 15 per cent.: how much did they cost him?

33. A bookseller sold a lot of books for $200, which was 12 per cent. more than the cost: what was the cost?

34. A wholesale druggist sold a quantity of medicines for $560, and made 50 per cent. profit on them: what was the cost of them?

35. A merchant sold a cargo of rice for $1500, which was $12\frac{1}{2}$ per cent. less than cost: what was the cost?

EXAMPLES FOR PRACTICE.

1. A merchant bought 25 boxes of raisins for $45: at what price per box must he retail them to gain 10 per cent. by his bargain?

Suggestion.—He must sell the whole for 10 per cent. more than the cost. Hence, if we add 10 per cent. to the cost, and divide the sum by the number of boxes, it will give the retail price per box. (Art. 217.)

2. A shopkeeper bought a piece of cotton containing 40 yards, at 6 cents a yard, and sold it for 7 cents a yard: what per cent. profit did he gain; and how much did he make by the bargain?

3. A merchant bought 60 yards of domestic flannel at 25 cents per yard, and sold it at 30 cents per yard: what per cent. was his profit; and how much did he clear by the operation?

4. A bookseller bought 100 Arithmetics at $31\frac{1}{4}$ cents

apiece, and retailed them at $37\frac{1}{2}$ cents apiece: what per cent.; and how much did he make by the operation.

5. A drover bought 175 sheep for $350, and sold them so as to gain 15 per cent.: how much did he sell them for per head?

6. A baker paid $2500 for 480 barrels of flour, and finding it damaged, sold it at a loss of 8 per cent.: how much did he sell it for per barrel?

7. A merchant bought 10 pieces of broadcloth, each piece contuing 30 yards, for $1400, and retailed the whole at a profit of 20 per cent.: at what price did he sell it per yard?

8. A grocer bought 500 lbs. of butter for $75, and sold it at a loss of 7 per cent.: how much did he get per pound?

9. A merchant bought 12 hogsheads of molasses at 25 cents per gallon: how must he sell it by the gallon in order to gain 20 per cent.; and how much was his profit?

10. A farmer raises 750 bushels of wheat at an expense of $675: how must he sell it per bushel, in order to make 18 per cent.?

11. A provision merchant bought 1500 barrels of pork at $10.25 per barrel, and sold it at a loss of 9 per cent.: how much did he lose; and what did he get per barrel?

12. An inn-keeper bought 150 bushels of oats, at 25 cents a bushel, and retailed them at the rate of $12\frac{1}{2}$ cents a peck: what per cent.; and how much did he make on the oats?

13. A miller bought 500 bushels of wheat, at 75 cents per bushel: how much must he sell the whole for in order to gain 20 per cent.?

14. A grocer bought 1630 pounds of tea, at $62\frac{1}{2}$ cents per pound, and sold it at 10 per cent. loss: how much did he sell it at per pound?

15. A merchant bought a bale of calico prints containing 750 yards and paid $75: how must he retail it per yard, in order to gain 20 per cent.; and how much would he make on a yard?

16. A bookseller purchased 1000 geographies, at 84

cents apiece: how must he retail them to gain 20 per cent.?

17. A milliner bought 1200 yards of ribbon, at 30 cents per yard: how must she sell it per yard to gain 50 per cent.?

18. A grocer bought 5000 lbs. of sugar for $350, and retailed it, at 6 cents per pound: what per cent. loss did he sustain?

19. A man purchased goods amounting to $1635: what per cent. profit must he gain, in order to make $350?

20. A speculator bought 10000 acres of land for $12500, and afterwards sold it, at 25 per cent. loss: for how much per acre did he sell it; and how much did he lose by the operation?

DUTIES.

271. DUTIES, in commerce, signify a *sum of money* required by Government to be paid on *imported* goods.

Duties are of two kinds, *specific* and *ad valorem.* A *specific duty* is a certain sum imposed on a ton, hundred weight, hogshead, gallon, square yard, foot, &c. without regard to the value of the article.

Ad valorem duties are those which are imposed on goods, at a certain per cent. on their *value* or *purchase price.*

Note.—The term *ad valorem* is a Latin phrase, signifying *according to*, or *upon the value.*

272. Before specific duties are imposed, it is customary to make certain deductions called *tare*, *draft*, or *tret*, *leakage*, &c.

Tare, in commerce, is an allowance of a certain

QUEST.—271. What are duties in commerce? Of how many kinds are they? What are specific duties? Ad valorem duties? *Note.* What is the meaning of the term ad valorem? 272. What deductions are made before specific duties are imposed? What is tare? Draft or tret? Leakage?

number of pounds made for the box, cask, &c., which contains the article under consideration.

Draft or *Tret* is an allowance of a certain per cent. (usually 4 per cent.) on the weight of goods for waste, or refuse matter.

Leakage is an allowance of a certain per cent. (usually 2 per cent.) for the waste of liquors contained in casks, &c.

OBS. 1. All duties, both specific and ad valorem, are regulated by the Government, and have been different at different times and in different countries.

2. The allowance or deductions [illegible] tare, leakage, &c., are also different on different articles, [illegible] lated by law.

3. In buying and selling groceri[illegible] quantities, allowances are sometimes made for draft, tare, [illegible] &c., similar to those in reckoning duties.

CASE I.

Ex. 1. What is the specific duty on 10 pipes of wine, at 15 cents per gallon, reckoning the leakage at 2 per cent.?

Suggestion.—First deduct the leakage. In 1 pipe there are 2 hogsheads or 126 gallons; in 10 pipes there are 10 times 126, or 1260 gallons. But 2 per cent. of 1260 gallons, is $1260 \times .02 = 25.20$ gallons; (Art. 225;) and 25.2 gallons subtracted from 1260 gallons leaves 1234.8 for the number of *net* gallons. Now if the duty on 1 gallon is 15 cents, on 1234.8 gallons it is $1234.8 \times .15 =$ $185.22, the duty required. Hence,

273. To find the *specific* duty on any given merchandise.

First deduct the legal draft, tare, leakage, &c. from the given quantity of goods; then multiply the remainder by the given duty per gallon, pound, yard, &c., and the product will be the duty required.

QUEST.—*Obs.* How are duties regulated? Are the allowances for draft, tare, &c. the same for all articles? Are allowances ever made in buying and selling groceries for draft, &c.? 273. How are specific duties calculated?

2. What is the specific duty, at 2 cents per pound, on 12 boxes of sugar, weighing 900 lbs. apiece, allowing 20 pounds per box for draft?

Ans. { The draft is 240 pounds.
And 2 per ct. on 10560 lbs. is $211.20.

3. At 3 cents a pound, what is the duty on 25 casks of nails, each weighing 125 lbs. allowing 8 pounds on a cask for tare?

4. At 5 cents a pound, what is the specific [illegible] on 75 boxes of raisins, w[illegible] 60 lbs. apiece, allowing 6 pounds a box for d[illegible]

5. At 4 cents p[illegible] wh[illegible] specific duty on 110 chests of cinna[illegible] weig[illegible]g 230 lbs. allowing 16 lbs. per chest fo[illegible]

6. At 15 cents a [illegible] what is the specific duty on 300 bags of indigo, [illegible] weighing 200 lbs., allowing 4 per cent. for tret?

CASE II.

7. What is the ad valorem duty, at 15 per cent. on an invoice of calico prints, which cost $150 in Liverpool?

Suggestion.—When duties are imposed upon the actual cost of merchandise, there are of course no deductions to be made; consequently we have only to find 15 per cent. of $150, the amount of the given invoice, or cost of the goods, and it will be the duty required.

Solution.—$150×.15=$22.50. *Ans.* Hence,

274. To find the *ad valorem* duty on any given merchandise.

Multiply the given invoice by the given or legal per cent., and the product will be the duty required. (Art. 225.)

OBS. 1. An *invoice* is a written statement of merchandise, with the value or prices of the articles annexed.

QUEST.—274. How are ad valorem duties calculated? *Obs.* What is an invoice? What does the law require respecting the invoice of imported goods?

2. The law requires that the invoice shall be verified by the owner, or one of the owners of the goods, wares, or merchandise, certifying that the invoice annexed contains a *true and faithful account of the actual costs* thereof, and of all charges thereon, and no other different discount, bounty, or drawback, but such as has been actually allowed on the same; which oath shall be administered by a consul, or commercial agent of the United States, or by some public officer duly authorized to administer oaths in the country where the goods were purchased, and the same shall be duly certified by the said consul, &c. Fraud on the part of the owners, or the consul, &c. who administers the oath, is visited with a heavy penalty.—*Laws of the United States.*

8. What is the ad valorem duty, at 30 per cent., on a box of books invoiced at $250?

9. What is the ad valorem duty, at 20 per cent., on a quantity of Java coffee, which cost $356.12?

10. What is the amount of ad valorem duty, at 25 per cent., on a quantity of Turkey carpeting, which cost $526.61.

11. What is the duty on a quantity of bombazines, invoiced at $310, at 30 per cent.?

12. What is the duty on a quantity of beeswax, the invoice of which is $460.25, at 15 per cent.?

13. At 25 per cent., what is the duty on an invoice of bleached linens, amounting to $745.85.

14. At 20 per cent., what is the duty on an invoice of jewelry, amounting to $4250?

15. What is the duty on a bale of goods, invoiced at $2500, at 40 per cent.?

16. What is the duty on an invoice of silks, amounting to $5650, at 30 per cent.?

17. What is the duty on a quantity of cutlery, invoiced at $4560, at 33 per cent.?

18. What is the duty on an invoice of broadcloths, which amounts to $8280, at 35 per cent.?

19. What is the duty on an invoice of wines, amounting to $10265, at 35 per cent.?

20. What is the duty on a quantity of cotton fabrics, invoiced at $13637.50, at 33 per cent.?

21. What is the duty on a quantity of ready-made clothing, amounting to $5638.25, at 50 per cent.?

ASSESSMENT OF TAXES.

275. A TAX is a sum imposed or levied on individuals for the support or benefit of the Government, a corporation, parish, district, &c. Taxes levied by the Government, are assessed either on the *person* or *property* of the *citizens*. When assessed on the *person*, they are called *poll taxes*, and are usually a *specific sum*. Those assessed on the *property* are usually apportioned at a *certain per cent.* on the amount of *real estate* and *personal property* of each citizen or taxable individual.

OBS. Property is divided into two kinds, viz: *real estate* and *personal* property. The *former* denotes possessions that are *fixed;* as houses, lands, &c. The *latter* comprehends *all other* property; as money, stocks, notes, mortgages, ships, furniture, carriages, cattle, tools, &c.

276. When a tax of any given amount is to be assessed, the first thing to be done is to obtain an inventory of the amount of *taxable* property, both personal and real, in the State, County, Corporation, or District, by which the tax is to be paid; also the amount of property of every citizen who is to be taxed, together with the number of Polls.

OBS. 1. By the *number of polls* is meant the number of *taxable individuals*, which usually includes every *native* or *naturalized freeman* over the age of 21, and under 70 years. In some States it also includes the young men over the age of eighteen years, who are subject to military duty.

2. When any part or the whole of a tax is assessed upon the polls, each citizen is taxed a *specific sum*, without regard to the amount of property he possesses.

Ex. 1. A certain town is taxed $325. The town contains 200 polls, which are assessed 25 cents apiece; and

---

QUEST.—275. What are taxes? Upon what are they assessed? When assessed upon the person, what are they called? When assessed upon the property, how are they apportioned? *Obs.* How is property divided? What does real estate denote? What is personal property? 276. When a tax is to be assessed, what is the first step? *Obs.* What is meant by the number of polls?

the whole amount of property both real and personal, is valued at \$13750. How much is the tax on a dollar; that is, what per cent. is the tax, and how much is a man's tax who pays for 1 poll, and whose property is valued at \$850?

Suggestion.—The tax on the polls is 200×.25=\$50. And \$50 subtracted from \$325 leaves \$275, which is to be assessed equally on the amount of property possessed by the citizens of the town. The next step is to find how much must be paid on a dollar. Now if \$13750 pay \$275, \$1 must pay $\frac{1}{13750}$ part of \$275. And \$275÷\$13750=\$.02, the tax on \$1, which is 2 per cent. Finally, at 2 per cent., or 2 cents on \$1, the tax on \$850, the amount of the man's property, is \$850×.02=\$17.00. And \$17+.25 (the poll)=\$17.25, the man's tax. Hence,

277. To assess a State, County, or other tax.

I. *First find the amount of tax on all the polls, if any, at the given rate, and subtract this sum from the whole tax to be assessed. Then dividing the remainder by the whole amount of taxable property in the State, County, &c., the quotient will be the per cent. or tax on 1 dollar.*

II. *Multiply the amount of each man's property by the per cent. or tax on one dollar, and the product will be the tax on his property.*

III. *Add each man's poll tax to the tax he pays on his property, and the amount will be his whole tax.*

278. PROOF.—*When a tax bill is made out, add together the taxes of all the individuals in the town, district, &c., and if the amount is equal to the whole tax assessed, the work is right.*

2. A certain parish is taxed \$237.50. The whole property of the parish is valued at \$8000; and there are 75 polls, which are assessed 50 cents apiece. What per cent. is the tax; and how much is a man's tax who pays for 3 polls, and whose property is valued at \$500?

QUEST.—277. How are taxes assessed? 278. When a tax bill is made out, how is its correctness proved?

Operation.

First multiply .50 cents, the tax on 1 poll,
By 75 the number of polls.
$37.50 amount on polls.

Then $237.50—$37.50=$200, the sum to be assessed on the property.

```
Property.   Tax.
$8000)$200.000(.025, the per cent. or tax on $1.
       160 00
       ------
        40 000
        40 000
```

And $500×.025=$12.50, the tax on the man's property.
.50×3= 1.50, tax for polls.
Ans. $14.00, his whole tax.

3. What amount of tax does a man living in the same parish pay, whose property is valued at $450, and pays for 2 polls?

4. A tax of $750 is assessed on a district to build a new school-house; the property of the district is valued at $15000. What is the tax on a dollar; and what is a man's tax whose property is $1150?

5. What is B's tax for erecting the same school-house, whose property is $1530?

6. A tax of $14752.50 is levied on a certain County, whose property is valued at $562875, and which has a list of 5825 polls, which are assessed at 60 cents apiece. What per cent. is the tax; and what is the amount of C's tax, who pays for 4 polls, and has property valued at $5000?

7. What is D's tax, who living in the same County, pays for 2 polls, and is worth $3500?

8. What is G's tax, who pays for 5 polls, and is worth $15300?

279. In making out a tax bill for a whole town, district, &c., assessors, having found the tax on $1, usually make a table, showing the amount of tax on any number of dollars from 1 to $10; then on 10, 20, 30, &c. to $100; then on 100, 200, &c. to $1000.

9. A tax of $3506.25 was levied on a corporation composed of 12 individuals, whose property was valued at $175000, and who were assessed for 25 polls at 25 cents apiece. What was the tax on a dollar?

Ans. 2 cents on a dollar.

Note.—Having found the tax on $1, we will make a table to aid us in making out the tax bill of the corporation. Since the tax on $1 is $.02, it is obvious that multiplying $.02 by 2 will be the tax on $2; multiplying it by 3, will be the tax on $3, &c.

TABLE.

$1	pays	$.02	$10	pays	$.20	$100	pays	$2.00
2	"	.04	20	"	.40	200	"	4.00
3	"	.06	30	"	.60	300	"	6.00
4	"	.08	40	"	.80	400	"	8.00
5	"	.10	50	"	1.00	500	"	10.00
6	"	.12	60	"	1.20	600	"	12.00
7	"	.14	70	"	1.40	700	"	14.00
8	"	.16	80	"	1.60	800	"	16.00
9	"	.18	90	"	1.80	900	"	18.00
10	"	.20	100	"	2.00	1000	"	20.00

10. In the above assessment, what was A's tax, whose property was valued at $1256, and who pays for 3 polls?

Operation.

$1000	pays	$20.00
200	"	4.00
50	"	1.00
6	"	.12
3 polls	"	.75
Amount,		$25.87.

$1256 is composed of 1000+200+50+6. Now, if we add the taxes paid on each of these sums together, the amount will be the tax paid on $1256.

A's tax, therefore, was $25.87.

11. What was B's tax, who paid for 4 polls, and had property to the amount of $1461?

12. C paid for 1 poll, and the valuation of his property was $5863. What was the amount of his tax?

QUEST.—279. When a tax bill is to be made out for a whole town, district, &c., what course do assessors usually take?

13. D paid for 1 poll, and the valuation of his property was $7961. What was his tax?

14. E paid for 2 polls, and his property was valued at $14236. What was his tax?

15. F paid for 2 polls, and his real estate was valued at $21000; his personal property at $4500. What was his tax?

16. G's property was valued at $20250, and he paid for 1 poll. What was his tax?

17. H paid for 2 polls, and the valuation of his estate was $15360. What was his tax?

18. J's property was valued at $33000, and he paid for 4 polls. What was his tax?

19. K paid for 1 poll, and his property was valued at $15013. What was his tax?

20. L paid for 3 polls, and his property was valued at $4500. What was his tax?

21. M paid for 1 poll, and the valuation of his property was $30600. What was his tax?

SECTION X.

PROPERTIES OF NUMBERS.*

DEFINITIONS.

ART. **280.** The progress as well as the pleasure of the pupil in the study of Arithmetic, depends very much upon the *accuracy* of his knowledge of the terms, which are employed in mathematical reasoning. Hence, particular care has been taken to *define* all the most important terms, as they have been introduced, and it is of the utmost importance for the pupil to *understand* their true import.

QUEST.—280. Upon what does the progress and pleasure of the student in Arithmetic very much depend?

* Barlow on the Theory of Numbers.

DEF. 1. Numbers are divided into two classes, *abstract* and *concrete.*

When they are applied to particular objects, as peaches, pounds, yards, &c., they are called *concrete.*

When they are not applied to any particular object, they are called *abstract.* (Art. 45. Obs. 1.) Thus, when it is said that two and three are five, the *two*, *three*, and *five* denote abstract numbers.

2. An *integer* signifies a *whole* number. (Art. 105.)

3. Whole numbers or integers are divided into *prime* and *composite* numbers.

4. A *prime* number is one which *cannot* be produced by multiplying any two or more numbers together. Thus, 1, 2, 3, 5, 7, 11, 13, 17, 19, 23, 29, 31, &c., are prime numbers.

OBS. 1. A prime number is exactly divisible only by *itself* and a *unit.*

2. One number is said to be *prime to another*, when a *unit* is the only number by which both can be divided without a remainder.

The number of prime numbers is unlimited. The first twelve are given above. The pupil can easily point out others.

5. A *composite* number is one which may be produced by multiplying two or more numbers together. (Art. 55. Obs. 1.) Thus, 4, 6, 8, 9, 10, 12, 14, 15, 16, &c. are composite numbers.

6. An *even* number is one which can be divided by 2 without a remainder; as, 4, 6, 8, 10.

7. An *odd* number is one which cannot be divided by 2 without a remainder; as, 1, 3, 5, 7, 9, 15.

OBS. All even numbers except 2, are *composite* numbers; an odd number is sometimes a *composite*, and sometimes a *prime* number.

QUEST.—Into how many classes are numbers divided? What is an abstract number? A concrete number? What is an integer? Into how many classes are whole numbers divided? What is a prime number? *Obs.* Are prime numbers divisible by other numbers? When is one number said to be prime to another? How many prime numbers are there? What is a composite number? What is an even number? An odd number? *Obs.* Are even numbers prime or composite? What is true of odd numbers in this respect?

8. One number is a *measure* of another, when the *former* is contained in the *latter* any number of times without a remainder. (Art. 93. Obs. 1.)

9. One number is a *multiple* of another, when the *former* can be divided by the *latter* without a remainder. (Art. 98.)

10. The *aliquot parts* of a number are the parts by which it can be measured, or into which it may be divided. Thus, 3 and 7 are the aliquot parts of 21.

11. The *reciprocal* of a number is the quotient arising from dividing a *unit* by that number. Thus, the reciprocal of 2 is 1÷2, or $\frac{1}{2}$; the reciprocal of 3 is 1÷3, or $\frac{1}{3}$.

PROPERTIES OF THE SUMS, DIFFERENCES, PRODUCTS, &c., OF NUMBERS.

281. By *properties* of numbers, is meant those qualities or elements of numbers which are inseparable from them.

1. The sum of any *two* or *more even* numbers, is an even number.

2. The difference of any *two even* numbers, is an even number.

3. The sum or difference of *two odd* numbers, is *even*; but the sum of *three odd* numbers, is *odd*.

4. The sum of any *even* number of odd numbers, is even; but the sum of any *odd* number of odd numbers, is odd.

5. The sum, or difference, of an *even* and an *odd* number, is an odd number.

6. The product of an *even* and an *odd* number, or of *two even* numbers, is even.

7. If an even number be divisible by an odd number, the *quotient* is an even number.

8. The product of any number of factors, is *even*, if any one of them be even.

9. An odd number cannot be divided by an even number without a remainder.

QUEST.—When is one number a measure of another? When is one number a multiple of another? What are aliquot parts? What is the reciprocal of a number? 281. What is meant by properties of numbers?

10. The product of any *two* or *more odd* numbers, is an odd number.

11. If an odd number divides an even number, it will also divide the *half* of it.

12. If an even number is divisible by an odd number, it will also be divisible by *double* that number.

13. The product of any two numbers is the *same*, whichever of the two numbers is the multiplier. (Art. 47.)

14. The *least* divisor of every number, is a *prime* number.

OBS. Hence, in obtaining the least common multiple, the *smallest* number which will divide any two or more of the given numbers, is always a *prime* number, and consequently we divide by a prime number. (Art. 102.)

15. Any number expressed by the decimal notation, divided by 9, will leave the *same remainder* as the sum of its *figures* or *digits* divided by 9. The same property belongs to the number 3, and to no other number. Thus, if 236 is divided by 9, the remainder is 2; so, if the sum of its digits, 2+3+6=11, is divided by 9, the remainder is also 2.

Note.—Upon this property of the number 9, is based a convenient method of proving multiplication and division.

PROOF OF MULTIPLICATION BY CASTING OUT THE NINES.

282. *First, cast the* 9s *out of the multiplicand and multiplier; multiply their remainders together, and cast the* 9s *out of their product, and set down the excess; then cast the* 9s *out of the answer obtained, and if this excess be the same as that obtained from the multiplier and the multiplicand, the work may be considered right.*

Note.—To cast out the 9s from a number, begin at the left hand, add the digits together, and as soon as the sum is 9 or over, drop the

---

QUEST.—What is the least divisor of every number? *Obs.* In obtaining the least common multiple of two or more numbers, by what kind of a number do we divide? 282. How is multiplication proved by casting out the 9s?

9, and add the remainder to the next digit, and so on. For example, to cast the 9s out of 4626357, we proceed thus: 4 and 6 are 10; drop the 9 and add the 1 to the next figure. 1 and 2 are 3, and 6 are 9; drop the 9 as above. 3 and 5 are 8, and 7 are 15; drop the 9, and we have 6 remainder.

Multiply 565 by 356.

Operation.	*Proof.*
565	The excess of 9s in the multiplicand is 7.
356	" " 9s " multiplier is 5.
3390	7×5=35; and the excess of 9s is 8.
2825	
1695	
Prod. 201140.	The excess of 9s in the *Ans.* is also 8.

PROOF OF DIVISION BY CASTING OUT THE NINES.

283. *First cast the* 9*s out of the divisor and quotient, and multiply the remainders together; to the product add the remainder, if any, after division; cast the* 9*s out of this sum, and set down the excess; finally cast the* 9*s out of the dividend, and if the excess is the same as that obtained from the divisor and quotient, the work may be considered right.*

AXIOMS.

284. In mathematics, there are certain propositions whose *truth* is so *evident* at sight, that no process of reasoning can make it plainer. These propositions are called *axioms.*

An *axiom*, therefore, is a *self-evident* proposition.

1. Quantities which are equal to the *same* quantity, are equal to each other.

2. If the same or equal quantities are *added* to equal quantities, the *sums* will be equal.

3. If the same or equal quantities are *subtracted* from equals, the *remainders* will be equal.

4. If the same or equal quantities are *added* to *unequals*, the *sums* will be unequal.

QUEST.—283. How is division proved by casting out the 9s?

5. If the same or equal quantities are *subtracted* from *unequals*, the *remainders* will be unequal.

6. If equal quantities are *multiplied* by the same or equal quantities, the *products* will be equal.

7. If equal quantities are *divided* by the same or equal quantities, the *quotients* will be equal.

8. If the same quantity is both *added* to and *subtracted* from another, the value of the latter will not be altered.

9. If a quantity is both *multiplied* and *divided* by the same or an equal quantity, its value will not be altered.

10. The *whole* of a quantity is *greater* than a *part*.

11. The *whole* of a quantity is equal to the sum of *all its parts*.

OBS. The term *quantity* signifies any thing which can be *multiplied*, *divided*, or *measured*. Thus, *numbers*, *yards*, *bushels*, *weight*, *time*, &c., are called quantities.

285. The following principles will at once be recognized by the pupil as *deductions* from the four Fundamental Rules of Arithmetic, viz: Addition, Subtraction, Multiplication, and Division.

286. When the *sum* of two numbers and *one* of the numbers are given, to find the *other* number.

From the given sum subtract the given number, and the remainder will be the other number.

Ex. 1. The sum of two numbers is 25, and one of them is 10; what is the other number?

Solution.—25—10=15, the other number. (Art. 40.)

PROOF.—15+10=25, the given sum. (Art. 284. Ax. 11.)

2. A and B together own 36 cows, 9 of which belong to A: how many does B own?

3. Two farmers bought 300 acres of land together, and one of them took 115 acres: how many acres did the other have?

QUEST.—284. What is an axiom? What is the first axiom? The second? Third? Fourth? Fifth? Sixth? Seventh? Eighth? Ninth? Tenth? Eleventh? *Obs.* What is meant by quantity? 286. When the sum of two numbers and one of them are given, how is the other found?

287. When the *difference* and the *greater* of two numbers are given, to find the *less.*

Subtract the difference from the greater, and the remainder will be the less number.

4. The greater of two numbers is 37, and the difference between them is 10: what is the less number?

Solution.—37—10=27, the less number. (Art. 40.)

PROOF.—27+10=37, the greater number. (Art. 39. Obs.)

5. A had 48 dollars in his pocket, which was 12 dollars more than B had: how many dollars had B?
6. D had 450 sheep, which was 63 more than E had: how many had E?

289. When the *difference* and the *less* of two numbers are given, to find the *greater.*

Add the difference and less number together, and the sum will be the greater number. (Art. 39.)

7. The difference between two numbers is 5, and the less number is 15: what is the greater number?

Solution.—15+5=20, the greater number.

PROOF.—20—15=5, the given difference. (Art. 40.)

8. A is 16 years old, and B is 8 years older: how old is B?
9. The number of male inhabitants in a certain town, is 935; and the number of females exceeds the number of males by 115: how many females does the town contain?

QUEST.—287. When the difference and the greater of two numbers are given, how is the less found? 289. When the difference and the less of two numbers are given, how is the greater found?

290. When the *sum* and *difference* of two numbers are given, to find the *two numbers.*

From the sum subtract the difference, and half the remainder will be the smaller number.

To the smaller number thus found, add the given difference, and the sum will be the larger number.

10. The sum of two numbers is 35, and their difference is 11: what are the numbers?

Solution.—35−11=24; and ½ of 24=12, the smaller number. And 12+11=23, the greater number.

PROOF.—23+12=35, the given sum. (Art. 284, Ax. 11.)

11. The sum of the ages of 2 boys is 25 years, and the difference between them is 5 years: what are their ages?

12. A man bought a chest of tea and a hogshead of molasses for $63; the tea cost $9 more than the molasses: what was the price of each?

291. When the *product* of two numbers and *one* of the numbers are given, to find the *other* number.

Divide the given product by the given number, and the quotient will be the number required. (Art. 74.)

13. The product of two numbers is 84, and one of the numbers is 7: what is the other number?

Solution.—84÷7=12, the required number. (Art. 72.)

PROOF.—12×7=84, the given product. (Art. 54.)

14. The product of A and B's ages is 120 years, and A's age is 12 years: how old is B?

15. A certain field contains 160 square rods, and the length of the field is 20 rods: what is its breadth?

QUEST.—290. When the sum and difference of two numbers are given, how are the numbers found? 291. When the product of two numbers and one of them are given, how is the other found?

Note.—The area of a field is found by multiplying its *length* and *breadth* together. (Art. 163.) Hence the area of a field may be considered as a product.

292. When the *divisor* and *quotient* are given to find the *dividend.*

Multiply the given divisor and quotient together, and the product will be the dividend. (Art. 73.)

16. If a certain divisor is 9, and the quotient is 12, what is the dividend?

Solution.—12×9=108, the dividend required.

PROOF.—108÷9=12, the given quotient. (Art. 72.)

17. A man having 11 children, gave them $75 apiece: how many dollars did he give them all?

18. A farmer divided a quantity of apples among 90 boys, giving each boy 15 apples: how many did he give them all?

293. When the *dividend* and *quotient* are given, to find the *divisor.*

Divide the given dividend by the given quotient, and the quotient thus obtained will be the number required. (Art. 73. Obs. 2.)

19. A certain dividend is 130, and the quotient is 10: what is the divisor?

Solution.—130÷10=13, the divisor required. (Art. 72.)

PROOF.—13×10=130, the given dividend. (Art. 73.)

20. A gentleman divided $120 equally among a company of sailors, giving them $10 apiece: how many sailors were there in the company?

QUEST.—292. When the divisor and quotient are given, how is the dividend found? 293. When the dividend and quotient are given, how is the divisor found?

21. A farmer having 600 sheep, divided them into flocks of 75 each: how many flocks had he?

294. When the *product* of three numbers and *two* of the numbers are given, to find the *other* number.

Divide the given product by the product of the two given numbers, and the quotient will be the other number.

22. There are three numbers whose product is 60; one of them is 3, and another 5: it is required to find the other number?

Solution.—$5\times3=15$; and $60\div15=4$, the number required.

Proof.—$5\times3\times4=60$, the given product.

23. The product of A, B, and C's ages, is 210 years; the age of A is 5 years, and that of B is 6 years: what is the age of C?

24. The product of three boys' marbles, is 1728; two of them have a dozen apiece: how many has the other?

SECTION XI.

ANALYSIS.

Art. **295.** Business men have a method of solving practical questions, which is frequently *shorter* and *more expeditious*, than that of arithmeticians fresh from the schools. If asked, by what rule they perform them, their reply is, "they do them *in their head*," or by the "*no rule method.*"

Their method consists in *Analysis*, and may, with propriety, be called the Common Sense Rule.

Quest—294. When the product of three numbers and two of them are given, how is the other found? 295. What is said of the method by which business men solve practical questions? In what does their method consist? What may it with propriety be called?

The term *analysis*, in physical science, signifies the *resolving* of a compound body into its *elements* or *component parts*.

ANALYSIS, in Arithmetic, signifies the *resolving of numbers* into the *factors* of which they are composed, and the *tracing of the relations* which they bear to each other.

OBS. In the preceding sections the student has become acquainted with the method of *analyzing particular examples* and *combinations* of numbers, and thence deducing general *principles* and *rules*. But analysis may be applied with advantage not only to the development of mathematical *truths*, but also to the *solution* of a great variety of problems both in arithmetic and practical life.

MENTAL EXERCISES.

Ex. 1. If 8 barrels of flour cost $40, how much will 5 barrels cost?

Analysis.—1 is 1 eighth of 8: therefore 1 barrel will cost 1 eighth as much as 8 barrels; and 1 eighth of $40 is $5. Now it is obvious that 5 barrels will cost 5 times as much as 1 barrel; and 5 times $5 are $25, the answer required.

Or, we may reason thus; 5 barrels are $\frac{5}{8}$ of 8 barrels; 5 barrels will therefore cost $\frac{5}{8}$ as much as 8 barrels. Now 1 eighth of $40 is $5, and 5 eighths is 5 times $5, which is $25. *Ans.*

2. If 7 lbs. of tea cost 42 shillings, what will 10 lbs. cost?

3. If 9 sheep are worth $27, how much are 15 sheep worth?

4. If 10 barrels of flour cost $60, what will 12 barrels cost?

5. Suppose 30 gallons of molasses cost $15, how many dollars will 7 gallons cost?

QUEST.—295. *a.* What is meant by analysis in physical science? What in arithmetic? *Obs.* To what may analysis be advantageously applied?

6. If a man earns 54 shillings in 6 days, how much can he earn in 15 days?

7. If 12 men can build 48 rods of wall in a day, how many rods can 20 men build in the same time?

8. A gentleman divided 90 shillings equally among 15 beggars: how many shillings did 7 of them receive?

9. Suppose 75 pounds of butter last a family of boarders 25 days, how many pounds will supply them for 12 days?

10. If 7 yards of cloth cost \$30, how much will 9 yards cost?

11. If 10 barrels of beef cost \$72, how much will 8 barrels cost?

12. If 7 acres of land cost \$50, what will 12 acres cost?

13. A farmer bought an ox cart, and paid \$15 down, which was $\frac{3}{10}$ of the price of it: what was the price of the cart; and how much does he owe for it?

Analysis.—The question to be solved is simply this: 15 is $\frac{3}{10}$ of what number? If 15 is $\frac{3}{10}$, $\frac{1}{10}$ is $\frac{1}{3}$ of 15, which is 5. Now if 5 is 1 tenth, 10 tenths is 10 times 5, which is 50.

Ans. { \$50 is the price of the cart, and \$50—\$15=35, the sum unpaid.

Note.—In solving examples of this kind, the learner is often perplexed in finding the value of $\frac{1}{10}$, &c. This difficulty arises from supposing that if $\frac{3}{10}$ of a certain number is 15, $\frac{1}{10}$ of it must be $\frac{1}{10}$ of 15. This mistake will be easily avoided by substituting in his mind the word *parts* for the given *denominator*.

Thus, if 3 parts cost \$15, 1 part will cost $\frac{1}{3}$ of \$15, which is \$5. But this part is a *tenth*. Now if 1 tenth cost \$5, then 10 tenths will cost 10 times as much.

14. A man bought a yoke of oxen, and paid \$56 cash down, which was $\frac{7}{8}$ of the price of them: what did they cost?

15. A merchant bought a quantity of wood and paid \$45 in goods, which was $\frac{5}{9}$ of the whole cost: how much did he pay for the wood?

16. A whale ship having been out 24 months, the cap-

tain found that his crew had consumed $\frac{4}{7}$ of his provisions: how many months' provision had he when he embarked; and how much longer would his provisions last?

17. How many times 7 in $\frac{4}{5}$ of 35?

Analysis.—$\frac{1}{5}$ of 35 is 7, and $\frac{4}{5}$ is 4 times 7, which is 28. Now 7 is contained in 28, 4 times. *Ans.* 4 times.

18. How many times 6 in $\frac{7}{9}$ of 45?
19. How many times 10 in $\frac{7}{6}$ of 60?
20. How many times 12 in $\frac{6}{7}$ of 84?
21. $\frac{5}{7}$ of 42 are how many times 6?
22. $\frac{7}{8}$ of 40 are how many times 5?
23. $\frac{6}{10}$ of 80 are how many times 12?
24. $\frac{5}{6}$ of 48 are how many times 4?
25. $\frac{7}{8}$ of 64 are how many times 7?
26. $\frac{9}{10}$ of 100 are how many times 12?
27. $\frac{7}{11}$ of 110 are how many times 8?
28. $\frac{8}{9}$ of 180 are how many times 10?
29. $\frac{5}{12}$ of 84 are how many times 9?
30. How many yards of cloth, at \$7 per yard, can be bought for $\frac{7}{9}$ of \$54?
31. How many barrels of flour, at \$5 per barrel, can be bought for $\frac{2}{3}$ of \$60?
32. A man had \$64 in his pocket, and paid $\frac{5}{8}$ of it for 10 barrels of flour: how much was that per barrel?
33. 40 is $\frac{5}{9}$ of how many times 6?

Analysis.—Since 40 is $\frac{5}{9}$, $\frac{1}{9}$ is $\frac{1}{5}$ of 40, or 8; and $\frac{9}{9}$ is 9 times 8, or 72. Now 6 is contained in 72, 12 times.

Ans. 12 times.

34. 56 is $\frac{8}{9}$ of how many times 7?
35. 81 is $\frac{9}{10}$ of how many times 30?
36. 72 is $\frac{8}{11}$ of how many times 9?
37. 96 is $\frac{8}{9}$ of how many times 12?
38. 64 is $\frac{8}{10}$ of how many times 20?
39. 54 is $\frac{6}{8}$ of how many times 24?
40. 108 is $\frac{9}{11}$ of how many times 12?

41. Frank sold 10 peaches, which was $\frac{2}{9}$ of all he had; he then divided the remainder equally among 5 companions: how many did they receive apiece?

42. Lincoln spent 60 cents for a book, which was $\frac{10}{12}$ of his money; the remainder he laid out for oranges, at 4 cents apiece: how many oranges did he buy?

43. A man paid away $35, which was $\frac{5}{7}$ of all he had; he then laid out the rest in cloth at $2 per yard: how many yards did he obtain?

44. A farmer bought a quantity of goods, and paid $20 down, which was $\frac{2}{5}$ of the bill: how many cords of wood, at $3 per cord, will it take to pay the balance?

45. A man bought a horse and paid $60 in cash, which was $\frac{5}{7}$ of the price: how many barrels of flour at $6 per barrel, will it take to pay the balance?

46. $\frac{3}{9}$ of 27 is $\frac{3}{4}$ of what number?

Analysis.—$\frac{3}{9}$ of 27 is 9. And if 9 is $\frac{3}{4}$ of a certain number, $\frac{1}{4}$ of that number is 3; and $\frac{4}{4}$ is 4 times 3, which is 12, the number required.

47. $\frac{4}{10}$ of 30 is $\frac{2}{7}$ of what number?

48. $\frac{3}{8}$ of 40 is $\frac{5}{7}$ of what number?

49. $\frac{4}{7}$ of 35 is $\frac{2}{10}$ of what number?

50. $\frac{7}{9}$ of 54 is $\frac{6}{15}$ of what number?

EXERCISES FOR THE SLATE.

296. It will be seen from the preceding examples, that no particular rules can be prescribed for solving questions by analysis. None in fact are requisite. The process will be easily suggested by the judgment of the pupil, and the conditions of the question.

OBS. The operation of solving a question by analysis, is called an *analytic solution*. In reciting the following examples, the pupil

---

QUEST.—296. Can any particular rules be prescribed for solving questions by analysis? How then will you know how to proceed? *Obs.* What is the operation of solving questions by analysis, called?

should be required to analyze each question, and give the reason for each step, as in the preceding mental exercises.

Ex. 1. If 40 barrels of beef cost $320, how much will 52 barrels cost?

Analytic Solution.—Since 40 bbls. cost $320, 1 bbl. will cost $\frac{1}{40}$ of $320. And $\frac{1}{40}$ of $320 is $320÷40=$8. Now if 1 bbl. cost $8, 52 bbls. will cost 52 times as much; and $8×52=$416, which is the answer required.

Or thus: 52 bbls. are $\frac{52}{40}$ of 40 bbls.; therefore 52 bbls. will cost $\frac{52}{40}$ of $320; (the cost of 40 bbls.;) and $\frac{52}{40}$ of $320 is $320×$\frac{52}{40}$=$416, the same result as before. (Arts. 132, 133.)

OBS.—1. Other solutions of this example might be given; but our present object is to show how this and similar examples may be solved by analysis. The former method is the simplest and most strictly analytic, though not so short as the latter. It contains two steps:

First, we separate the given price of 40 bbls. ($320) into 40 equal parts, to find the value of one part, or the cost of 1 bbl., which is $8.

Second, we multiply the price of 1 bbl. ($8) by 52, the number of barrels, whose cost is required, and the product is the answer sought.

2. In solving questions analytically, it may be remarked in general, that we reason from the *given number* to 1, then from 1 to the *required number*.

3. This and similar questions are usually placed under Simple Proportion, or the "Rule of Three;" but business men almost invariably solve them by *analysis*.

2. If 30 cows cost $360.90, how much will 47 cows cost, at the same rate?

3. If 25 barrels of apples cost $15, how much will 37 barrels cost?

4. If 15 hogsheads of molasses cost $450, how much will 21 hogsheads cost?

5. If 31 yards of cloth cost $127, how much will 89 yards cost?

6. If 55 tons of hay cost $660, what will 17 tons come to?

7. An agent paid $159 for 530 pounds of wool: how much was that per 100?

8. A man bought 30 cords of wood for $76.80: how much must he pay for 65 cords?

9. A gentleman bought 85 yards of carpeting for $106.25: how much would 38 yards cost?

10. A drover bought 350 sheep for $525: how much would 65 cost, at the same rate?

11. If $12\frac{1}{2}$ pounds of coffee cost $1.25, how much will 45 pounds cost?

12. If $16\frac{1}{2}$ bushels of corn are worth $8, how much are 25 bushels worth?

13. Paid $20 for 60 pounds of tea: how much would $12\frac{1}{2}$ pounds cost, at the same rate?

14. Bought 41 yards of flannel for $16.40: how much would $8\frac{3}{5}$ yards cost?

15. Bought 18 pounds of ginger for $4.50: how much will $10\frac{3}{5}$ pounds cost?

16. If a stage goes 84 miles in 12 hours, how far will it go in $15\frac{1}{2}$ hours?

17. If 8 horses eat 36 bushels of oats in a week, how many bushels will 25 horses eat in the same time?

18. If a Railroad car runs 120 miles in 5 hours, how far will it run in $12\frac{3}{4}$ hours?

19. If a steamboat goes 180 miles in 12 hours, how far will it go in $5\frac{3}{5}$ hours?

20. If 4 men can do a job of work in 12 days, how long will it take 6 men to do it?

Solution—Since the job requires 4 men 12 days, it will require 1 man 4 times as long; and 4 times 12 days are 48 days. Again, it requires 1 man 48 days, it will require 6 men $\frac{1}{6}$ as long; and 48 days$\div$6=8 days, which is the answer required.

21. If 6 men eat a barrel of flour in 24 days, how long will it last 10 men?

22. If a given quantity of corn lasts 9 horses 96 days, how long will the same quantity last 15 horses?

23. If 12 men can build a house in 90 days, how long will it take 20 men to build it?

24. If 100 barrels of pork last a crew of 20 men 45 months, how long will it last a crew of 28 men?

25. If 4 stacks of hay will keep 60 cattle 120 days, how long will they keep 25 cattle?

26. If $\frac{3}{8}$ of a bushel of wheat cost 30 cents, what will $\frac{7}{8}$ of a bushel cost?

27. If $\frac{5}{9}$ of a ton of hay cost $7, what will $\frac{3}{9}$ of a ton cost?

28. If $\frac{3}{5}$ of a pound of imperial tea cost 27 cents, how much will $\frac{4}{5}$ of a pound cost?

29. If $\frac{3}{7}$ of a ton of coal cost $2.61, how much will $\frac{6}{7}$ of a ton cost?

30. If $\frac{2}{3}$ of a yard of silk cost 6 shillings, how much will $\frac{7}{8}$ of a yard cost?

Solution.—Since $\frac{2}{3}$ of a yard cost 6s., $\frac{1}{3}$ will cost 3s., and $\frac{3}{3}$ or 1 yard will cost 9s. Again, if 1 yard costs 9s., $\frac{1}{8}$ yd. will cost $1\frac{1}{8}$s.; and $\frac{7}{8}$ yd. will cost $7\frac{7}{8}$ shillings, which is the answer required.

31. If $\frac{4}{5}$ of a cord of wood cost $1.80, how much will $\frac{3}{7}$ of a cord cost?

32. If $\frac{2}{9}$ of a yard of broadcloth cost 14 shillings, how much will $\frac{3}{4}$ of a yard cost?

33. A man bought $\frac{7}{8}$ of an acre of land for $56, and afterwards sold $\frac{2}{5}$ of an acre at cost: how much did he receive for it?

34. A grocer bought 7 barrels of vinegar for $28, and sold $\frac{3}{4}$ of a barrel at cost: how much did it come to?

35. A grocer bought a firkin of butter containing 56 lbs. for $11.20, and sold $\frac{3}{4}$ of it at cost: how much did he get a pound?

36. If $6\frac{1}{4}$ bushels of peas are worth $5.50, how much are $20\frac{1}{2}$ bushels worth?

37. If a man pays $47 for building $23\frac{1}{2}$ rods of ornamental fence, how much would it cost him to build $42\frac{3}{4}$ rods?

38. A farmer paid $45.42 for making $36\frac{2}{5}$ rods of stone wall: how much will it cost him to make $60\frac{7}{10}$ rods?

39. A man paid $\frac{20}{100}$ of a dollar for 4 pounds of veal; how much would a quarter of veal cost, which weighs 20 pounds?

40. If 5 pounds of butter cost $4\frac{2}{7}$ shillings, how much will 42 pounds cost?

Note.—It will be seen that $4\frac{2}{7}$s. $=\frac{30}{7}$s. (Art. 122.) Therefore 1 pound will cost $\frac{6}{7}$s.; and 42 lbs. will cost $\frac{6}{7}\times 42=\frac{252}{7}$, or 36s. *Ans.*

41. If 20 pounds of cheese cost \$$3\frac{6}{8}$, how much will 168 pounds cost?

42. If 30 yards of cotton cost \$$4\frac{1}{2}$, how much will a piece containing 19 yards cost?

43. If $\frac{5}{12}$ of a cord of wood costs $\frac{5}{8}$ of a dollar, how much will $\frac{3}{4}$ of a cord cost?

Solution.—Since $\frac{5}{12}$ of a cord cost \$$\frac{5}{8}$, $\frac{1}{12}$ will cost \$$\frac{1}{8}$; consequently $\frac{12}{12}$ or 1 cord will cost \$$\frac{12}{8}$. Again, if 1 cord cost \$$\frac{12}{8}$, $\frac{1}{4}$ of a cord will cost \$$\frac{3}{8}$; and $\frac{3}{4}$ will therefore cost $\frac{9}{8}$ or \$$1\frac{1}{8}$, which is the answer required.

44. If $\frac{3}{4}$ of a yard of cloth cost £$\frac{6}{7}$, how much will $\frac{7}{8}$ of a yard cost?

45. If $\frac{3}{16}$ of a ship cost \$16000, how much is $\frac{5}{8}$ of her worth?

46. A man bought a quantity of land and sold $\frac{9}{10}$ of an acre for \$63, which was only $\frac{3}{4}$ of the cost: how much did he give per acre?

47. If $7\frac{1}{2}$ yards of satinet cost \$$9\frac{3}{8}$, how much will $18\frac{1}{2}$ yards cost?

48. A ship's company of 30 men have 4500 pounds of flour: how long will it last them, allowing each man $2\frac{1}{2}$ lbs. per day?

49. How long will 56700 pounds of meat last a garrison of 756, allowing each man $\frac{3}{4}$ lb. per day?

50. How long will the same quantity of meat last the same garrison, allowing $1\frac{1}{2}$ lb. apiece per day?

51. A merchant sold 12 yards of silk at 7 shillings per yard, and took his pay in wheat at 6 shillings per bushel: how many bushels did he receive?

Solution.—First find the cost of the silk. If 1 yd. costs 7s., 12 yds. will cost 7 times 12s., which is 84s. Now the question is, how many bushels of wheat it will take to pay this 84s. But as the wheat is 6s. a bushel, it will manifestly take as many bushels as 6s. is contained times in 84s.; and $84\div 6=14$. *Ans.* 14 bushels.

297. The last and similar examples are frequently solved by a rule called ***Barter***.

Barter signifies an exchange of articles of commerce at prices agreed upon by the parties.

OBS. A specific rule for such operations seems to be *worse* than *useless;* for it burdens the memory of the learner with particular directions for the solution of questions which his *common sense*, if permitted to be exercised, will solve more expeditiously by the *principles of analysis*.

52. A shoemaker sold 6 pair of thick boots at 32 shillings a pair, and took his pay in corn at 3 shillings per bushel: how many bushels did he receive?

53. A man bought 50 pounds of sugar at 12½ cents a pound, and was to pay for it in wood at $3.12½ per cord: how many cords did it take?

54. How many pair of hose at 3 shillings a pair, will it take to pay for 135 pounds of tea at 6 shillings a pound?

55. How many pounds of butter at 17½ cents a pound, must be given in exchange for 186 yards of calico at 18¾ cents per yard?

56. How many pounds of tobacco at 16¼ cents a pound, must be given in exchange for 256 pounds of sugar at 6¼ cents a pound?

57. A farmer bought 325 sheep at $2¼ apiece, and paid for them in hay at $10½ per ton: how many tons did it take?

58. A man bought a hogshead of molasses worth 37½ cents per gallon, and gave 331½ pounds of cheese in exchange: how much was the cheese a pound?

59. Bought 74 bushels of salt at 42½ cents per bushel, and paid in oats at ¼ of a dollar per bushel: how many oats did it require?

60. A bookseller exchanges 400 dictionaries worth 87½ cents apiece for 700 grammars: how much did the grammars cost apiece?

61. What cost 680 tons of chalk, at 10 shillings sterling per ton?

QUEST.—297. What is meant by Barter? *Obs.* Is a specific rule necessary for such operations?

Solution.—10s.=£$\frac{1}{2}$. Now, if 1 ton costs £$\frac{1}{2}$, 680 tons will cost 680 times as much; that is, 680 tons will cost *half* as many pounds as there are tons; (Art. 132;) and 680÷2=340. *Ans.* £340.

298. Examples like the preceding one are often classed under the Rule called *Practice.*

TABLE OF ALIQUOT PARTS OF $1, £1, AND 1s.

Parts of $1.	N. E. Currency.	Parts of £1.	Parts of 1s.
50cts.=$$\frac{1}{2}$.	3 shil.=$$\frac{1}{2}$.	10 shil.=£$\frac{1}{2}$.	6d. = $\frac{1}{2}$ shil.
33$\frac{1}{3}$ c.=$$\frac{1}{3}$.	2 shil.=$$\frac{1}{3}$.	6s. 8d. =£$\frac{1}{3}$.	4d. = $\frac{1}{3}$ shil.
25 c.=$$\frac{1}{4}$.	1s. 6d.=$$\frac{1}{4}$.	5 shil.=£$\frac{1}{4}$.	3d. = $\frac{1}{4}$ shil.
20 c.=$$\frac{1}{5}$.	1 shil.=$$\frac{1}{6}$.	4 shil.=£$\frac{1}{5}$.	2d. = $\frac{1}{6}$ shil.
16$\frac{2}{3}$ c.=$$\frac{1}{6}$.	N. Y. Currency.	3s. 4d. =£$\frac{1}{6}$.	1$\frac{1}{2}$d.= $\frac{1}{8}$ shil.
12$\frac{1}{2}$ c.=$$\frac{1}{8}$.	4 shil.=$$\frac{1}{2}$.	2s. 6d. =£$\frac{1}{8}$.	1d. =$\frac{1}{12}$ shil.
10 c.=$$\frac{1}{10}$.	2s. 8d.=$$\frac{1}{3}$.	2 shil.=£$\frac{1}{10}$.	$\frac{1}{2}$d. =$\frac{1}{24}$ shil.
8$\frac{1}{3}$ c.=$$\frac{1}{12}$.	2 shil.=$$\frac{1}{4}$.	1s. 8d. =£$\frac{1}{12}$.	9d. = $\frac{3}{4}$ shil.
6$\frac{1}{4}$ c.=$$\frac{1}{16}$.	1s. 4d.=$$\frac{1}{6}$.	1 shil. =£$\frac{1}{20}$.	8d. = $\frac{2}{3}$ shil.
5 c.=$$\frac{1}{20}$.	1 shil.=$$\frac{1}{8}$.	11s.=£$\frac{1}{2}$+£$\frac{1}{20}$.	7d.=$\frac{1}{2}$s.+ $\frac{1}{6}$s.

62. What cost 720 bushels of corn, at 2 shillings and 6 pence per bushel?

Solution.—2s. 6d.=£$\frac{1}{8}$. And 720×$\frac{1}{8}$=£90. *Ans.*

63. What cost 840 chairs, at 3s. N. E. cur. apiece?

64. What cost 360 melons, at 1s. 6d. N. E. cur. apiece?

65. What cost 360 knives, at 2s. 8d. N. Y. cur. apiece?

66. What cost 760 brooms, at 1s. N. Y. cur. apiece?

67. At 10s. 6d. sterling per barrel, what will 350 barrels of mackerel come to?

68. At 17s. 6d. st. apiece, what will 540 hats come to?

69. What cost 33750 sheep, at 6s. 8d. st. apiece?

70. At $20.60 per ton, what cost 12 tons and 5 hundred weight of hay?

71. What is the cost of 480 yards of ribbon, at 6$\frac{1}{4}$ cents per yard?

72. What cost 750 bushels of potatoes, at 33$\frac{1}{3}$ cents per bushel?

73. What cost 360 barrels of cider, at 66$\frac{2}{3}$ cents per barrel?

74. What cost 450 chaldrons of coal, at 15s. per chaldron?

75. What will 150 acres of land cost, at £8, 10s. per acre?

76. Three men, A, B, and C join in an adventure; A puts in $200; B, $300; and C, $400; and they gain $72: how much is each man's share of the gain?

Analysis.—The whole sum invested is $200+$300+$400=$900. Now, since $900 gain $72, $1 will gain $\frac{1}{900}$ of $72; and $72÷900=$.08.

If $1 gains 8c., $200 will gain $200×.08=$16, A's sh.
" 1 " " 300 " 300×.08= 24, B's "
" 1 " " 400 " 400×.08= 32, C's "

Or, we may reason thus: since the sum invested is $900,
A's part of the investment is $\frac{200}{900}=\frac{2}{9}$;
B's " " is $\frac{300}{900}=\frac{3}{9}$;
C's " " is $\frac{400}{900}=\frac{4}{9}$. Hence,

A must receive $\frac{2}{9}$ of $72 (the gain)=$16
B " $\frac{3}{9}$ " 72 " = 24
C " $\frac{4}{9}$ " 72 " = 32
PROOF.—The whole gain is $72. (Ax. 11.)

299. When two or more individuals associate themselves together for the purpose of carrying on a joint business, the union is called a *partnership* or *copartnership*.

OBS. The process by which examples like the last one are solved, is often called *Fellowship*.

77. A and B entered into partnership; A furnished $400, and B $500; they gained $300: how much was each man's share of the gain?

78. A, B, and C hired a farm together, for which they paid $175 rent: A advanced $75; B, $60; and C, $40. They raised 250 bushels of wheat: what was each man's share?

79. A, B, and C together spent $1000 in lottery tickets. A put in $400; B, $250; and C, $350; they drew a prize of $1500: how much was each man's share?

80. A, B, C, and D fitted out a whale ship; A advanced $10000; B, $12000; C, $15000; and D, $8000; the ship brought home 3000 bbls. of oil: what was each man's share?

81. A, B, and C formed a partnership; A furnished $900; B, $1500; and C, $1200: they lost $1260; what was each man's share of the loss?

82. X, Y, and Z entered into a joint speculation, on a capital of $20000, of which X furnished $5000; Y, $7000; and Z the balance; their net profits were $5000 per annum: what was the share of each?

83. A bankrupt owes one of his creditors $300; another $400; and a third $500; his property amounts to $800: how much can he pay on a dollar; and how much will each of his creditors receive?

Note.—The solution of this example is the same in principle as example seventy-sixth.

300. A *bankrupt* is a person who is insolvent, or unable to pay his just debts.

OBS. Examples like the preceding one are sometimes arranged under a rule called *Bankruptcy*.

84. A bankrupt owes $2000, and his property is appraised at $1600: how much can he pay on a dollar?

85. A man failing in business, owes A $156.45; B $256.40; and C $360.40; and his effects are valued at $317: how much will each man receive?

86. The whole effects of a man failing in business amounted to $3560, he owed $35600: how much can he pay on a dollar; and how much will B receive, who has a claim on him of $5000?

87. A man died insolvent, owing $55645; and his property was sold at auction for $2350: how much will his estate pay on a dollar?

88. How much can a bankrupt, who has $6540 real estate and owes $56000, pay on a dollar?

301. It often happens in storms and other casualties at sea, that masters of vessels are obliged to throw portions of their cargo overboard, or sacrifice the ship and their crew. In such cases, the law requires that the loss shall be divided among the owners of the vessel and cargo, in proportion to the amount of each one's property at stake.

The process of finding each man's loss, in such instances, is called *General Average.*

OBS. The operation is the same as that in solving questions in bankruptcy and partnership.

89. A, B, and C freighted a sloop with flour from New York to Boston; A had on board 600 barrels; B, 400; and C, 200. On her passage 200 barrels were thrown overboard in a gale, and the loss was shared among the owners according to the quantity of flour each had on board: what was the loss of each?

90. A Liverpool packet being in distress, the master threw goods overboard to the amount of $10000. The whole cargo was valued at $72000, and the ship at $28000: what per cent. loss was the general average; and how much was A's loss, who had goods aboard to the amount of $15000?

91. A coasting vessel being overtaken in a gale, the master was obliged to throw overboard part of his cargo valued at $15500. The whole cargo was worth $85265, and the vessel $17000: what per cent. was the general average; and what was the loss of the master, who owned $\frac{1}{4}$ of the vessel?

92. A farmer mixed 15 bushels of oats worth 2 shillings per bushel, with 5 bushels of corn worth 4 shillings per bushel: what is the mixture worth per bushel?

Solution.—15 bu. at 2s.=30s., value of oats.
5 bu. at 4s.=20s., value of corn.
20 bu. mixed. 50s., value of whole mixture.

Now, if 20 bu. mixture are worth 50s., 1 bu. is worth $\frac{1}{20}$ of 50s., which is $2\frac{1}{2}$s., the answer required.

PROOF.—20 bu.×$2\frac{1}{2}$s.=50s. the value of the whole mixture.

93. A miller has a quantity of rye worth 6s. per bushel, and wheat worth 9s. per bushel; he wishes to make a mixture of them which shall be worth 8s. per bushel: what part of each must the mixture contain?

Analysis.—The difference in their prices per bushel is 3s.; hence, the difference in the price of 1 third of a bushel of each is 1s. Now if 1 third of a bushel is taken from a bushel of rye, the remaining 2 thirds will be worth 4s.; and if 1 third of a bushel of wheat which is worth 3s., be added to the rye, the mixture will be worth 7s. Again, if $\frac{2}{3}$ of a bushel is taken from a bushel of rye, the remaining third will be worth 2s., and if $\frac{2}{3}$ of a bushel of wheat, which is worth 6s., be added to the rye, the mixture will be worth 8s.; therefore, $\frac{1}{3}$ of a bushel of rye added to $\frac{2}{3}$ of wheat, will make a mixture of 1 bushel, which is worth 8 shillings; consequently the mixture must be $\frac{1}{3}$ rye and $\frac{2}{3}$ wheat; or 1 part rye to 2 parts wheat.

PROOF.—Since 1 bushel of rye is worth 6s., $\frac{1}{3}$ bu. is worth $\frac{1}{3}$ of 6s., or 2s.; and as 1 bu. of wheat is worth 9s., $\frac{2}{3}$ bu. is worth $\frac{2}{3}$ of 9s., or 6s.; and 6s.+2s.=8s.

Note.—If we make the difference between the less price and the price of the mixture, the numerator, and the difference betweeen the prices of the commodities to be mixed, the denominator, the fraction will express the part to be taken of the higher priced article; and if we place the difference between the higher price and the price of the mixture over the same denominator, the fraction will express the part to be taken of the lower priced article.

94. A goldsmith has a quantity of gold 16 carats fine, and another quantity 22 carats fine; he wishes to make a mixture 20 carats fine: what part of each will the mixture contain?

Ans. $\frac{2}{6}$ of 16 carats fine, and $\frac{4}{6}$ of 22 carats fine.

302. Examples requiring a mixture of commodities of different values, like the last three, are commonly classed under a rule called *Alligation.*

Note. Alligation is usually divided into *medial* and *alternate.* The

92d example is an instance of Medial Alligation; the 93d and 94th are instances of Alternate Alligation. Questions in the latter very seldom occur in practical life.

95. A grocer mixes 50 pounds of tea worth 4 shillings a pound, with 100 lbs. worth 7s. a pound: what is a pound of the mixture worth?

96. A milk-man mixed 30 quarts of water with 120 quarts of milk, worth 5 cents per quart: what is a quart of the mixture worth?

97. A farmer made a mixture of provender containing 30 bushels of oats, worth 25 cents per bushel; 10 bushels of peas, worth 75 cents per bushel, and 15 bushels of corn, worth 50 cents per bushel: what is the value of the whole mixture; and what is it worth per bushel?

98. An oil dealer mixed 60 gallons of whale oil, worth $31\frac{1}{4}$ cents per gallon, with 85 gallons of sperm oil, worth 90 cents per gallon: what is the mixture worth per gallon?

99. A grocer had three kinds of sugar, worth 6, 8, and 12 cents per pound; he mixed 112 lbs. of the first, 150 lbs. of the second, and 175 of the third together: what was the mixture worth per pound?

100. A goldsmith melted 10 oz. of gold 20 carats fine, with 8 oz. 22 carats fine, and 4 oz. of alloy: how many carats fine was the mixture?

101. If 4 men reap 12 acres in 2 days, how long will it take 9 men to reap 36 acres?

Analysis.—If 4 men can reap 12 acres in 2 days, 1 man can reap $\frac{1}{4}$ of 12 acres in the same time; and $\frac{1}{4}$ of 12 acres is 3 acres. But if 1 man can reap 3 acres in 2 days, in 1 day he can reap $\frac{1}{2}$ of 3 acres, and $\frac{1}{2}$ of 3 is $1\frac{1}{2}$ acre. Again, if $1\frac{1}{2}$ acre requires a man 1 day, 36 acres will require him as many days as $1\frac{1}{2}$ is contained times in 36; and $36 \div 1\frac{1}{2} = 24$ days. Now if 1 man can reap the given field in 24 days, 9 men will reap it in $\frac{1}{9}$ of the time: and $24 \div 9 = 2\frac{2}{3}$.

Ans. 9 men can reap 36 acres in $2\frac{2}{3}$ days.

OBS.—This and similar examples are usually placed under Compound Proportion, or "Double Rule of Three." If the analysis of them is found too difficult for beginners, they can be deferred till review.

102. If 7 men can reap 42 acres in 6 days, how many men will it take to reap 100 acres in 5 days?

103. If 14 men can build 84 rods of wall in 3 days, how long will it take 20 men to build 300 rods?

104. If 1000 barrels of provisions will support a garrison of 75 men for 3 months, how long will 3000 barrels support a garrison of 300?

105. If a man travels 320 miles in 10 days, traveling 8 hours per day, how far can he go in 15 days, traveling 12 hours per day?

106. If 24 horses eat 126 bushels of oats in 36 days, how many bushels will 32 horses eat in 48 days?

107. A lad returning from market being asked how many peaches he had in his basket, replied that $\frac{1}{2}$, $\frac{1}{3}$, and $\frac{1}{4}$ of them made 52: how many peaches had he?

Analysis.—The sum of $\frac{1}{2}$, $\frac{1}{3}$, and $\frac{1}{4}=\frac{13}{12}$. (Art. 127.) The question then resolves itself into this: 52 is $\frac{13}{12}$ of what number? Now if 52 is $\frac{13}{12}$, $\frac{1}{12}$ is $\frac{1}{13}$ of 52, which is 4; and $\frac{12}{12}$ is $4\times12=48$. *Ans.* 48 peaches.

PROOF.—$\frac{1}{2}$ of 48 is 24; $\frac{1}{3}$ is 16; and $\frac{1}{4}$ is 12. Now, $24+16+12=52$.

303. This and similar examples are often placed under a rule called ***Position.***

OBS. The shortest and easiest method of solving them is by *Analysis.*

108. A farmer lost $\frac{1}{4}$ of his sheep by sickness; $\frac{1}{5}$ were destroyed by wolves; and he had 72 sheep left: how many had he at first?

109. A person having spent $\frac{1}{2}$ and $\frac{1}{3}$ of his money, finds he has $48 left: what had he at first?

110. After a battle a general found that $\frac{1}{6}$ of his army had been taken prisoners, $\frac{1}{8}$ were killed, $\frac{1}{12}$ had deserted, and he had 900 left: how many had he at the commencement of the action?

111. What number is that $\frac{1}{3}$ and $\frac{1}{4}$ of which is 84?

112. What number is that $\frac{1}{2}$ and $\frac{1}{3}$ of which being added to itself, the sum will be 110?

113. A certain post stands $\frac{1}{3}$ in the mud, $\frac{1}{4}$ in the water, and 10 feet above the water: how long is the post?

114. Suppose I pay \$85 for $\frac{5}{8}$ of an acre of land: what is that per acre?

115. A man paid \$2700 for $\frac{3}{16}$ of a vessel: what is the whole vessel worth?

116. A gentleman spent $\frac{1}{3}$ of his life in Boston, $\frac{1}{4}$ of it in New York, and the rest of it, which was 30 years, in Philadelphia: how old was he?

117. What number is that $\frac{7}{8}$ of which exceeds $\frac{2}{3}$ of it by 10?

118. In a certain school $\frac{1}{3}$ of the scholars were studying arithmetic, $\frac{1}{4}$ algebra, $\frac{1}{6}$ geometry, and the remainder, which was 18, were studying grammar: how many scholars were there in the school?

119. A owns $\frac{1}{3}$, and B $\frac{1}{12}$ of a ship; A's part is worth \$650 more than B's: what is the value of the ship?

120. In a certain orchard $\frac{1}{3}$ are apple-trees, $\frac{1}{4}$ peach-trees, $\frac{1}{6}$ plumb-trees, and the remaining 15 were cherry-trees: how many trees did the orchard contain?

SECTION XII.

RATIO AND PROPORTION.

ART. **305.** RATIO is that relation between two numbers or quantities, which is expressed by the *quotient* of the one divided by the other. Thus, the ratio of 6 to 2 is 6÷2, or 3; for 3 is the quotient of 6 divided by 2.

MENTAL EXERCISES.

Ex. 1. What is the ratio of 14 to 7? *Ans.* 2.

2. What is the ratio of 10 to 2? Of 16 to 4?

QUEST.—305. What is ratio?

3. What is the ratio of 18 to 9? Of 18 to 6?

4. What is the ratio of 24 to 3? Of 24 to 4? Of 24 to 6? Of 24 to 8? Of 24 to 12?

5. What is the ratio of 30 to 6? Of 25 to 5? Of 27 to 9? Of 40 to 8? Of 56 to 7? Of 84 to 12?

6. What is the ratio of 3 to 7? *Ans.* $\frac{3}{7}$.

7. What is the ratio of 5 to 8? Of 7 to 10? Of 9 to 13? Of 10 to 17? Of 21 to 43?

306. The two given numbers thus compared, when spoken of together, are called a *couplet;* when spoken of separately, they are called the *terms* of the ratio.

The *first* term is the *antecedent;* and the *last,* the *consequent.*

307. Ratio is expressed in two ways:

First, in the form of a fraction, making the *antecedent* the *numerator,* and the *consequent* the *denominator.* Thus, the ratio of 8 to 4 is written $\frac{8}{4}$; the ratio of 12 to 3, $\frac{12}{3}$, &c.

Second, by placing two points or a colon (:) between the numbers compared. Thus, the ratio of 8 to 4, is written 8 : 4; the ratio of 12 to 3, 12 : 3, &c.

OBS. 1. The expressions $\frac{8}{4}$, and 8 : 4 are equivalent to each other, and one may be exchanged for the other at pleasure.

2. The English mathematicians put the antecedent for the numerator and the consequent for the denominator, as above; but the French put the consequent for the numerator and the antecedent for the denominator. The English method appears to be equally simple, and is confessedly the most in accordance with reason.

3. In order that concrete numbers may have a ratio to each other, they must necessarily express objects so far of the same nature, that one can be properly said to be *equal* to, or *greater,* or *less* than the other. (Art. 280.) Thus a foot has a ratio to a yard; for one is *three times* as long as the other; but a foot has not properly a ratio to an hour, for one cannot be said to be *longer* or *shorter* than the other.

QUEST.—306. What are the two given numbers called when spoken of together? What, when spoken of separately? 307. In how many ways is ratio expressed? What is the first? The second? *Obs.* Which of the terms do the English mathematicians put for the numerator? Which do the French? In order that concrete numbers may have a ratio to each other, what kind of objects must they express?

8. What is the ratio of 15 lbs. to 3 lbs.? Of 21 lbs. to 7 lbs.? Of 35 bu. to 7 bu.? Of 36 yds. to 12 yds?

9. What is the ratio of £1 to 10s.?

Note.—£1 is 20s. The question then is simply this: what is the ratio of 20s. to 10s.? *Ans.* 2.

10. What is the ratio of £2 to 5s.? Of £3 to 12s.?

308. A *direct* ratio is that which arises from dividing the antecedent by the consequent, as in Art. 305.

309. An *inverse* or *reciprocal* ratio, is the ratio of the *reciprocals* of two numbers. (Art. 280. Def. 11.) Thus, the direct ratio of 9 to 3, is 9 : 3, or $\frac{9}{3}$; the reciprocal ratio is $\frac{1}{9}$: $\frac{1}{3}$, or $\frac{1}{9}\div\frac{1}{3}=\frac{3}{9}$; (Art. 139;) that is, the consequent 3, is divided by the antecedent 9. Hence,

A reciprocal ratio is expressed by inverting the fraction which expresses the direct ratio; or when the notation is by points, by inverting the order of the terms. Thus, 8 is to 4, inversely, as 4 to 8.

309. *a.* A *simple* ratio is a ratio which has but one antecedent and one consequent, and may be either direct or inverse; as 9 : 3, or $\frac{1}{9}$: $\frac{1}{3}$.

310. A *compound* ratio is the ratio of the *products* of the corresponding terms of two or more simple ratios. Thus,

The simple ratio of	9 : 3 is 3;
And " " of	8 : 4 is 2;
The ratio compounded of these is	72 : 12 = 6;

OBS. A compound ratio is of the *same nature* as any other ratio. The term is used to denote the *origin* of the ratio in particular cases.

311. From the definition of ratio and the mode of expressing it in the form of a fraction, it is obvious that the *ratio* of two numbers is the same as the *value* of a fraction whose numerator and denominator are respectively equal to the antecedent and consequent of the giv-

QUEST.—308. What is a direct ratio? 309. What is an inverse or reciprocal ratio? How is a reciprocal ratio expressed by a fraction? How by points? 309. *a.* What is a simple ratio? 310. What is a compound ratio? *Obs.* Does it differ in its nature from other ratios? 311. What is the ratio of two numbers equal to?

en couplet; for, *each* is the *quotient* of the numerator divided by the denominator. (Arts. 305, 110.)

OBS. From the principles of fractions already established, we may, therefore, deduce the following truths respecting ratios.

312. *To multiply the antecedent of a couplet by any number, multiplies the ratio by that number; and to divide the antecedent, divides the ratio:* for, multiplying the numerator, multiplies the value of the fraction by that number, and dividing the numerator, divides the value. (Arts. 111, 112.)

Thus, the ratio of 16 : 4 is 4;
The ratio of 16×2 : 4 is 8, which equals 4×2;
And " 16÷2 : 4 is 2, " " 4÷2.

313. *To multiply the consequent of a couplet by any number, divides the ratio by that number; and to divide the consequent, multiplies the ratio:* for, multiplying the denominator, divides the value of the fraction by that number, and dividing the denominator, multiplies the value. (Arts. 113, 114.)

Thus the ratio of 16 : 4 is 4;
The " 16 : 4×2 is 2, which equals 4÷2;
And " 16 : 4÷2 is 8, which equals 4×2.

314. *To multiply or divide both the antecedent and consequent of a couplet by the same number, does not alter the ratio:* for, multiplying or dividing both the numerator and denominator by the same number, does not alter the value of the fraction. (Art. 116.)

Thus the ratio of 12 : 4 is 3;
The " 12×2 : 4×2 is 3;
And " 12÷2 : 4÷2 is 3.

315. If the two numbers compared are *equal*, the *ratio* is a *unit* or 1: for, if the numerator and denomina-

---

QUEST.—312. What is the effect of multiplying the antecedent of a couplet by any number? Of dividing the antecedent? How does this appear? 313. What is the effect of multiplying the consequent by any number? Of dividing the consequent? Why? 314. What is the effect of multiplying or dividing both the antecedent and consequent by the same number? Why?

tor are equal, the *value* of the fraction is a unit, or 1. (Art. 117.) Thus the ratio of 6×2 : 12 is 1; for the value of $\frac{12}{12}$=1. (Art. 121.)

OBS. This is called a ratio of *equality.*

316. If the antecedent of a couplet is *greater* than the consequent, the ratio is *greater* than a unit: for, if the numerator is greater than the denominator, the *value* of the fraction is greater than 1. (Art. 117.) Thus the ratio of 12 : 4 is 3.

OBS. This is called a ratio of *greater inequality.*

317. If the antecedent is *less* than the consequent, the ratio is *less* than a unit: for, if the numerator is *less* than the denominator, the value of the fraction is less than 1. (Art. 117.) Thus, the ratio of 3 : 6 is $\frac{3}{6}$, or $\frac{1}{2}$; for $\frac{3}{6}=\frac{1}{2}$. (Art. 120.)

OBS. This is called a ratio of *less inequality.*

11. What is the direct ratio of 3 : 9, expressed in the lowest terms? What the inverse ratio?

Ans. $\frac{1}{3}$; and $\frac{1}{3}\div\frac{1}{9}=3$. (Arts. 308, 309.)

12. What is the inverse ratio of 4 to 12? Of 6 to 18? Of 9 to 24? Of 21 to 25? Of 40 to 56?

13. What is the direct ratio of 15s. to £2? Of 13s. 6d. to £1? Of £2, 10s. to £3, 5s.?

14. What is the direct ratio of 6 inches to 3 feet?

15. What is the direct ratio of 15 oz. to 1 cwt.?

PROPORTION.

318. PROPORTION is an equality of ratios. Thus, the two ratios 6 : 3 and 4 : 2 form a proportion; for $\frac{6}{3}=\frac{4}{2}$, the ratio of each being 2.

QUEST.—315. When the two numbers compared are equal, what is the ratio? *Obs.* What is it called? 316. When the antecedent is greater than the consequent, what is the ratio? *Obs.* What is it called? 317. If the antecedent is less than the consequent, what is the ratio? *Obs.* What is it called? 318. What is proportion?

OBS. The terms of the two couplets, that is, the numbers of which the proportion is composed, are called *proportionals.*

319. *Proportion* may be expressed in two ways.

First, by the sign of equality (=) placed between the two ratios.

Second, by four points or a double colon (: :) placed between the two ratios.

Thus, each of the expressions, 12 : 6=4 : 2, and 12 : 6 : : 4 : 2, is a proportion, one being equivalent to the other.

OBS. The latter expression is read, "the ratio of 12 to 6 equals the ratio of 4 to 2," or simply; "12 is to 6 as 4 is to 2."

320. The number of *terms* in a proportion must at least be *four*, for the equality is between the ratios of *two couplets,* and each couplet must have an antecedent and a consequent. (Art. 306.) There may, however, be a proportion formed from *three numbers,* for one of the numbers may be repeated so as to form *two terms.* Thus the numbers 8, 4, and 2, are proportional: for the ratio of 8 : 4= 4 : 2. It will be seen that 4 is the consequent in the first couplet, and the antecedent in the last. It is therefore a *mean proportional* between 8 and 2.

OBS. 1. In this case, the number repeated is called the *middle term* or *mean proportional* between the other two numbers.

The *last* term is called a *third* proportional to the other two numbers. Thus 2 is a third proportional to 8 and 4.

2. Care must be taken not to confound *proportion* with *ratio.* (Arts. 305, 318.) In a simple ratio there are but *two* terms, an antecedent and a consequent; whereas in a proportion there must at least be *four* terms, or *two couplets.*

Again, one *ratio* may be *greater* or *less* than another; the ratio of 9 to 3 is greater than the ratio of 8 to 4, and less than 18 to 2. One *proportion,* on the other hand, cannot be *greater* or *less* than another; for *equality* does not admit of degrees.

QUEST—*Obs.* What are the numbers of which a proportion is composed, called? 319. In how many ways is proportion expressed? What is the first? The second? 320. How many terms must there be in a proportion? Why? Can a proportion be formed of three numbers? How? Will there be four terms in it? *Obs.* What is the number repeated called? What is the last term called in such a case? What is the difference between proportion and ratio?

321. The *first* and *last* terms of a proportion are called the *extremes;* the other two, the *means.*

OBS. *Homologous* terms are either the two antecedents, or the two consequents. *Analogous* terms are the antecedent and consequent of the same couplet.

322. *Direct* proportion is an equality between two *direct* ratios. Thus, 12 : 4 : : 9 : 3 is a direct proportion.

OBS. In a direct proportion, the first term has the same ratio to the second, as the third has to the fourth.

323. *Inverse* or *reciprocal* proportion is an equality between a *direct* and a *reciprocal* ratio. Thus, 8 : 4 : : $\frac{1}{3}$: $\frac{1}{6}$; or 8 is to 4, reciprocally, as 3 is to 6.

OBS. In a reciprocal or inverse proportion, the first term has the same ratio to the second, as the fourth has to the third.

324. *If four numbers are proportional, the product of the extremes is equal to the product of the means.* Thus, 8 : 4 : : 6 : 3 is a proportion : for $\frac{8}{4}=\frac{6}{3}$. (Art. 318.)

Now $8\times3=4\times6$.

Again, 12 : 6 : : $\frac{1}{3}$: $\frac{1}{6}$ is a proportion. (Art. 323.)

And $12\times\frac{1}{6}=6\times\frac{1}{3}$.

OBS. 1. The truth of this proposition may also be illustrated in the following manner:

The numbers 2 : 3 : : 6 : 9 are obviously proportional. (Art. 318.)
For, $\frac{2}{3}=\frac{6}{9}$. (Art. 120.) Now,

Multiplying each ratio by 27, (the product of the denominators,)
The proportion becomes $\frac{2\times27}{3}=\frac{6\times27}{9}$ (Art. 284. Ax. 6.)

QUEST.---321. Which terms are the extremes? Which the means? *Obs.* What are homologous terms? Analogous terms? 322. What is direct proportion? *Obs.* In direct proportion what ratio has the first term to the second? 323. What is inverse proportion? *Obs.* What ratio has the first term to the second in this case? 324. If four numbers are proportional, what is the product of the extremes equal to? *Obs.* If the product of the extremes is equal to the product of the means, what is true of the four numbers? If the products are not equal, what is true of the numbers?

Dividing both the numerator and the denominator of the first couplet by 3; (Art. 116;) or canceling the denominator 3 and the same factor in 27; (Art. 136;) also canceling the 9, and the same factor in 27, we have $2\times9=6\times3$. But 2 and 9 are the extremes of the given proportion, and 3 and 6 are the means; hence, the product of the extremes $2\times9=6\times3$, the product of the means.

2. Conversely, if the product of the extremes is equal to the product of the means, the four numbers are proportional; and if the products are not equal, the numbers are not proportional.

325. *Proportion* in arithmetic, is usually divided into *Simple* and *Compound*.

SIMPLE PROPORTION.

326. SIMPLE PROPORTION is an equality between two *simple* ratios. (Art. 309. *a.*) It may be either *direct* or *inverse*. (Arts. 322, 323.)

The most important application of simple proportion, is the *solution* of that class of examples in which *three terms are given to find a fourth*.

326. *a.* We have seen that, if four numbers are in proportion, the product of the extremes is equal to the product of the means. (Art. 324.) Hence,

If the product of the means is divided by one of the extremes, the quotient will be the other extreme; and if the product of the extremes is divided by one of the means, the quotient will be the other mean. For, if the product of two factors is divided by one of them, the quotient will be the other factor. (Art. 291.)

Take the proportion 8 : 4 : : 6 : 3.
Now the product $8\times3\div4=6$, one of the means;
So the product $8\times3\div6=4$, the other mean.
Again, the product $4\times6\div8=3$, one of the extremes;
And the product $4\times6\div3=8$, the other extreme.

QUEST.—325. Into what is proportion usually divided? 326. What is simple proportion? What is the most important application of it? 326. *a.* If the product of the means is divided by one of the extremes, what will the quotient be? If the product of the extremes is divided by one of the means, what will the quotient be?

326. *b. If, therefore, any three terms of a proportion are given, the fourth may be found by dividing the product of two of them by the other term.*

OBS. Simple Proportion is often called the *Rule of Three*, from the circumstance that three terms are given to find a fourth. In the older arithmetics, it is also called the *Golden Rule*. But the fact that these names convey no idea of the nature or object of the rule, seems to be a strong objection to their use, not to say a sufficient reason for discarding them.

EX. 1. If the first three terms of a proportion are 4, 6, 8, what is the fourth term?

Solution.—6×8=48 and 48÷4=12, which is the number required; that is, 4 : 6 : : 8 : 12.

PROOF.—4×12 is equal to 6×8. (Art. 324. Obs. 2.)

2. If 12 bbls. of flour cost $72, what will 4 bbls. cost, at the same rate?

Solution.—It is evident 12 bbls. has the same ratio to 4 bbls., as the cost of 12 bbls. ($72) has to the cost of 4 bbls., which is required. That is, 12 bbls. : 4 bbls : : $72 is to the cost of 4 bbls. Now, 72×4=288; and 288÷12 =24. *Ans.* $24.

3. If 6 men can dig a cellar in 12 days, how many men will it take to dig it in 4 days?

Note.—Since the answer is men, we put the given number of men for the third term. Then, as it will require more men to dig the cellar in 4 days than it will to dig it in 12 days, we put the larger number of days for the second term, and the smaller for the first term.

Operation.

4d. : 12d. : : 6 m. : to the men required.

$$\begin{array}{r} 6 \\ 4\overline{)72} \\ \hline 18 \end{array}$$ men. *Ans.*

QUEST.—*Obs.* What is simple proportion often called? Do these terms convey an idea of the nature or object of the rule?

327. From the preceding illustrations and principles, we deduce the following general

RULE FOR SIMPLE PROPORTION.

I. *Place that number for the third term, which is of the same kind as the answer or number required.*

II. *Then, if by the nature of the question the answer must be greater than the third term, place the greater of the other two numbers for the second term; but if it is to be less, place the less of the other two numbers for the second term, and the other for the first.*

III. *Finally, multiplying the second and third terms together, divide the product by the first, and the quotient will be the answer in the same denomination as the third term.*

PROOF.—*Multiply the first term and the answer together, and if the product is equal to the product of the second and third terms, the work is right.* (Art. 324.)

OBS. 1. If the first and second terms are compound numbers, reduce them to the lowest denomination mentioned in either, before the multiplication or division is performed.

When the third term contains different denominations, it must also be reduced to the lowest denomination mentioned in it.

2. The process of arranging the terms of a question for solution, that is, putting it into the form of a proportion, is called *stating the question.*

3. We have seen that questions in Simple Proportion may easily be solved by *Analysis.* (Art. 296.) After solving the following examples by proportion, it will be an excellent exercise for the pupil to solve each by analysis.

4. If 6 yards of broadcloth cost 30 dollars, how much will 20 yards cost? *Ans.* $100.

5. If 8 bbls. of flour cost $40, what will 15 bbls. cost?

6. If 16 lbs. of tea cost $12, what will 41 lbs. cost?

QUEST.--327. In arranging the terms in simple proportion, which number is put for the third term? How arrange the other two numbers? Having stated the question, how is the answer found? Of what denomination is the answer? How is Simple Proportion proved? *Obs.* If the first and second terms contain different denominations, how proceed? When the third term contains different denominations, what is to be done? What is meant by stating the question?

7. If 12 acres of land produce 240 bushels of wheat, how much will 57 acres produce?

8. If a man can travel 400 miles in 15 days, how far can he travel in 9 days?

9. If 63 barrels of beef cost $504, how much will 7 barrels cost?

Common Method.

```
bbls. bbls.   dolls.
63  :  7  ::  504 : Ans.
                7
       63)3528($56. Ans.
          315
          ---
           378
           378
```

Multiplying the second and third terms together and dividing the product by the first, we have $56 for the answer.

By Cancelation.

```
bbls. bbls.   dolls.
 63  :  7  ::  504 : Ans.
  9  :  1
```

(63 and 7 are printed struck through.)

Now 504÷9=$56. *Ans.*

By canceling the factor 7, which is common to the first two terms; that is, which is common to the divisor and dividend, we avoid the necessity of multiplying by it. (Art. 91. *a.*)

PROOF.—63×56=504×7. (Art. 327.) Hence,

328. When the first term has one or more factors common to either of the other two terms.

CANCEL *the factors which are common, then proceed according to the rule above.* (Art. 91. *a.*, 136.)

OBS. 1. The question should be *stated*, before attempting *to cancel* the common factors.

2. When the terms are of different denominations, the reduction of them may sometimes be shortened by cancelation.

10. If 12 yds. of lace cost £1, what will 1 qr. of a yard cost?

Operation.

```
 yds.     qr.   £
12×4  :  1  :.  1×20×12 : Ans. (Art. 327. Obs. 1.)
```

$$\text{Then, } \frac{1\times1\times20\times12}{12\times4}=\frac{\overset{5}{\not{20}}\times\not{12}}{\not{12}\times\not{4}}=5 \text{ d. } Ans.$$

QUEST.—328. When the first term has factors common to either of the other two terms, how may the operation be shortened?

11. If 6 men can build a wall in 36 days, how long will it take 18 men to build it?

12. If 10 quintals of fish cost $35, how much will 17 quintals cost?

13. If a ship has water sufficient to last a crew of 25 men for 8 months, how long will it last 15 men?

14. If 12 lbs. sugar cost $1, how much will 84 lbs. cost?

15. If 15 lbs. lard cost $1.15, how much will 80 lbs. cost?

16. If $\frac{5}{8}$ of an acre of land cost £$\frac{3}{7}$, how much will $\frac{7}{8}$ of an acre cost?

Solution.—$\overset{A.}{\frac{5}{8}} : \overset{A.}{\frac{7}{8}} :: \overset{£}{\frac{3}{7}}$: to the answer.

Hence, $\frac{8}{5}\times\frac{7}{8}\times\frac{3}{7}=$Answer. (Arts. 327, 139.)

By cancelation, (Art. 136,) $\frac{\not{7}}{\not{8}}\times\frac{3}{\not{7}}\times\frac{\not{8}}{5}=$ £$\frac{3}{5}$. *Ans.*

17. If $\frac{4}{7}$ of a hogshead of molasses cost $28, how much will 16 hogsheads cost?

18. If $2\frac{1}{4}$ yds. of broadcloth cost $18, how much will 27 yds. cost?

19. If 6 acres and 40 rods of land cost $125, how much will 25 acres and 120 rods cost?

20. If 15 yds. of silk cost £4, 10s., how much will 75 yds. cost?

21. If a Railroad car goes 35 m. in 1 hr. 45 min., how far will it go in 3 days?

22. If $4\frac{1}{2}$ lbs. of chocolate cost 9s., how much will $22\frac{1}{2}$ lbs. cost?

23. If $35\frac{2}{3}$ lbs. of butter cost $4, how much will $15\frac{1}{6}$ lbs. cost?

24. If 84 lbs. of cheese cost $$5\frac{2}{5}$, how much will 60 lbs. cost?

25. If $\frac{3}{5}$ of a ship is worth $6000, how much is $\frac{5}{16}$ of her worth?

26. If $4\frac{1}{2}$ bu. of wheat make 1 barrel of flour, how many barrels will 84 bu. make?

27. If the interest of $1500 for 12 mo. is $90, what will be the interest of the same sum for 8 mo.?

28. If a tree 20 ft. high, casts a shadow 30 ft. long, how long will be the shadow of a tree 50 ft. high?

29. How long will it take a steam ship to sail round the globe, allowing it to be 25000 miles in circumference, if she sails at the rate of 3000 miles in 12 days?

30. How many acres of land can a man buy for $840, if he pays at the rate of $56 for every 7 acres?

31. How much will 85 cwt. of iron cost, at the rate of $91 for 13 cwt.?

32. At the rate of $45 for 6 cwt. of beef, how much can be bought for $980?

33. If 9 ounces of silver will make 4 tea spoons, how many spoons will 25 pounds of silver make?

34. If 15 tons of wool are worth $90000, how much is 5 cwt. worth?

35. If $5\frac{1}{2}$ yds. of cloth are worth $$27\frac{1}{2}$, how much are $50\frac{1}{4}$ yards worth?

36. If 60 men can build a house in $90\frac{1}{2}$ days, how long will it take 15 men to build it?

37. A bankrupt owes $25000, and his property is worth $20000: how much can he pay on a dollar?

38. At 7s. 6d. per week, how long can a man board for £24, 10s.?

39. What cost 94 tons of coal, if 141 tons cost £85?

40. What cost 291 yds. of cambric, if 13 yds. cost £8, 6s. $3\frac{1}{2}$d.?

41. What cost 3 lbs. of raisins, at £6, 7s. 6d. per 100 lbs.?

42. If 20 sheep cost £37, $12\frac{1}{2}$s., what will 311 cost?

43. At 7s. 6d. per ounce, what is the value of a silver pitcher weighing 9 oz. 13 pwt. 8 grs.?

44. If 405 yards of linen cost £69, 7s. 6d., what will 243 yards cost?

45. If A can saw a cord of wood in 6 hours, and B in 9 hours, how long will it take both together to saw a cord?

46. A cistern has 3 cocks, the first of which will empty it in 10 min.; the second, in 15 min.; and the third, in 30 min.: how long will it take all of them together to empty it?

47. A man and a boy together can mow an acre of grass in 4 hours; the man can mow it alone in 6 hours: how long will it take the boy to mow it?

COMPOUND PROPORTION.

329. COMPOUND PROPORTION is an equality between a *compound* ratio and a *simple* one. (Arts. 309. *a*, 310.)

Thus, 8 : 4 }
Into 6 : 3 } : : 12 : 3, is a compound proportion.

That is, 8×6 : 4×3 : : 12 : 3; for, 8×6×3=4×3×12.

OBS. Compound proportion is chiefly applied to the solution of examples which would require *two or more statements* in simple proportion. It is sometimes called *Double Rule of Three.*

Ex. 1. If 4 men can earn $24 in 6 days, how much can 8 men earn in 10 days?

Suggestion.—When stated in the form of a *compound proportion*, the question will stand thus:

4m. : 8m. }
6d. : 10d. } : : $24 : to the answer required. That is, "the product of the antecedents 4×6, has the same ratio to the product of the consequents, 8×10, as $24 has to the answer."

Operation.
24×8×10=1920,
and 4×6=24.
Now 1920÷24=80.
Ans. 80 dollars.

We divide the product of all the numbers standing in the 2d and 3d places of the proportion, by the product of those standing in the first place.

Note.—1. The learner will observe, that it is not the ratio of 4 to 8 alone, nor that of 6 to 10, which is equal to the ratio of 24 to the answer, as it is sometimes stated; but it is the ratio *compounded* of 4 to 8 and 6 to 10, which is equal to the ratio of 24 to the answer. Thus, 4×6 : 8×10 : : 24 : 80, the answer.

2. A compound proportion, when stated as above, is read, "the ratio of 4 into 6 is to 8 into 10 as 24 to the answer."

2. If 5 men can mow 20 acres of grass in 4 days, working 10 hours per day, how much can 8 men mow in 5 days, working 12 hours per day?

Operation.
5m. : 8m. }
4d. : 5d. } : : 20 : Ans. (Acres.)
10hr. : 12hr. }

State the question, then multiply and divide as before.

8×5×12×20=9600; and 5×4×10=200. Now 9600÷200=48. *Ans.* 48 acres.

QUEST.—329. What is compound proportion? *Obs.* To what is it chiefly applied? What is it sometimes called?

330. From the foregoing illustrations we derive the following general

RULE FOR COMPOUND PROPORTION.

I. *Place that number which is of the same kind as the answer required for the third term.*

II. *Then take the other numbers in pairs, or two of a kind, and arrange them as in simple proportion.* (Art. 327.)

III. *Finally, multiply together all the second and third terms, divide the result by the product of the first terms, and the quotient will be the fourth term or answer required.*

PROOF.—*Multiply the answer into all of the first terms or antecedents of the first couplets, and if the product is equal to the continued product of all the second and third terms multiplied together, the work is right.* (Art. 324.)

OBS. 1. Among the given numbers there is but one which is of the same kind as the answer. This is sometimes called the *odd* term, and is always to be placed for the *third* term.

2. Questions in Compound Proportion may be solved by *Analysis;* also by *Simple Proportion,* by making *two* or *more* separate statements. (Art. 302. Obs. 327.)

3. If 8 men can clear 30 acres of land in 63 days, working 10 hours a day, how many acres can 10 men clear in 72 days, working 12 hours a day?

Statement.

$$\left.\begin{array}{l} 8\text{m.} : 10\text{m.} \\ 63\text{d.} : 72\text{d.} \\ 10\text{hr.} : 12\text{hr.} \end{array}\right\} :: \overset{\text{Acres.}}{30} : \text{to the answer.}$$ That is,

$8\times63\times10 : 10\times72\times12 :: 30 :$ to the answer.

$$\begin{array}{l}\text{But the prod.}\\ \text{Divided by}\end{array} \quad \frac{10\times72\times12\times30}{8\times63\times10} = \text{the Ans. (Art. 330.)}$$

QUEST.—330. In arranging the numbers in compound proportion, which number do you put for the third term? How arrange the other numbers? Having stated the question, how is the answer found? How are questions in compound proportion proved? *Obs.* Among the given numbers, how many are of the same kind as the answer? Can questions in compound proportion be solved by simple proportion? How?

Now by *canceling* equal factors, (Art. 116,) we have

$$\frac{\cancel{1}\times\cancel{72}\times12\times30}{\cancel{8}\times\underset{7}{\cancel{63}}\times\cancel{1}}=\frac{360}{7},$$ or $51\frac{3}{7}$ acres. *Ans.* Hence,

331. *After stating the question according to the rule above, if the antecedents or first terms have factors common to the consequents or second terms, or to the third term, they should be* CANCELED *before performing the multiplication and division.*

Note.—Instead of placing points between the first and second terms, that is, between the antecedents and consequents of the left hand couplets of the proportion, it is sometimes more convenient to put a perpendicular line between them, as in division of fractions. (Art. 140.) This will bring all the terms whose product is to be the dividend on the right of the line, and those whose product is to form the divisor, on the left. In this case the third term should be placed below the second terms, with the sign of proportion (::) before it, to show its *origin*, and its *relation* to the answer.

4. If a man can walk 192 miles in 4 days, traveling 12 hours a day, how far can he go in 24 days, traveling 8 hours a day?

Operation.

$\cancel{4}$ d.	$\cancel{24}$ d. 2
$\cancel{12}$ hr.	$\cancel{8}$ hr. 2
	:: 192m.
Ans.	192×2×2×2=768 miles.

The product of the antecedents, 4×12, has the same ratio to the product of the consequents, 24×8, as 192 has to the answer required.

5. If 8 men can make 9 rods of wall in 12 days, how many men will it require to make 36 rods in 4 days?

6. If 5 men make 240 pair of shoes in 24 days, how many men will it require to make 300 pair in 15 days?

7. If 60 lbs. of meat will supply 8 men 15 days, how long will 72 lbs. last 24 men?

8. If 12 men can reap 80 acres of wheat in 6 days, how long will it take 25 men to reap 200 acres?

9. If 18 horses eat 128 bushels of oats in 32 days, how many bushels will 12 horses eat in 64 days?

10. If 8 men can build a wall 20 ft. long, 6 ft. high,

QUEST.—331. When the antecedents have factors common to the consequents, what should be done with them?

and 4 ft. thick, in 12 days, how long will it take 24 men to build one 200 ft. long, 8 ft. high, and 6 ft. thick?

11. If 8 men reap 36 acres in 9 days, working 9 hours per day, how many men will it take to reap 48 acres in 12 days, working 12 hours per day?

12. If $100 gain $6 in 12 months, how long will it take $400 to gain $18?

13. If $200 gain $12 in 12 months, what will $400 gain in 9 months?

14. If 8 men spend £32 in 13 weeks, how much will 24 men spend in 52 weeks?

15. If 6 men can dig a drain 20 rods long, 6 feet deep, and 4 feet wide, in 16 days, working 9 hours each day, how many days will it take 24 men to dig a drain 200 rods long, 8 ft. deep, and 6 ft. wide, working 8 hours per day?

SECTION XIII.

DUODECIMALS.

ART. **332.** DUODECIMALS are a species of compound numbers, the *denominations* of which increase and decrease uniformly in a *twelvefold ratio.* Its denominations are *feet, inches* or *primes, seconds, thirds, fourths, fifths, &c.*

Note.—The term *duodecimal* is derived from the Latin numeral *duodecim,* which signifies *twelve.*

TABLE.

12 fourths	('''')	make	1 third,	marked '''
12 thirds		"	1 second,	" ''
12 seconds		"	1 inch or prime,	" *in.* or '
12 inches or primes		"	1 foot,	" *ft.*

Hence $1' = \frac{1}{12}$ of 1 foot.

$1'' = \frac{1}{12}$ of 1 in. or $\frac{1}{12}$ of $\frac{1}{12}$ of 1 ft. $= \frac{1}{144}$ of 1 ft.

$1''' = \frac{1}{12}$ of $1''$, or $\frac{1}{12}$ of $\frac{1}{12}$ of $\frac{1}{12}$ of 1 ft. $= \frac{1}{1728}$ of 1 ft.

QUEST.—332. What are duodecimals? What are its denominations? *Note.* What is the meaning of the term duodecimal? Repeat the Table. *Obs.* What are the accents called, which are used to distinguish the different denominations?

OBS. The accents used to distinguish the different denominations below feet, are called *Indices.*

333. Duodecimals may be added and subtracted in the same manner as other compound numbers. (Arts. 168, 169.)

MULTIPLICATION OF DUODECIMALS.

334. Duodecimals are principally applied to the measurement of *surfaces* and *solids.* (Arts. 153, 154.)

Ex. 1. How many square feet are there in a board 8 ft. 9 in. long, and 2 ft. 6 in. wide?

Operation.

8 ft. 9′ length.	
2 ft. 6′ width.	
17 ft. 6′	
4 ft. 4′ 6″	
21 ft. 10′ 6″. *Ans.*	

We first multiply each denomination of the multiplicand by the feet in the multiplier, beginning at the right hand. Thus, 2 times 9′ are 18′, equal to 1 ft. and 6′. Set the 6′ under inches, and carry the 1 ft. to the next product. 2 times 8 ft. are 16 ft. and 1 to carry makes 17 ft. Again, since $6'=\frac{6}{12}$ of a ft. and $9'=\frac{9}{12}$ of a ft., 6′ into 9′ is $\frac{54}{144}$ of a ft.=54″, or 4′ and 6″. Write the 6″ one place to the right of inches, and carry the 4′ to the next product. Then 6′ or $\frac{6}{12}$ of a foot multiplied into 8 ft.$=\frac{48}{12}$ of a ft., or 48′, and 4′ to carry make 52′; but 52′=4 ft. and 4′. Now adding the partial products, the sum is 21 ft. 10′ 6″.

OBS. It will be seen from this operation, that feet multiplied into feet, produce feet; feet into inches, produce inches; inches into inches, produce seconds, &c. Hence,

335. To find the denomination of the product of any two factors in duodecimals.

Add the indices of the two factors together, and the sum will be the index of their product.

Thus, feet into feet, produce feet; feet into inches, produce inches; feet into seconds, produce seconds; feet into thirds, produce thirds, &c.

QUEST.—333. How are duodecimals added and subtracted? 334. To what are duodecimals chiefly applied? 335. How find the denomination of the product in duodecimals? What do feet into feet produce? Feet into inches? Feet into seconds?

Inches into inches, produce seconds; inches into seconds, produce thirds; inches into fourths, produce fifths, &c.

Seconds into seconds, produce fourths; seconds into thirds, produce fifths; seconds into sixths, produce eighths, &c.

Thirds into thirds, produce sixths; thirds into fifths, produce eighths; thirds into sevenths, produce tenths, &c.

Fourths into fourths, produce eighths; fourths into eighths, produce twelfths, &c.

Note.—The foot is considered the *unit*, and has no *index.*

336. From these illustrations we have the following

RULE FOR MULTIPLICATION OF DUODECIMALS.

I. *Place the several terms of the multiplier under the corresponding terms of the multiplicand.*

II. *Multiply each term of the multiplicand by each term of the multiplier separately, beginning with the lowest denomination in the multiplicand, and the highest in the multiplier, and write the first figure of each partial product one or more places to the right, under its corresponding denomination.* (Art. 335.)

III. *Finally, add the several partial products together, carrying* 1 *for every* 12 *both in multiplying and adding, and the sum will be the answer required.*

OBS. It is sometimes asked whether the inches in duodecimals are *linear, square,* or *cubic.* The answer is, they are *neither.* An *inch* is 1 *twelfth* of a foot. Hence, in measuring surfaces an inch is $\frac{1}{12}$ of a *square* foot; that is, a surface 1 foot *long* and 1 inch *wide.* In measuring solids, an inch denotes $\frac{1}{12}$ of a *cubic* foot. In measuring lumber, these inches are commonly called *carpenter's inches.*

2. How many square feet are there in a board 18 feet 9 inches long, and 2 feet 6 inches wide?

3. How many square feet are there in a board 14 feet 10 inches long, and 11 inches wide?

QUEST.—What do inches into inches produce? Inches into thirds? Inches into fourths? Seconds into seconds? Seconds into thirds? Seconds into eighths? Thirds into thirds? Thirds into sixths? 336. What is the rule for multiplication of duodecimals? *Obs.* What kind of inches are those spoken of in measuring surfaces by duodecimals? In measuring solids? In measuring lumber what are they called?

4. How many square feet in a gate 12 feet 5 inches wide, and 6 feet 8 inches high?

5. How many square feet in a floor 16 feet 6 inches long, and 12 feet 9 inches wide?

6. How many square feet in a ceiling 53 feet 6 inches long, and 25 feet 6 inches wide?

7. How many square feet are there in a stock of 6 boards 17 feet 7 inches long, and 1 foot 5 inches wide?

8. How many feet in a stock of 10 boards 12 feet 8 inches long, and 1 foot 1 inch wide?

9. How many cubic feet in a stick of timber 12 feet 10 inches long, 1 foot 7 inches wide, and 1 foot 9 inches thick?

10. How many cubic feet in a block of marble 8 feet 4 inches long, 2 feet 6 inches wide, and 1 foot 10 inches thick?

11. How many cubic feet in a load of wood 6 feet 7 inches long, 3 feet 5 inches high, and 3 feet 8 inches wide?

12. How many feet in a load of wood 7 feet 2 inches long, 4 feet high, and 3 feet wide?

13. How many feet in a load of wood 9 feet long, 4 feet 3 inches wide, and 5 feet 6 inches high?

14. How many feet in a pile of wood 100 feet long, $5\frac{1}{2}$ feet high, and 4 feet wide?

15. How many feet in a pile of wood 150 feet long, $8\frac{1}{2}$ feet high, and 5 feet wide?

16. How many cubic feet in a wall 40 feet 6 inches long, 5 feet 10 inches high, and 2 feet thick?

17. How many solid feet in a vat 10 feet 8 inches long, 7 feet 2 inches wide, and 6 feet 4 inches deep?

18. How many bricks 8 inches long, 4 inches wide, and 2 inches thick, are there in a wall 20 feet long, 10 feet high, and $1\frac{1}{2}$ feet thick?

19. How much will the flooring of a room which is 20 feet long, and 18 feet wide come to, at $6\frac{1}{4}$ cents per square foot?

20. How much will the plastering of a wall 16 feet square come to, at $12\frac{1}{2}$ cents per square yard?

SECTION XIV.

INVOLUTION.

MENTAL EXERCISES.

ART. **337.** Ex. 1. What is the product of 5 multiplied by 5? *Ans.* 5×5=25.

2. What is the product of 3 multiplied into 3 *twice?* *Ans.* 3×3×3=27.

3. What is the product of 2 into itself *three* times? *Ans.* 2×2×2×2=16.

338. When any number or quantity is multiplied into *itself*, the *product* is called a *power*. Thus, in the examples above, the products 25, 27, and 16 are powers.

The *original* number, that is, the number which being multiplied into itself, produces a power, is called the *root* of all the powers of that number; because they are derived from it.

339. *Powers* are divided into *different orders;* as the *first, second, third, fourth, fifth power*, &c. They take their name from the *number of times* the given number is used as a *factor*, in producing the *given power.*

Note.—1. The *first* power of a number is said to be the number itself. Strictly speaking, it is not a *power*, but a root. (Art. 338.)

1. The *second* power of a number is also called the *square;* (Art. 153. Obs. 1;) for, if the side of a square is 3 yards, then the product of 3×3=9 yards, will be the area of the given square. (Art. 163.) But 3×3=9 is also the *second* power of 3; hence, it is called the *square.*

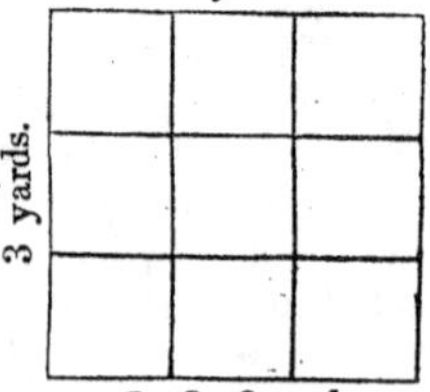

QUEST.—338. What is a power? 339. How are powers divided? From what do they take their name? *Note.* What is said to be the first power? What is the second power called? The third? The fourth?

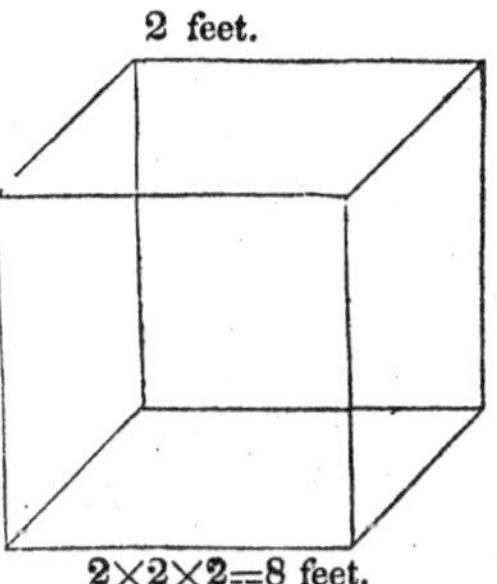

3. The *third* power of a number is also called the *cube;* (Art. 154. Obs. 2;) for, if the side of a cube is 2 feet, then the product of 2×2×2= 8 feet, will be the solidity of the given cube. (Art. 164.) But 2×2×2=8, is also the *third* power of 2; hence it is called the *cube.*

4. The *fourth* power of a number is called the *biquadrate.*

4. What is the square of 4? *Ans.* 16.
5. What is the cube of 3? The fourth power of 3?
6. The fourth power of 2? The fifth power of 2?
7. What is the square of 5? Of 6? Of 7? Of 9? Of 8? Of 10? Of 11? Of 12?
8. What is the cube of 3? Of 4? Of 5? Of 6?

340. Powers are frequently denoted by a *small figure* placed above the given number at the right hand.

This figure is called the *index* or *exponent.* It shows how many times the given number is employed as a factor to produce the required power. Thus,

The index of the *first* power is 1, but this is omitted; for, $(2)^1=2$.

The index of the *second* power is 2;
The index of the *third* power is 3;
The index of the *fourth* power is 4;
The index of the *fifth* power is 5; &c. That is,

$2^1=2$, the first power of 2;
$2^2=2\times2$, the square, or 2d power of 2;
$2^3=2\times2\times2$, the cube, or 3d power of 2;
$2^4=2\times2\times2\times2$, the biquadrate, or 4th power of 2;
$2^5=2\times2\times2\times2\times2$, the fifth power of 2;
$2^6=2\times2\times2\times2\times2\times2$, the 6th power of 2; &c.

QUEST.—340. How are powers denoted? What is this figure called? What does it show? What is tne index of the first power? Of the second? The third? Fourth? Fifth? Sixth?

EXERCISES FOR THE SLATE.

9. Express the third power of 6; the 4th power of 12.

10. Express the square of 16; the cube of 20; the fourth power of 25; the fifth power of 72; the sixth power of 100; the tenth power of 500.

341. The process of finding a *power* of a given number by multiplying it into itself, is called INVOLUTION.

342. Hence, to *involve* a number to any required power.

Multiply the given number into itself, till it is taken as a factor, as many times as there are units in the index of the power to which the number is to be raised. (Art. 339.)

OBS. 1. The *number of multiplications* in raising a number to any given power, is *one* less than the index of the required power. Thus, the square of 3 is written 3^2, and $3\times3=9$, the 3 is taken twice as a factor, but there is but one multiplication.

2. A *Fraction* is raised to a power by multiplying it into itself. Thus, the square of $\frac{2}{3}$ is $\frac{2}{3}\times\frac{2}{3}=\frac{4}{9}$.

Mixed numbers should be reduced to improper fractions, or the common fraction may be reduced to a decimal.

3. All powers of 1 are the same, viz: 1; for $1\times1\times1\times1$, &c.$=1$.

11. What is the square of 24?

Common Operation.	*Analytic Operation.*
24	24=2 tens or 20+4 units.
24	24=2 tens or 20+4 units.
96	80+16
48	400+80
576. *Ans.*	And 400+160+16=576.

It will be seen from this operation that the square of 20+4, contains the square of the first part, viz: $20\times20=400$, added to twice the product of the two parts, viz: $20\times4+20\times4=160$, added to the square of the last part, viz: $4\times4=16$. Hence,

QUEST.—341. What is involution? 342. How is a number involved to any required power? *Obs.* How many multiplications are there in raising a number to a given power? How is a fraction involved? A mixed number? What are all powers of 1?

342. *a. The square of any number which consists of two figures, is equal to the square of the tens, added to twice the product of the tens into the units, added to the square of the units.*

OBS. 1. The product of any two factors cannot have more figures than both factors, nor but one less than both. For example, take 9, the greatest number which can be expressed by one figure. (Art. 7.) And $(9)^2$, or $9 \times 9 = 81$, has two figures, the same number which both factors have. 99 is the greatest number which can be expressed by two figures; (Art. 7;) and $(99)^2$, or $99 \times 99 = 9801$, has four figures, the same as both factors have.

Again, 1 is the smallest number expressed by one figure, and $(1)^2$, or $1 \times 1 = 1$, has but one figure less than both factors. 10 is the smallest number which can be expressed by two figures; and $(10)^2$, or $10 \times 10 = 100$, has one figure less than both factors. Hence,

2. *Any square number cannot have more figures than double the number of the root or first power, nor but one less.*

3. *A cube cannot have more figures than triple the number of the root or first power, nor but two less.*

12. What is the square of 45? 50? 75? 100? 540?
13. What is the cube of 5? Of 8? 10? 12? 60?
14. What is the fourth power of 3? Of 4? 16? 20?
15. What is the fifth power of 2? Of 3? 4? 5? 6?
16. What is the square of $\frac{1}{2}$? Of $\frac{1}{3}$? $\frac{1}{4}$? $\frac{2}{3}$? $\frac{3}{4}$? $\frac{5}{8}$?
17. What is the cube of $\frac{2}{3}$? Of $\frac{4}{5}$? Of $\frac{7}{8}$? Of $\frac{10}{20}$?
18. What is the square of $2\frac{1}{2}$? Of $3\frac{1}{4}$? $5\frac{2}{3}$? $10\frac{3}{4}$?
19. What is the square of 1.5? Of 3.25? Of 10.25?

EVOLUTION.

343. If we resolve 25 into *two equal factors*, viz: 5 and 5, each of these equal factors is called a *root* of 25. So if we resolve 27 into *three equal factors*, viz: 3, 3, and 3, each factor is called a *root* of 27; if we resolve 16 into *four* equal factors, viz: 2, 2, 2, and 2, each factor is called a *root* of 16. And, universally, when a number is resolved into *any number* of *equal factors*, each of those factors is said to be a *root* of that number. Hence,

---

QUEST.—342. *a.* What is the square of any number consisting of two figures equal to? OBS. How many figures are there in the product of any two factors? How many figures will the square of a number contain? The cube? 343. When a number is resolved into any number of equal factors, what is each of those factors called?

344. A *root* of a number is a *factor*, which, being *multiplied into itself* a certain number of times, will produce that number. (Art. 338.)

OBS. When a number is resolved into *two equal* factors, each of these factors is called the *second* or *square* root; when resolved into *three equal* factors, each of these factors is called the *third* or *cube* root; when resolved into *four equal* factors, each factor is called the *fourth* root; &c. Hence,

The name of the root expresses the number of equal factors into which the given number is to be resolved.

For example, the *second* or *square* root, shows that the number is to be resolved into *two equal* factors; the *third* or *cube* root, into *three equal* factors; the fourth root, into *four equal* factors, &c. Thus,

The *square* root of 16 is 4; for $4\times4=16$.

The *cube* root of 27 is 3; for $3\times3\times3=27$.

The *fourth* root of 16 is 2; for $2\times2\times2\times2=16$, &c.

MENTAL EXERCISES.

Ex. 1. Resolve 25 into two equal factors.

Solution.—$25=5\times5$. *Ans.* 5, and 5.

2. Resolve 8 into three equal factors.

Solution.—$8=2\times2\times2$. *Ans.* 2, 2, and 2.

345. The process of resolving numbers into *equal factors* is called EVOLUTION, or *the Extraction of Roots.*

OBS. 1. Evolution is the *opposite* of involution. (Art. 341.) One is finding a *power* of a number by multiplying it into itself; the other is finding a *root* by resolving a number into equal factors. *Powers* and *roots* are therefore *correlative* terms. If one number is a *power* of another, the latter is a *root* of the former. Thus, 27 is the cube of 3; and 3 is the cube root of 27.

2. The learner will be careful to remember, that

In *subtraction*, a number is resolved into *two parts;*

In *division*, a number is resolved into *two factors;*

In *evolution*, a number is resolved into *equal factors.*

3. What is the square root of 16? *Ans.* 4.

4. What is the square root of 36? Of 49?

QUEST.—344. What then is a root? *Obs.* What does the name of the root express? What does the square root show? The cube root? The fourth root? 345. What is evolution? *Obs.* Of what is it the opposite? Into what are numbers resolved in subtraction? In division? In evolution?

5. What is the square root of 64? Of 81? Of 100? Of 121? Of 144?

6. What is the cube or third root of 8?

Solution.—If we resolve 8 into three equal factors, each of these factors is 2: for $2\times2\times2=8$. The cube root of 8 therefore, is 2.

7. What is the cube root of 27?
8. What is the cube root of 64?
9. What is the cube root of 125?
10. What is the fourth root of 16?
11. What is the square root of $\frac{9}{16}$?

Solution.—The square root of the numerator 9, is 3; and the square root of the denominator 16, is 4. Therefore $\frac{3}{4}$ is the square root of $\frac{9}{16}$; for $\frac{3}{4}\times\frac{3}{4}=\frac{9}{16}$.

12. What is the square root of $\frac{1}{9}$? *Ans.* $\frac{1}{3}$.
13. What is the square root of $\frac{16}{25}$? Of $\frac{25}{49}$?
14. What is the square root of $\frac{36}{81}$? Of $\frac{64}{100}$?
15. What is the cube root of $\frac{1}{8}$? *Ans.* $\frac{1}{2}$.
16. What is the cube root of $\frac{1}{27}$? Of $\frac{27}{64}$?

346. Roots are expressed in *two* ways; one by the *radical sign* ($\surd$) placed before a number; the other by a *fractional index* placed above the number on the right hand. Thus, $\sqrt{4}$, or $4^{\frac{1}{2}}$ denotes the *square* or 2d root of 4; $\sqrt[3]{27}$, or $27^{\frac{1}{3}}$ denotes the *cube* or 3d root of 27; $\sqrt[4]{16}$, or $16^{\frac{1}{4}}$ denotes the 4th root of 16.

OBS. 1. The figure placed over the radical sign, denotes the *root*, or the number of equal factors into which the given number is to be resolved. The figure for the *square* root is usually omitted, and simply the radical sign $\surd$ is placed before the given number. Thus, the square root of 25 is written $\sqrt{25}$.

2. When a root is expressed by a *fractional* index, the *denominator* like the figure over the radical sign, denotes the *root* of the given number. Thus, $(25)^{\frac{1}{2}}$ denotes the s[illegible] 25; $(27)^{\frac{1}{3}}$ denotes the cube root of 27.

QUEST.—346. In how many ways are roots expressed? What are they? *Obs.* What does the figure over the radical sign denote? What the denominator of the fractional index?

EXERCISES FOR THE SLATE.

17. Express the cube root of 45 both ways.

18. Express the cube root of 64 both ways. Of 125.

19. Express the fourth root of 181 both ways. Of 576.

20. Express the 5th root of 32; the 6th root of 64.

21. Express the 7th root of 84; the 8th root of 91; the 9th root of 105; the 10th root of 256.

22. Express the cube root of 576; the fourth root of 675; the fifth root of 1000; the twelfth root of 840.

347. A number which can be resolved into *equal* factors, or whose root can be *exactly* extracted, is called a *perfect power*, and its root is called a *rational number*. Thus, 16, 25, 27, &c., are perfect powers, and their roots 4, 5, 3, are rational numbers.

348. A number which *cannot* be resolved into *equal* factors, or whose root *cannot* be *exactly* extracted, is called an *imperfect power;* and its root is called a *Surd*, or *irrational number*. Thus, 15, 17, 45, &c., are imperfect powers, and their roots 3.8+; 4.1+; 6.7+, &c., are surds, for their roots cannot be exactly extracted.

OBS. A number may be a perfect power of one degree and an imperfect power of another degree. Thus, 16 is a perfect power of the second degree, but an imperfect power of the third degree; that is, it is a perfect *square* but not a perfect *cube*. Indeed numbers are seldom perfect powers of more than one degree. 16 is a perfect power of the 2d and 4th degrees: 64 is a perfect power of the 2d, 3d and 6th degrees.

349. Every *root*, as well as every power of 1, is 1. (Art. 342. Obs. 3.) Thus, $(1)^2$, $(1)^3$, $(1)^6$, and $\sqrt{1}$, $\sqrt[3]{1}$, $\sqrt[6]{1}$, &c., are all equal.

QUEST.—347. What is a perfect power? What is a rational number? 348. What is an imperfect power? What is a surd? *Obs.* Are numbers ever perfect powers of one degree and imperfect powers of another degree? Are they often perfect powers of more than one degree? 349. What are all roots and powers of 1?

EXTRACTION OF THE SQUARE ROOT.

350. *To extract the square root, is to resolve a given number into two equal factors; or, to find a number which being multiplied into itself, will produce the given number.* (Art. 344. Obs.)

Ex. 1. What is the side of a square room which contains 16 square yards?

Solution.—Let the room be represented by the adjoining figure. It is divided into 16 equal squares, which we will call square yards.

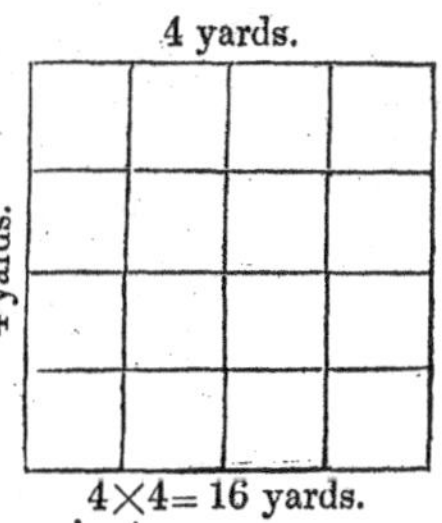

Since the room is square, the question is simply this: What is the square root of 16? Now if we resolve 16 into two equal factors, each of those factors will be the square root of 16. But 16=4 ×4. The square root of 16, therefore, is 4.

2. What is the length of one side of a square room which contains 576 square feet?

Operation.

```
 5̇76̇(24
 4
44)176
   176
```

Since we may not see what the root of 576 is at once, as in the last example, we will separate it into periods of two figures each, by putting a point over the 5, and also over the 6; that is, over the units' figure and over the hundreds. This shows us that the root is to have two figures; (Art. 342. *a.* Obs. 2;) and thus enables us to find the root of part of the number at a time. Now the greatest square of 5, the left hand period, is 4, the root of which is 2. We place the 2 on the right hand of the number for the first part of the root; then subtract its square from 5, the period under consideration, and to the right of the remainder bring down 76, the next period, for a dividend. To find the next figure in the root, we double the 2, the part of the root already found, and placing it on the left of the dividend for a partial divisor, we find how many times

QUEST.—350. What is it to extract the square root of a number?

it is contained in the dividend, omitting the right hand figure. Now 4 is contained in 17, 4 times. Placing the 4 on the right of the root, also on the right of the partial divisor, we multiply 44, the divisor thus completed, by 4, the last figure in the root, and subtracting the product 176 from the dividend, find there is no remainder. The answer therfore is 24.

Note.—Since the root is to contain two figures, the 2 stands in tens' place; hence the first part of the root found is properly 20; which being doubled, gives 40 for the divisor. For convenience we omit the cipher on the right; and to compensate for this, we omit the right hand figure of the dividend. This is the same as dividing both the divisor and dividend by 10, and therefore does not alter the quotient. (Art. 88.)

PROOF.—24=2 tens, or 20+4 units.
24=2 " 20+4 "

96 80+16
48 400+80

$(24)^2=576 = 400+160+16$. (Art. 342. *a.*)

ILLUSTRATION BY GEOMETRICAL FIGURE.

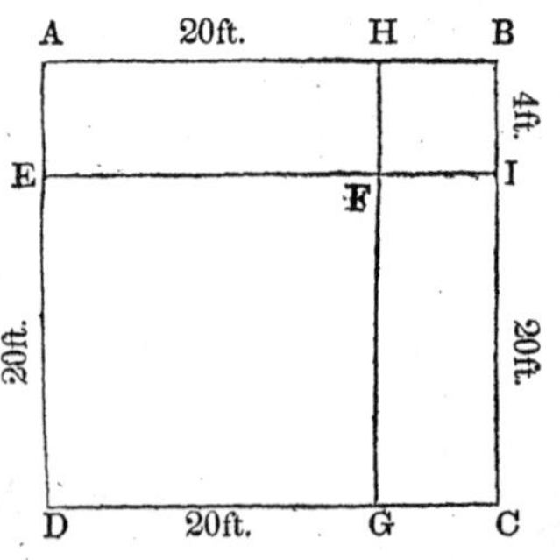

Let the large square ABCD, represent the room in the last example; then the square DEFG will be the greatest square of the left hand period, the root of which is 20 ft., and 20×20=400, the number of feet in its area. (Art. 163.) But this square 400 ft. taken from 576 ft. leaves a remainder of 176 ft. Now it is plain, if this remaining space is all added to one side of this square, its sides will become unequal; consequently it will cease to be a square. (Art. 153. Obs. 1.) But if it is equally enlarged on two sides it will obviously continue to be a

QUEST.—*Note.* What place does the first figure of the root occupy in the example above? Why is the right hand figure of the dividend omitted?

square. For this reason the root is doubled for a divisor in the operation. The parallelograms AEFH and GFIC will therefore represent the additions made to the two sides, each of which is 4 ft. wide; consequently the area of each is $20\times4=80$ ft., and the area of both is $40\times4=160$ ft.

But having made these additions to two sides of the square, there is a vacancy at the corner. The square BIFH represents this vacancy, the side of which is 4 ft., or the same as the width of the additions; and its area is $4\times4=16$ ft. For convenience of finding the area of this vacancy, it is customary in the operation to place the last figure of the root on the right of the divisor, and thus it is multiplied into itself. The figure is now a perfect square, the length of whose side, is $20+4=24$ ft.

351. From these principles and illustrations we derive the following general

RULE FOR EXTRACTING THE SQUARE ROOT.

I. *Separate the given number into periods of two figures each, by placing a point over the units' figure, another over the hundreds, and so on over each alternate figure.*

II. *Find the greatest square number in the first or left hand period, and place its root on the right of the number for the first figure in the root. Subtract the square of this figure of the root from the period under consideration; and to the right of the remainder bring down the next period for a dividend.*

III. *Double the root just found and place it on the left of the dividend for a partial divisor, find how many times it is contained in the dividend; omitting its right hand figure; place the quotient on the right of the root, also on the right of the partial divisor; multiply the divisor thus completed by the last figure of the root; subtract the product from the dividend, and to the remainder bring down the next period for a new dividend as before.*

IV. *Double the root already found for a new partial di-*

QUEST.—351. What is the first step in extracting the square root? The second? Third? Fourth?

visor, divide, &c. as before, and thus continue the operation till the root of all the periods is extracted.

PROOF.—*Multiply the root into itself; and if the product is equal to the given number, the work is right.* (Art. 344.)

OBS. The product of the divisor completed into the figure last placed in the root, cannot *exceed* the *dividend.* Hence, in finding the figure to be placed in the root, some allowance must be made for carrying, when the product of this figure into itself *exceeds* 9.

351. *a. Demonstration.*—The *reason* for the several steps in the rule may easily be inferred from the preceding illustrations. The following is a summary of them:

1. Separating the given number into *periods of two figures* each shows how many figures the root is to contain, and thus enables us to find *part* of the *root* at a time. (Art. 342, *a.* Obs. 2.)
2. The *square* of the first figure of the root, is the number of feet, yards, &c. disposed of by the first figure of the root; it is subtracted from the period to find how many feet, yards, &c., *remain* to be added.
3. The root thus found is *doubled* for a partial divisor, because the addition must be made on *two sides* of the square already found, or it will cease to be a *square.*
4. In dividing, the *right hand figure* of the dividend is *omitted,* because the cipher on the right of the divisor is omitted; otherwise the quotient would be 10 times too large for the next figure in the root.
5. The last figure of the root is placed on the right of the divisor for convenience of multiplying. The divisor is then multiplied by the last figure of the root to find the area of the several additions thus made.

3. What is the square root of 625?
4. What is the square root of 900?
5. What is the square root of 1225?
6. What is the square root of 1764?
7. What is the square root of 2916?
8. What is the square root of 4761?
9. What is the square root of 8649?
10. What is the square root of 12321?

QUEST.—How is the square root proved? *Dem.* Why do we separate the given number into periods of two figures each? Why subtract the square of the first figure in the root from the first period? Why double the root thus found for a divisor? Why omit the right hand figure of the dividend? Why place the last figure of the root on the right of the divisor? Why multiply the divisor by the last figure in the root?

11. What is the square root of 53824?
12. What is the square root of 531441?

352. If there are decimals in the given sum, they must be separated into periods like whole numbers, by placing a point over *units*, then over *hundredths*, and so on, over every alternate figure towards the right.

If there is a remainder after all the periods are brought down, the operation may be continued by annexing *periods of ciphers.*

OBS. 1. There will always be as many decimal figures in the root, as there are periods of decimals in the given number.

2. The square root of a common fraction is found by extracting the root of the numerator and denominator.

3. A mixed number should be reduced to an improper fraction. When either the numerator or denominator of a common fraction is not a *perfect square*, the fraction may be reduced to a decimal, and the *approximate* root be found as above.

13. What is the square root of 6.25? *Ans.* 2.5.
14. What is the square root of 1.96?
15. What is the square root of 29.16?
16. What is the square root of 234.09?
17. What is the square root of .1225?
18. What is the square root of .776161?
19. What is the square root of 2?
20. What is the square root of 17?
21. What is the square root of 175?
22. What is the square root of 116964?
23. What is the square root of 10316944?
24. What is the square root of $\frac{36}{49}$?
25. What is the square root of $\frac{144}{576}$?
26. What is the square root of $6\frac{1}{4}$?
27. What is the square root of $52\frac{9}{16}$?

QUEST.—352. When there are decimals in the given number, how are they pointed off? When there is a remainder, how proceed? *Obs.* How do you determine how many decimal figures there should be in the root? How is the square root of a common fraction found? Of a mixed number?

APPLICATIONS OF THE SQUARE ROOT.

353. The principles of the square root may be applied to the solution of questions in which two sides of a right-angled triangle are given, and it is required to find the other side.

354. A *triangle* is a figure which has *three sides* and *three angles*, as in the adjoining diagram.

When one of the sides of a triangle is *perpendicular* to another side, the angle between them is called a *right-angle.* (Legendre, B. I. Def. 12.)

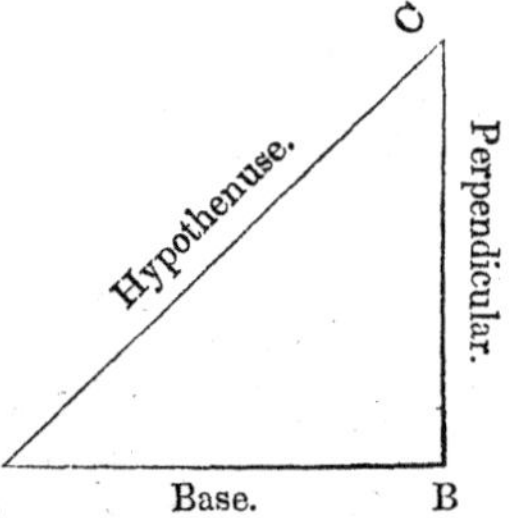

355. A *right-angled triangle* is a triangle which has a right-angle. (Leg. B. I. Def. 17.)

The side opposite the right-angle is called the *hypothenuse,* and the other two sides, the *base* and *perpendicular.* The triangle ABC is right-angled at B, and the side AC is the hypothenuse.

356. It is an established principle in geometry, that the square described on the *hypothenuse* of a right-angled triangle, is equal to the *sum* of the squares described on the *other two* sides. (Leg. IV. 11., Euc. I. 47.) Thus, if the base of the triangle ABC, is 4 feet, and the perpendicular 3 feet; then the square of 4 added to the square of 3 is equal to the square of the hypothenuse BC; that is, $(4)^2+(3)^2$, or $16+9=25$, the square of the hypothenuse; therefore the square root of 25, which is 5, must be the hypothenuse itself. Hence, when any two sides of a right-angled triangle are given, the third side may be easily found.

QUEST.—354. What is a triangle? What is a right-angle? 355. What is a right-angled triangle? Draw a right-angled triangle upon the black-board. What is the side opposite the right-angle called? What are the other two sides called? 356. What is the square described on the hypothenuse equal to? Draw a right-angled triangle, and describe a square on each of its sides?

357. When the base and perpendicular are given, to find the *hypothenuse.*

Add the square of the base to the square of the perpendicular, and the square root of the sum will be the hypothenuse.

Thus, in the right-angled triangle ABC, if the base is 4 and the perpendicular is 3, then $(4)^2+(3)^2=25$, and $\sqrt{25}=5$, the hypothenuse.

358. When the hypothenuse and base are given, to find the *perpendicular.*

From the square of the hypothenuse subtract the square of the base, and the square root of the remainder will be the perpendicular.

Thus, if the hypothenuse is 5, and the base 4, then $(5)^2-(4)^2=9$, and $\sqrt{9}=3$, the perpendicular.

359. When the hypothenuse and the perpendicular are given, to find the *base.*

From the square of the hypothenuse subtract the square of the perpendicular, and the square root of the remainder will be the base.

Thus, if the hypothenuse is 5, and the perpendicular 3, then $(5)^2-(3)^2=16$, and $\sqrt{16}=4$, the base.

28. What is the length of a ladder which will just reach to the top of a house 32 feet high, when its foot is placed 24 feet from the house?

Operation.

Perpendicular $(32)^2=32\times32=1024$
Base $(24)^2=24\times24=\underline{\ 576}$

The square root of their sum $\overline{1600}=40$. *Ans.*

29. The side of a certain school-room having square corners, is 8 yards, and its width 6 yards: what is the distance between two of its opposite corners?

QUEST.—357. When the base and perpendicular are given, how is the hypothenuse found? 358. When the hypothenuse and base are given, how is the perpendicular found? 359. When the hypothenuse and perpendicular are given, how is the base found?

30. Two men start from the same place, one goes exactly south 40 miles a day, the other goes exactly west 30 miles a day: how far apart will they be at the close of the first day?

31. How far apart will the same travelers be at the end of 4 days?

32. A line 75 feet long fastened to the top of a flag staff reaches the ground 45 feet from its base: what is the height of the flag staff?

33. Suppose a house is 40 feet wide, and the length of the rafters is 32 feet: what is the distance from the beam to the ridge pole?

34. The side of a square field is 30 rods: how far is it between its opposite corners?

35. If a square field contains 10 acres, what is the length of its side, and how far apart are its opposite corners?

EXTRACTION OF THE CUBE ROOT.

360. *To extract the cube root, is to resolve a given number into three equal factors; or, to find a number which being multiplied into itself twice, will produce the given number.* (Art. 345.)

1. What is the side of a cubical block containing 27 solid feet?

Solution.—Let the given block be represented by the adjoining cubical figure, each side of which is divided into 9 equal squares, which we will call square feet. Now, since the length of a side is 3 feet, if we multiply 3 into 3 into 3, the product 27, will be the solid contents of the cube. (Art. 164.)

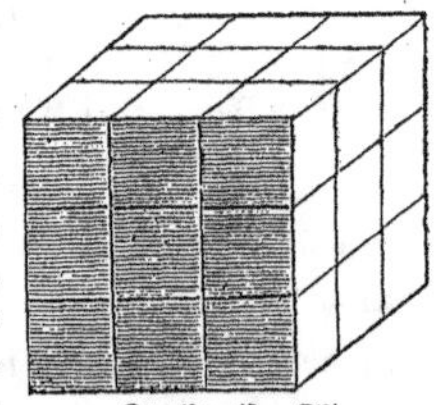

$3 \times 3 \times 3 = 27.$

Hence, if we reverse the process, i. e. if we resolve 27 into three equal factors, one of these factors will be the side of the cube. (Art. 344. Obs.) *Ans.* 3 ft.

2. A man wishes to form a cubical mound containing 15625 solid feet of earth: what is the length of its side?

QUEST.—360. What is it to extract the cube root?

Operation.

```
          15625(25
           8
Divisor. Dividend.
 1200  |  7625
  300  |
   25  |
 1525  |  7625
```

1. We first separate the given number into periods of three figures each, by placing a point over the units' figure, then over thousands. This shows us that the root must have two figures, (Art. 342. *a.* Obs. 3,) and thus enables us to find part of it at a time.

2. Beginning with the left hand period, we find the greatest cube of 15 is 8, the root of which is 2. Placing the 2 on the right of the given number for the first figure in the root, we subtract its cube from the period, and to the remainder bring down the next period for a dividend. This shows that we have 7625 solid feet to be added to the cubical mound already found.

3. We square the root already found, which in reality is 20, for since there is to be another figure annexed to it, the 2 is tens; then multiplying its square 400 by 3, we write the product on the left of the dividend for a divisor; and finding it is contained in the dividend 5 times, we place the 5 in the root.

4. We next multiply 20, the root already found, by 5, the last figure placed in the root; then multiply this product by 3 and place it under the divisor. We also place the square of 5, the last figure placed in the root, under the divisor, and adding these three results together, multiply their sum 1525 by 5, and subtract the product from the dividend. The answer is 25.

PROOF. $(25)^3 = 25 \times 25 \times 25 = 15625$. (Art. 360.)

DEMONSTRATION BY CUBICAL BLOCKS.

361. The simplest method of illustrating the process of extracting the cube root to those unacquainted with algebra and geometry, is by means of *cubical blocks.**

1. Dividing the number into *periods* of *three figures*, shows how

* A set of these blocks contains 1st, a *cube*, the side of which is usually about 1½ in. square; 2d, *three side* pieces about ⅓ in. thick, the upper and lower base of which is just the size of a side of the cube; 3d, *three corner* pieces, whose ends are ⅓ in. square, and whose length is the same as that of the side pieces: 4th, a *small cube*, the side of which is equal to the end of the corner pieces. It is desirable for every teacher and pupil to have a set. If not conveniently procured at the shops, any one can easily make them for himself.

many figures the *root* will contain, and also enables us to find *part* of it at a time. Now, placing the large cube upon a table or stand, let it represent the greatest cube in the left hand period, which in the example above is 8, the root of which is 2. We subtract this cube from the left hand period, and to the remainder bring down the next period, in order to find how many feet remain to be added. In making this addition, it is plain the cube must be equally increased on *three* sides; otherwise its sides will become *unequal*, and it will then cease to be a *cube*. (Art. 154. Obs. 2.)

2. The object of squaring the root already found is to find the area of one side of this cube; (Art. 163;) we then multiply its square by 3, because the additions are to be made to three of its sides; and, dividing the dividend by this product shows the thickness of these additions. Now placing one of the side pieces on the top, and the other two on two adjacent sides of the cube, they will represent these additions.

3. But we perceive there is a vacancy at three corners, each of which is of the same length as the root already found, or the side of the cube, viz: 20 ft., and the breadth and thickness of each is 5 ft., the thickness of the side additions. Placing the corner pieces in these vacancies, they will represent the additions necessary to fill them. The object of multiplying the root already found by the figure last placed in it, is to obtain the area of a side of one of these additions; we then multiply this area by 3, to find the area of a side of each of them.

4. We find also another vacancy at one corner, whose length, breadth, and thickness are each 5 ft., the same as the thickness of the side additions. This vacancy therefore is cubical. It is represented by the small cube, which being placed in it, will render the mound an exact cube again. The object of squaring 5, the figure last placed in the root, is to find the area of a side of this cubical vacancy. We now have the area of one side of each of the side additions, the area of one side of each of the corner additions, and the area of one side of the cubical vacancy, the sum of which is 1525. We next multiply the sum of these areas by the figure last placed in the root, in order to find the cubical contents of the several additions. (Art. 164.) These areas are added together, and their sum multiplied by the last figure placed in the root, for the sake of finding the solidity of all the additions at once. The result would obviously be the same, if we multiplied them separately, and then subtracted the sum of their products from the dividend.

362. From the preceding illustrations we derive the following general

QUEST.—362. What is the first step in extracting the cube root? The second? Third? Fourth? Fifth? How is the cube root proved? *Dem.* Why separate the given number into periods of three figures each? Why subtract the greatest cube from the left hand period? Why square the root already found? Why multiply its square by 3? Why divide the dividend by this product? Why multiply the root already found by the last figure placed in it? Why multiply this product by 3? Why square the figure last placed in the root? Why multiply the sum of these areas, by the last figure placed in the root?

RULE FOR EXTRACTING THE CUBE ROOT.

I. *Separate the given number into periods of three figures each, placing a point over units, then over every third figure towards the left in whole numbers, and over every third figure towards the right in decimals.*

II. *Find the greatest cube in the first period on the left hand; then placing its root on the right of the number for the first figure of the root, subtract its cube from the period, and to the remainder bring down the next period for a dividend.*

III. *Square the root already found, regarding its local value; multiply this square by* 3, *and place the product on the left of the dividend for a divisor; find how many times it is contained in the dividend, and place the result in the root.*

IV. *Multiply the root previously found, regarding its local value, by this last figure placed in it, then multiply this product by* 3, *and write the result on the left of the dividend under the divisor; under this result write also the square of the last figure placed in the root.*

V. *Finally, add these results to the divisor; multiply the sum by the last figure placed in the root, and subtract the product from the dividend. To the right of the remainder bring down the next period for a new dividend; find a new divisor, and proceed with the operation as above.*

PROOF.—*Multiply the root into itself twice, and if the last product is equal to the given number, the work is right.*

OBS. 1. When there is a remainder, *periods of ciphers* may be added, and the operation continued as in square root.

2. If the right hand period of decimals is deficient, this deficiency must be supplied by ciphers.

3. When there are *decimals* in the given example, find the root as in whole numbers; then point off as many decimal figures in the answer, as there are *periods of decimals* in the given number.

4. The cube root of a *common fraction* is found by extracting the root of its numerator and denominator.

A *mixed* number should be reduced to an improper fraction.

5. When there are more than *two periods* in the given example, it is sufficient to annex *one cipher* to the root previously found, before squaring it for the divisor.

3. What is the cube root of 1728?

4. What is the cube root of 13824?

5. If a box in the form of a cube, contains 373248 solid inches, what is the length of one side?

6. What is the side of a cubical vat, which contains 571787 solid feet?

7. What is the side of a cubical mound which contains 1953125 solid yards?

8. What is the cube root of 2?

9. What is the cube root of 2357947691?

10. What is the cube root of 12.167?

11. What is the cube root of 91.125?

12. What is the cube root of $\frac{27}{64}$?

13. What is the cube root of $\frac{125}{729}$?

SECTION XV.

EQUATION OF PAYMENTS.

ART. **363.** EQUATION OF PAYMENTS is the process of finding the *equalized* or *average time* when two or more payments due at different times, may be made *at once*, without loss to either party.

OBS. The *equalized* or *average* time for the payment of several debts, due at different times, is often called the *mean* time.

364. From principles already explained, it is manifest, when the *rate* is fixed, the *interest* depends both upon the *principal* and the *time.* (Art. 241.) Thus, if a given principal produces a certain interest in a given time,

Double that principal will produce *twice* that interest;

Half that principal will produce *half* that interest; &c.

In *double* that time the same principal will produce *twice* that interest;

In *half* that time the same principal will produce *half* that interest; &c.

QUEST.—363. What is Equation of Payments? *Obs.* What is the average time for the payment of several debts sometimes called? 364. When the rate is fixed, upon what does the interest depend?

365. Hence, it is evident that any given principal will produce the same interest in any given time, as

One half that prin. will produce in *double* that time;
One third that prin. will " " *thrice* that time;
Twice that principal will " " *half* that time;
Thrice that principal will " " *a third* of that time, &c.

For example, at any given per cent.,

The int. of $2 for 1 year, is the same as the int. of $1 for 2 years;
The int. of $3 for 1 year, " " " $1 for 3 years; &c.
The int. of $4 for 1 mo. " " " $1 for 4 mos.;
The int. of $5 for 1 mo. " " " $1 for 5 mos; &c.

366. *The interest, therefore, of any given principal for 1 year, or 1 month, &c., is the same, as the interest of 1 dollar for as many years, or months, &c. as there are dollars in the given principal.*

1. Suppose you owe a man $15 and are to pay him $5 in 8 months, and $10 in 2 months, at what time may both payments be made without loss to either party?

Analysis.—Since the interest of $5 for 1 month is the same as the interest of $1 for 5 months, (Art. 365,) the interest of $5 for 8 months must be equal to the interest of $1 for 8 times 5 months. And 5 mo. ×8=40 mo. In like manner the interest of $10 for 1 month is equal to the interest of $1 for 10 months, and the interest of $10 for 2 months is equal to the interest of $1 for 2 times 10 months. And 10 mo. ×2=20 months. Now 40 months added to 20 months make 60 months; that is, you are entitled to the use of $1 for 60 months. But $1 is $\frac{1}{15}$ of $15, consequently you are entitled to the use of $15, $\frac{1}{15}$ part of 60 months, and 60 months ÷15=4. *Ans.* 4 months.

Proof.

The interest of $5 at 6 per ct. for 8 mo. is $5 ×.04=$.20
The interest of $10 " " " 2 mo. is $10×.01= .10
Sum of both $.30

The interest of $15 at 6 per ct. for 4 mo. is 15×.02=$.30

367. Hence, we derive the following general

RULE FOR EQUATION OF PAYMENTS.

First multiply each debt by the time before it becomes due; then divide the sum of the products thus obtained by the sum of the debts, and the quotient will be the average time required.

OBS. 1. If one of the debts is to be *paid down*, its product will be *nothing;* but in finding the *sum* of the debts, this payment must be added in with the others.

2. This rule is based upon the supposition that *discount* and *interest paid in advance* are *equal.* But this is not exactly true; (Art. 261. Obs. 1;) consequently, the rule, though in general use, is not strictly accurate.

2. If I owe a man $20, payable in 4 months, $40 payable in 6 months, and $60 in 3 months, at what time may I justly pay the whole at once?

Operation.

$20×4=$80, the same as $1 for 80 mo. (Art. 366.)
$40×6=240, " " " $1 for 240 "
$60×3=180, " " " $1 for 180 "

$120 debts. 500 sum of products.
120)500(4⅙ months. *Ans.*

3. A merchant bought three lots of goods amounting to $300; for the first he gave $100, payable in 5 months; for the second $150, payable in 8 months; for the third $50, payable in 2 months: what is the average time of all the payments?

4. A farmer has 3 notes; one of $50, due in 2 months; another of $100, due in 5 months; and the third of $150, due in 8 months: what is the average time of the whole?

5. A merchant buys goods amounting to $1200, and agrees to pay $400 down, $400 in 4 months, and $400 in 8 months; he finally concluded to give his note for the whole: at what time must the note be made payable?

6. A man borrows $600, and agrees to pay $100 in 2 months, 200 in 5 months, and the balance in 8 months: when can he justly pay the whole at once?

QUEST.—367. What is the rule for equation of payments?

7. A man buys a house for $1600, and agrees to pay $400 down, and the rest in 3 equal annual instalments: what is the average credit for the whole?

8. I have $1200 owing to me, $\frac{1}{2}$ of which is now due; $\frac{1}{4}$ of it will be due in 4 months, and the remainder in 8 months: what is the average time of the whole?

9. A grocer bought goods amounting to $1500, for which he was to pay $250 down; $300 in 4 mo.; and $950 in 9 mo.: when may he pay the whole at once?

10. A young man bought a farm for $2000, and agrees to pay $500 down, and the balance in 5 equal annual instalments: what is the average time of the whole?

PARTNERSHIP.

368. PARTNERSHIP is the associating of two or more individuals together for the transaction of business. (Art. 299.) The persons thus associated are called *partners;* and the association is termed a *company* or *firm.* The money employed is called the *capital* or *stock;* and the profit or loss to be shared among the partners, the *dividend.*

CASE I.

Ex. 1. A and B formed a partnership; A furnished $300 capital, and B $500; they gained $200: what was each partner's share of the gain?

Solution.—Since the whole stock is $\$300+\$500=\$800$, A's part of it was $\frac{300}{800}=\frac{3}{8}$, and B's part was $\frac{500}{800}=\frac{5}{8}$. Now since A put in $\frac{3}{8}$ of the stock, he must have $\frac{3}{8}$ of the gain; and $\$200\times\frac{3}{8}=\$75$. For the same reason B must have $\frac{5}{8}$ of the gain; and $\$200\times\frac{5}{8}=\$125$.

PROOF.—$\$75+125=\$200$, the whole gain. (Art. 284. Ax. 11.) Hence,

---

QUEST.—368. What is partnership? What are the persons thus associated called? What is the association called? What is the money employed called? What the profit or loss?

369. To find each partner's share of the gain or loss, when the *stock* of each is employed for the *same time.*

Make each man's stock the numerator, and the whole stock the denominator of a common fraction; multiply the gain or loss by the fraction which expresses each man's share of the stock, and the product will be his share of the gain or loss.

Or, multiply each man's stock by the whole gain or loss; divide the product by the whole stock, and the quotient will be his share of the gain or loss.

PROOF.—*Add the several shares of the gain or loss together, and if the sum is equal to the whole gain or loss, the work is right.* (Art. 284. Ax. 11.)

OBS. 1. This rule is applicable to questions in Bankruptcy, General Average, and all other operations in which there is to be a division of property in specified proportions.

2. The preceding case is often called *Single* Fellowship. But since a *partnership* is always composed of *two* or *more* individuals, it is somewhat difficult to see the propriety of calling it *single.*

2. A, B, and C entered into partnership; A put in $\frac{5}{24}$ of the capital, B $\frac{9}{24}$, and C $\frac{10}{24}$; they gain $4800: what was each man's share of the gain?

3. A, B, and C form a partnership; A furnishes $600, B $800, and C $1000; they gain $480: what is each man's share of the gain?

4. A Bankrupt owes A $1200, B $2300, C $3400, and D $4500; his whole effects are worth $5600: how much will each creditor receive?

5. A, B, C, and D make up a purse to buy lottery tickets; A puts in $30, B $40, C $60, and D $70; they draw a prize of $2000: what is each man's share?

6. A, B, and C freight a vessel with a cargo worth $30000; of which A owned $8000, B $10000, and C $12000; in a gale the master throws $\frac{1}{3}$ of the cargo overboard: what was each man's loss?

QUEST.—369. How is each man's share of the gain or loss found, when the stock of each is employed for the same time? How is the operation proved? *Obs.* To what is this rule applicable? What is it sometimes called?

CASE II.

7. A and B formed a partnership; A put in $300, and B $200. At the end of 2 months A took out his stock, while B's was employed 6 months; they gained $150: what was each man's just share of the gain?

Note.—It is obvious that the gain of each depends both upon the *capital* he furnished, and the *time* it was employed. (Art. 364.)

Solution.—Since A's capital $300, was employed 2 mo., his share of the gain is the same as if he had put in $600 for 1 mo.; (Art. 365;) for $\$300 \times 2 = \$600$. Also, B's capital $200, being employed 6 mo., his share of the gain is the same as if he had put in $1200 for 1 mo.; for $\$200 \times 6 = \$1200$. The sum of $600 and $1200 is $1800.
A's share of the gain must therefore be $\frac{600}{1800} = \frac{1}{3}$.
B's " " " " " " $\frac{1200}{1800} = \frac{2}{3}$.
Now $\$150 \times \frac{1}{3} = \$50$, A's share.
And $\$150 \times \frac{2}{3} = \$100$, B's share. Hence,

370. To find each partner's share of the gain or loss, when the *stock* of each is employed for *different periods*.

Multiply each partner's stock by the time it is employed; make each man's product the numerator, and the sum of the products the denominator of a common fraction; then multiply the whole gain or loss by each man's fractional share of the stock, and the product will be his share of the gain or loss.

OBS. This case is often called *Compound* or *Double* Fellowship.

8. A, B, and C enter into business together; A puts in $500 for 4 months, B $400 for 6 months, and C $800 for 3 months; they gain $340: what is each man's share of the gain?

9. A and B hire a pasture together for $60; A put in 120 sheep for 6 months, and B put in 180 sheep for 4 months: what should each pay?

QUEST.—370. When the stock of each partner is employed for different periods, how is each man's share found? *Obs.* What is this case sometimes called?

10. The firm A, B, and C lost $246; A had put in $85 for 8 mo., B $250 for 6 mo., and C $500 for 4 mo.: what is each man's share of the loss?

EXCHANGE OF CURRENCIES.

371. The term *currency* signifies *money*, or the *circulating medium* of trade.

372. The *intrinsic value* of the coins of different nations, depends upon their *weight* and the *purity* of the metal of which they are made. (Art. 203. Obs. 1.)

Note.—1. The present *standard gold* coins of Great Britain are 22 parts of pure gold and 2 parts of copper, i. e. 22 carats fine. The *standard silver* coins are 37 parts of pure silver and 3 parts of copper. A Pound Sterling or Sovereign weighs 123.274 grs., and a shilling, (silver,) 3 pwts. $15\frac{3}{11}$ grs.*

2. For the *present standard weight* and *purity* of gold and silver coins of the United States, see Art. 203. Obs. 2.

373. The *relative value* of foreign coins is determined by the laws of the country. By act of Congress, 1842,

The value of a Pound Sterling, or Sovereign,	is	$4.84
" " " Guinea, English,	is	5.075
" " " Franc, French,	is	.185
" " " Five-franc piece, (Act of 1843,)	is	.93
" " " Doubloon of Spain, Mexico, &c., of standard weight and purity,	is	15.535

OBS. 1. The *legal* value of a Pound Sterling has been changed several times. By the law of 1842, its value was fixed at $4.84, and it now passes for this sum in all payments to or from the Treasury, and in reckoning duties on imported goods invoiced in Sterling money. The *intrinsic* value of a £ Sterling or sovereign, is $4.861.

2. In 1799, the value of a Pound Sterling was fixed at $4.44\frac{4}{9}$, which is now called its *nominal value.*

QUEST.—371. What is currency? 372. On what does the intrinsic value of the coins of different countries depend? 373. How is the relative value of foreign coins determined? What is the value of a Pound Sterling? Of a guinea? A franc? Five-franc piece? A doubloon?

* Hind's Arithmetic.

374. The process of changing money expressed in the denominations of one country to its equivalent value in the denominations of another country, is called *Exchange of Currencies.*

Ex. 1. Change £20 sterling to Federal money.

Suggestion.—Since £1 is worth $4.84, £20 are worth 20 times as much; and $4.84×20=$96.80. *Ans.*

2. Change £5, 13s. 6d. to Federal money.

Operation.

£5, 13s. 6d.=£5.675. (Art. 200.)
Value of £1=$ 4.84
Ans. $27.467. (Art. 215.)

Reduce 13s. 6d, to the decimal of a pound, and multiply the sum by $4.84.

375. Hence to reduce Sterling to Federal money.

Set down the pounds as whole numbers, and reduce the given shillings, pence, and farthings to the decimal of a pound; then multiply the whole sum by $4.84, (*the value of* £1,) *point off the product as in multiplication of decimals, and it will be the answer required.*

OBS. 1. Guineas, Francs, Doubloons, and all foreign coins, may be reduced to Federal currency, by multiplying the *given number* by the *value* of *one* expressed in Federal money.

2. The rule usually given for reducing Sterling to Federal money, is to reduce the shillings, pence, and farthings to the decimal of a pound, and placing it on the right of the given pounds, divide the whole sum by $\frac{9}{40}$. This rule is based on the law of 1798, which fixed the value of a pound at 4.44\frac{4}{9}$, and that of a dollar at 4s. 6d. But 4.44\frac{4}{9}$ is 9 per cent. of itself, or 40 cents, less than $4.84, which is the present *legal* value of a pound; consequently, the result or answer obtained by it, must be 9 per cent. too *small.* A dollar is now equal to 49.6d. very nearly, instead of 54d. as formerly.

3. What is the value of £100 in Federal money?
4. What is the value of £275, 15s. in Federal money?
5. Change £450, 7s. 6d. to Federal money.
6. Change $27.467 to Sterling money.

Solution.—Since there is £1 in $4.84, in $27.467 there

QUEST.—374. What is meant by exchange of currencies? 375. How is Sterling money reduced to Federal? *Obs.* How may any foreign coins be reduced to Federal money?

are as many pounds, as \$4.84 is contained times in it; and \$27.467÷4.84=5.675; that is, £5.675. Reducing the decimal .675 to shillings and pence, (Art. 201,) we have £5, 13s. 6d. for the answer. Hence,

376. To reduce Federal to Sterling money.

Divide the given sum by \$4.84, (*the value of* £1,) *and point off the quotient as in division of decimals. The figures on the left hand of the decimal point will be pounds; those on the right, decimals of a pound, which must be reduced to shillings, pence, and farthings.* (Art. 201.)

7. Change \$486.42 to Sterling money.
8. Change \$1452 to Sterling money.

376. *a.* In buying and selling Bills of Exchange on England, the *premium* or *discount* is commonly reckoned at a certain per cent. on the *nominal* value of a Pound Sterling, which is \$4.44$\frac{4}{9}$. (Art. 373. Obs.)

9. What is the worth of a bill of exchange of £100 on London, at 9 per cent. premium?

Solution.—£100×\$4.44$\frac{4}{9}$=\$444.44$\frac{4}{9}$, the nominal value.
Then, \$444.44$\frac{4}{9}$×.09=\$40.00, the premium.
And \$444.44+\$40=\$484.44. *Ans.*

10. What is the value of £1325, 10s., at 8½ per cent. premium.

377. Previous to the adoption of Federal money in 1786, accounts in the United States were kept in pounds, shillings, pence, and farthings.

OBS. At the time Federal money was adopted, the *colonial currency*, or *bills of credit* issued by the colonies, had more or less *depreciated* in value: that is, a colonial pound was worth less than a pound Sterling; a colonial shilling, than a shilling Sterling, &c. This depreciation being greater in some of the colonies than in others, gave rise to the *different State currencies.* Thus,

In New England currency, Va., Ky., and Tenn, 6s. or £$\frac{3}{10}$=\$1.
In New York currency, North Carolina, and Ohio, 8s. or £$\frac{2}{5}$=\$1.
In Penn. cur., New Jer., Del., and Md., 7s. 6d. (7½s.) or £$\frac{3}{8}$=\$1.
In Georgia cur., and South Carolina, 4s. 8d. (4$\frac{2}{3}$s.) or £$\frac{7}{30}$=\$1.
In Canada currency, and Nova Scotia, 5s. or £¼=\$1.

QUEST.—376. How is Federal money reduced to Sterling? 377. Previous to the adoption of Federal money, in what were accounts kept?

11. Reduce $45 to New England currency.

Solution.—Since there are 6s. in $1, in $45 there are 45 times 6s. And 6s.×45=270s. Now 270s.÷20=£13, 10s. *Ans.* Hence,

378. To reduce Federal money to either of the State currencies.

Multiply the given sum by the number of shillings which, in the required currency, make $1, and the product will be the answer in shillings, and decimals of a shilling. The shillings should be reduced to pounds, and the decimals to pence and farthings. (Art. 201.)

12. Reduce $378 to New England currency.
13. Reduce $465.45 to New York Currency.
14. Reduce $640 to Pennsylvania currency.
15. Reduce $1000 to Canada currency.
16. Reduce £15, 7s. 6d., N. E. cur. to Federal money.

Solution.—£15, 7s. 6d.=307.5s. (Art. 200.) Now since 6s. make $1, 307.5s. will make as many dollars, as 6 is contained times in 307.5. And 307.5÷6=$51.25. *Ans.* Hence,

379. To reduce either of the State currencies to Federal money.

Reduce the pounds to shillings, and the given pence and farthings to the decimal of a shilling; then divide the sum by the number of shillings which, in the given currency, make $1, and the quotient will be the answer in dollars and cents.

17. Reduce £48, 15s., N. E. cur., to Federal Money.
18. Reduce £73, 4s., N. E. cur., to Federal Money.
19. Reduce £100, 18s., N. Y. cur., to Federal Money.
20. Reduce £256, 5s., N. Y. cur., to Federal Money.
21. Reduce £296, 12s., Penn. cur., to Federal Money.
22. Reduce £430, 8s., Penn. cur., to Federal Money.
23. Reduce £568, 10s., Ga. cur., to Federal Money.
24. Reduce £1000, 15s., Canada cur., to Federal Money.

QUEST.—378. How is Federal Money reduced to the State currencies? 379. How are the several State currencies reduced to Federal Money?

SECTION XVI.

MENSURATION.

ART. **380.** MENSURATION is the art of measuring *magnitudes.*

OBS. The term *magnitude*, denotes that which has one or more of the three dimensions, *length, breadth,* and *thickness.*

381. In measuring *surfaces,* it is customary to assume a *square* as the *measuring unit,* as a square inch, a square foot, a square rod, &c.; that is, a *square* whose side is a *linear unit* of the same name. (Thomson's Legendre, IV. 4. Sch. Art. 153. Obs. 1.)

Note.—For the demonstration of the following principles, see references.

382. To find the area of a *parallelogram,* and a *square.* (Art. 163. Obs.)

Multiply the length by the breadth. (Leg. IV. 5.)

OBS. When the *area* and *one side* of a rectangle are given, *the other side* is found by dividing the *area* by the *given side.* (Art. 291. Note.)

1. How many acres are there in a field 120 rods long, and 90 rods wide? *Ans.* $67\frac{1}{2}$ acres.
2. How many acres in a field 800 rods long, and 128 rods wide?
3. Find the area of a square field whose sides are 65 rods in length.
4. A man fenced off a rectangular field containing 3750 sq. rods, the length of which was 75 rods: what was its breadth?
5. One side of a rectangular field is 1 mile in length, and the field contains 160 acres: what is the length of the other side?

383. To find the area of a *rhombus.* (Leg. I. Def. 18.)

Multiply the length by the altitude. (Leg. IV. 5.)

Note.—The term *altitude*, denotes perpendicular height.

6. The length of a rhombus is 17 ft., and its perpendicular height 12 ft.: what is its area? *Ans.* 204 sq. ft.
7. What is the area of a rhombus whose altitude is 25 rods, and its length 28.6 rods?

384. To find the area of a *trapezium.* (Leg. IV. 7.)

Multiply half the sum of the parallel sides by the altitude.

8. The parallel sides of a trapezium are 15 ft. and 21 ft., and its altitude 12 ft.: what is its area? *Ans.* 216 ft.
9. Find the area of a trapezium whose parallel sides are 25 rods and 37 rods, and its altitude 18 rods.

385. To find the area of a *triangle.* (Leg. IV. 6.)

Multiply the base by half the altitude.

OBS. 1. The *base* of a triangle is found by dividing the area by *half* the *altitude.*

2. The *altitude* of a triangle is found by dividing the area by *half* the *base.*

10. What is the area of a triangle whose base is 45 ft., and its altitude 20 ft.? *Ans.* 450 sq. ft.

11. What is the area of a triangle whose base is 156 ft., and its altitude 63 ft.?

386. To find the area of a *triangle, the three sides* being given.

From half the sum of the three sides subtract each side respectively; then multiply together half the sum and the three remainders, and extract the square root of the product.

12. What is the area of a triangle whose sides are 10 ft., 12 ft., and 16 ft.? *Ans.* 59.92+ft.

13. What is the area of a triangle whose sides are each 12 yds.?

387. To find the *circumference* of a circle, when the diameter is given. (Leg. V. 11. Sch.)

Multiply the given diameter by 3.14159.

Note.—The *circumference* of a circle is a curve line, all the points of which are equally distant from a point within, called the *centre.*

The *diameter* of a circle is a straight line which passes through the centre, and is terminated on both sides by the circumference.

The *radius* or *semi-diameter* is a straight line drawn from the centre to the circumference.

14. What is the circumference of a circle whose diameter is 15 ft.? *Ans.* 47.12385 ft.

15. What is the circumference of a circle whose diameter is 100 rods?

388. To find the *diameter* of a circle, when the circumference is given.

Divide the given circumference by 3.14159.

OBS. The *diameter* of a circle may also be found by dividing the *area* by .7854, and extracting the *square root* of the quotient.

16. What is the diameter of a circle whose circumference is 94.2477 ft.? *Ans.* 30 ft.

17. What is the diameter of a circle whose circumference is 628.318 yards?

389. To find the area of a *circle.* (Leg. V. 11.)

Multiply half the circumference by half the diameter; or, multiply the circumference by a fourth of the diameter.

Note.—The area of a circle may also be found by multiplying the *square* of its diameter by the decimal .7854.

18. What is the area of a circle whose diameter is 100 ft.? *Ans.* 7854 sq. ft.

19. What is the area of a circle whose diameter is 120 rods?

20. How many square yards in a circle whose circumference is 160 yards?

21. Required the diameter of a circle containing 50.2656 sq. rods.

22. Required the diameter of a circle containing 201.0624 sq. ft.

390. *The side of a square equal in area to any given surface, is found by extracting the square root of the given surface.* (Arts. 350, 339. Obs. 2.)

OBS. When it is required to find the dimensions of a rectangular field, equal in *area* to a given surface, and whose length is double, triple, or quadruple, &c., of its breadth, the square root of $\frac{1}{2}$, $\frac{1}{3}$, $\frac{1}{4}$, of the given surface, will be the *width;* and this being *doubled, tripled,* or *quadrupled,* as the case may be, will be the *length.*

23. What is the side of a square, whose area is equal to that of a circle which contains 225 sq. yds.? *Ans.* 15 yds.

24. What is the side of a square, whose area is equal to that of a triangle containing 576 sq. ft.?

25. The length of a rectangular field containing 80 acres, is twice its breadth: what are its length and breadth?

391. *A mean proportional between two numbers is found by multiplying the given numbers together, and extracting the square root of the product.* (Art. 320. Obs. 1.)

26. What is the mean proportional between 9 and 16?

27. What is the mean proportional between 49 and 144?

28. What is the mean proportional between $\frac{1}{4}$ and $\frac{1}{9}$?

392. In measuring *solids,* it is customary to assume a *cube* as the *measuring unit,* whose sides are *squares* of the same name. Thus, the sides of a cubic inch, are square inches; of a cubic foot, are square feet, &c. (Art. 154. Obs. 2.)

OBS. To find the *capacity, solidity,* or *cubical contents* of a body, is to find the number of *cubic inches, feet,* &c., contained in the body.

393. To find the *solidity* of bodies whose sides are *perpendicular* to each other. (Art. 164. Leg. VII. 11. Sch.)

Multiply the length, breadth, and thickness together.

OBS. When the *contents* of a solid body and *two* of its *sides* are given, the *other side* is found by dividing the *contents* by the *product* of the two given sides. (Art. 294.)

29. How many cubic feet are there in a stick of timber 60 ft. long, $3\frac{1}{3}$ ft. wide, and 2 ft. thick? *Ans.* 400 cu. ft.

30. How many cubic feet in a wall 100 ft. long, $15\frac{1}{2}$ ft. high, and $3\frac{1}{2}$ ft. thick?

31. A gentleman wishes to construct a cubical bin, which shall contain 19683 solid feet: what must be the length of its side?

32. If a stick of timber containing 400 cu. ft., is 60 ft. long, and $3\frac{1}{3}$ ft. thick, what is its width? *Ans.* 2 ft.

394. To find the solidity of a *prism*.

Multiply the area of the base by the height. (Leg. VII. 12.)

OBS. 1. This rule is applicable to *all prisms*, triangular, quadrangular, pentagonal, &c.; also to *all parallelopipedons*, whether rectangular or oblique. (Leg. VII. Def. 4, 8, 9.)

2. The *height* of a prism is the perpendicular distance between the planes of the bases. Hence, in a right prism, the height is equal to the length of one of the sides.

33. What is the solidity of a prism whose base is 5 ft. square, and its height 15 ft.? *Ans.* 375 cu. ft.

34. What is the solidity of a triangular prism whose height is 20 ft., and the area of whose base is 460 sq. ft.?

395. To find the *lateral surface* of a right prism.

Multiply the length by the perimeter of the base.

OBS. If we add the areas of both ends to the lateral surface, the sum will be the whole surface of the prism.

35. Required the lateral surface of a triangular prism whose perimeter is $4\frac{1}{2}$ in., and its length 12 in. *Ans.* 54 sq. in.

36. Required the lateral surface of a quadrangular prism whose sides are each 2 ft., and its length 19 ft.

396. To find the solidity of a *pyramid*, or *cone*. (Leg. VII. 18. VIII. 4.)

Multiply the area of the base by $\frac{1}{3}$ of the altitude.

37. Required the solidity of a square pyramid, the side of whose base is 25 ft., and whose height is 60 ft. *Ans.* 12500 cu. ft.

38. Required the solidity of a cone, the diameter of whose base is 30 ft., and whose height is 90 ft.

397. To find the *lateral* or *convex surface* of a regular pyramid, or cone. (Leg. VII. 16. VIII. 3.)

Multiply the perimeter of the base by $\frac{1}{2}$ the slant-height.

OBS. The *slant-height* of a regular pyramid, is the distance from the vertex or summit to the middle of one of the sides of the base.

39. What is the lateral surface of a regular triangular pyramid whose slant-height is 10 ft., and whose sides are each 8 ft.? *Ans.* 120 sq. ft.

40. What is the convex surface of a cone, the perimeter of whose base is 500 yds., and whose slant-height is 120 yds.?

398. To find the solidity of a *frustum* of a pyramid, or cone. (Leg. VII. 19. Sch., VIII. 6.)

To the sum of the areas of the two ends, add the square root of the product of these areas; then multiply this sum by $\frac{1}{3}$ of the perpendicular height.

41. The areas of the two ends of a frustum of a cone are 9 sq. ft., and 4 sq. ft., and its height is 15 ft.: what is its solidity?
Ans. 95 cu. ft.

42. The two ends of a frustum of a pyramid are 4 ft. and 3 ft. square, and its height is 10 ft.: what is its solidity?

399. The convex surface of a *frustum* of a pyramid, or cone, is found *by multiplying half the sum of the circumferences of the two ends by the slant-height.* (Leg. VII. 17.)

43. The circumferences of the two ends of a frustum of a pyramid are 12 ft. and 8 ft., and its slant-height 7 ft.: what is its convex surface?
Ans. 70 sq. ft.

44. The circumferences of the two ends of a frustum of a cone are 15 yds. and 9 yds., and its slant-height, 7 yds.: what is its convex surface?

400. To find the solidity of a *cylinder.* (Leg. VIII. 2.)

Multiply the area of the base by the height or length.

45. Required the solidity of a cylinder 6 ft. in diameter, and 20 ft. high.
Ans. 565.488 cu. ft.

46. Required the solidity of a cylinder 30 ft. in diameter, and 65 ft. long.

401. To find the *convex surface* of a cylinder.

Multiply the circumference of the base by the height.

47. What is the convex surface of a cylinder 16 inches in circumference and 40 in. long?
Ans. 640 sq. in.

48. What is the convex surface of a cylinder, the diameter of whose base is 20 ft., and whose height is 65 ft?

402. To find the surface of a *sphere or globe.*

Multiply the circumference by the diameter. (Leg. VIII. 9.)

49. Required the surface of a globe 13 inches in diameter.
Ans. 531 sq. in. nearly.

50. Required the surface of the earth, allowing its diameter to be 8000 miles.

403. To find the *solidity* of a sphere or globe.

Multiply the surface by $\frac{1}{6}$ of the diameter.

51. What is the solidity of a globe 12 in. in diameter?

52. What is the solidity of the earth, reckoning its diameter at 8000 miles?

404. *The solid contents of similar bodies are to each other, as the cubes of their homologous sides, or like dimensions.* (Leg. VII. 20. VIII. 11. Cor.)

53. If a ball 4 inches in diameter weighs 32 lbs., what is the weight of a ball whose diameter is 5 inches?

Solution.—$4^3 : 5^3 :: 32$ lbs. : to the weight. *Ans.* 62.5 lbs.

54. If a ball 3 inches in diameter weighs 4 lbs., what is the diameter of a ball which weighs 32 lbs.?

405. To find the side of a cube whose solidity shall be *double, triple,* &c., that of a cube whose side is given.

Cube the given side, multiply it by the given proportion, and the cube root of the product will be the side of the cube required.

55. What is the side of a cubical mound, which contains 8 times as many solid feet as one whose side is 3 ft. *Ans.* 6 ft.

56. Required the side of a cubical vat, which contains 16 times as many solid feet as one whose side is 5 ft.

GAUGING OF CASKS.

406. To find the contents or capacity of casks.

Multiply the square of the mean diameter into the length in inches; then this product multiplied into .0034 *will be the wine gallons required, or multiplied into* .0028 *will be the beer gallons.*

OBS. The *mean* diameter of a cask is found by adding to the head diameter .7 of the difference between the head and bung diameters when the staves are *very much* curved; or by adding .5 when *very little* curved; and by adding .65 when they are of a *medium* curve.

57. How many wine gallons does a cask contain whose length is 35 inches, its bung diameter 30 in., and its head diameter 26 in., it being but little curved? *Ans.* 93.296 gals.

58. How many beer gallons in a cask 54 in. long, whose bung diameter is 42 in., and head diameter 36 in., its staves being much curved?

MISCELLANEOUS EXAMPLES.

Ex. 1. How much will 500 sheep cost, at $\$2\frac{1}{2}$ apiece?

2. How much can a man earn in 240 days, at $37\frac{1}{2}$ cts. per day?

3. What will 690 bushels of apples cost, at $18\frac{3}{4}$ cts. per bushel?

4. What cost 476 cows, at $\$12\frac{3}{4}$ apiece?

5. What cost $685\frac{3}{4}$ gallons of oil, at $87\frac{1}{2}$ cts. per gal.?

6. What cost $325\frac{1}{8}$ acres of land, at $\$10\frac{1}{4}$ per acre?

7. How much flour, at $\$4\frac{1}{2}$ per bbl., can be bought for \$525?

8. How many yards of cloth, at $\$5\frac{1}{8}$ per yard, can be bought for \$1230? *Ans.* 240 yds.

9. How many saddles, at $\$11\frac{1}{4}$, can be bought for \$5625?

10. How many horses, at $\$75\frac{3}{5}$, can be bought for \$3780?

11. A man bought $\frac{7}{8}$ of a ship, and sold $\frac{4}{5}$ of it: how much had he left? *Ans.* $\frac{7}{40}$.

12. A broker negotiated a bill of exchange of \$10360, at $1\frac{3}{8}$ per cent.: what was his commission?

13. What is the interest of \$2345 for 1 year and 6 months, at 6 per cent.?

14. What is the int. of \$1356.25 for 90 days, at 6 per ct.?

15. What is the int. of \$533.11 for 6 months, at 7 per ct.?

16. What is the amount of \$925 for 1 yr. and 4 mo., at 8 per ct.?

17. What is the amount of \$4635 for 30 days, at 7 per ct.?

18. What is the amount of \$10360 for 60 days, at 5 per ct.?

19. What is the present worth of \$1365, payable in 6 months, when money is worth 7 per cent. per annum?

20. At 6 per ct. discount, what is the present worth of \$1623.28, due in 1 year?

21. What is the bank discount on a note of \$730, payable in 4 months, at $6\frac{1}{2}$ per ct.? *Ans.* \$16.212.

22. What is the bank discount on a note of \$1575, payable in 60 days, at 7 per ct.?

23. What will 35 shares of Railroad stock cost, at $10\frac{1}{2}$ per ct. advance? *Ans.* \$3867.50.

24. What cost 63 shares of bank stock, at $3\frac{1}{3}$ per ct. discount?

25. What premium must a man pay annually for insuring \$8500 on his store and goods, at $1\frac{1}{4}$ per ct.?

26. If I obtain insurance on goods, worth \$16265, at $2\frac{1}{2}$ per ct., and the goods are lost, how much shall I lose?

27. What is the insurance on \$925.68, at $1\frac{1}{2}$ per ct.?

28. What is the insurance on \$63460, at $\frac{5}{8}$ per ct.?

29. What is the insurance on \$48256, at $1\frac{1}{5}$ per ct.?

30. A man bought a farm for \$5640, and afterwards sold it for 15 per ct. more than it cost: how much did he make by his bargain?

31. A merchant bought a stock of goods for \$4390, and retailed them at a profit of $22\frac{1}{2}$ per ct.: how much did he make?

32. An oil merchant bought 15000 gallons of oil for $8500, and sold it at 15 per ct. advance: how did he sell it per gal.?

33. If I buy 1675 yards of flannel for $368.50, how must I retail it per yard to gain 25 per ct.? *Ans.* 27½ cts.

34. A grocer bought 2500 lbs. of coffee for $250, and sold it at 6 per ct. loss: what did he get per pound?

35. A merchant bought 1824 yds. of cloth, at $2.50 per yd., and retailed it at $3 per yd.: what per ct. was his profit, and how much did he make?

36. A shop-keeper bought 100 pieces of lace, for $250, and sold them for $375: what per ct. did he make?

37. If a grocer buys 3680 lbs. of cheese, at 4½ cts. per lb., and sells it at 6½ cents, what per ct. is his profit?

38. What is the ad valorem duty, at 33⅓ per ct., on a quantity of cloths which cost $10436?

39. What is the ad valorem duty, at 15½ per ct., on a cargo of tea invoiced at $35856?

40. At 37½ per ct., what is the duty on a quantity of silks which cost $23265?

41. The sum of two numbers is 856, and their difference is 75: what are the numbers?

42. The sum of two numbers is 5643, and their difference is 125: what are the numbers?

43. The difference of two numbers is 63, and the smaller number is 365: what is the greater number?

44. The product of two numbers is 3750, and one of the numbers is 75: what is the other?

45. What number is that $\frac{4}{9}$ of which is 265? *Ans.* 477.

46. What number is that $\frac{3}{5}$ of $\frac{5}{8}$ of which is 120?

47. How long will it take a person to count a billion, if he counts 50 a minute, and works 6 hours per day, for 5 days a week, and 52 weeks a year?

48. How many dollars, each weighing 412½ grains, can be made from 7 lbs. 1 oz. 18 pwt. 18 grs. of silver?

49. How many pounds of silk will it take to spin a thread which will reach round the earth, allowing its circumference to be 25000 miles, and 2½ oz. to make 160 rods of thread?

50. How many times will the hind wheel of a carriage, 7 ft. 6 in. in circumference, turn round in 7 miles, 1 furlong, 30 rods?

51. How many times will the fore wheel of a carriage, 5 ft. 7½ in. in circumference, turn round in the same distance?

52. What cost 645 bushels of salt, at 4s. N. Y. currency per bu.?

53. What cost 744 yards of muslin, at 1s. 4d. N. Y. cur. per yd.?

54. What cost 241 melons, at 2s. 8d. N. Y. cur. apiece?

55. What cost 1536 yards of calico, at 1s. N. E. cur. per yd.?

56. What cost 873 baskets of peaches, at 3s. N. E. cur. a basket?

57. What cost 632 bushels of oats, at 1s. 6d. N. E. cur. a bushel?

58. What cost 848 lambs, at 5s. sterling apiece?

59. What cost 258 yards of cloth, at 15s. sterling per yard?

60. What cost 912 bushels of rye, at 2s. 6d. sterling per bu.?

61. What cost 657 yards of silk, at 6s. 8d. ster. per yard?

62. What cost 735 bushels of apples, at 1s. 8d. ster. per bushel?

63. What cost 3 pieces of cloth, each containing 27 yards, at 3s. 4d. per yard? *Ans.* £13, 10s.

64. What cost 248 pair of boots, at 12s. 6d. sterling a pair?

65. If 156 lbs. of butter cost $15.60, what will 730 lbs. cost?

66. If 48 yards of cloth cost $480, what will 125 yards cost?

67. If 96 horses eat 192 tons of hay in a winter, how many tons will 150 horses eat?

68. If 10 lbs. of sugar cost $9\frac{3}{8}$s., what will 240 lbs. cost?

69. If 25 lbs. of veal cost \$$\frac{3}{4}$, how much will 872 lbs. cost?

70. If 50 lbs. of ginger cost \$$7\frac{1}{7}$, how much will 460 lbs. cost?

71. What cost 260 cords of wood, if 45 cords cost \$$87\frac{3}{5}$?

72. A man sold a sheep for £$1\frac{1}{2}$, and a pig for $\frac{5}{8}$s. $\frac{3}{4}$d.: what did he get for both?

73. A goldsmith melted up $\frac{3}{4}$ lb. $10\frac{1}{2}$ pwts. of gold, at one time, and $3\frac{1}{2}$ oz. 10 grs. at another: how much did he melt in all?

74. A man having $2\frac{1}{4}$ oz. of silver, sold $6\frac{3}{4}$ pwts.: how much had he left?

75. A man owing £$\frac{4}{5}$, $2\frac{1}{2}$s., paid $7\frac{1}{4}$s. $2\frac{1}{2}$d.: how much does he still owe?

76. If 50 lbs. of rice cost £$\frac{4}{7}$, what will 840 lbs. cost?

77. If 13 yards of edging cost \$$\frac{19}{20}$, what will 200 yds. cost?

78. If $\frac{7}{9}$ of a ton of iron cost $35, what will 381 tons cost?

79. If I owe a man £6950, and can pay him but 13s. 4d. on a pound, how much will he receive for his debt?

80. If 385 yards of linen cost £63, how much can be bought for £18?

81. How much brandy can be bought for £396, if 90 gallons cost £18?

82. If $15\frac{1}{2}$ yards silk cost \$$18\frac{3}{4}$, what will $56\frac{3}{4}$ yards cost?

83. A grocer used a false weight of $13\frac{1}{2}$ oz. for a pound: what was the amount of his fraud in weighing 500 pounds?

84. If $\frac{2}{7}$ of a barrel of apples costs \$$\frac{4}{5}$, how much will $\frac{7}{8}$ of a barrel cost? *Ans.* $2.45.

85. If $\frac{7}{15}$ of a pound of lard costs $\frac{14}{17}$ of a shilling, how much will $\frac{29}{30}$ of a pound cost?

86. If $\frac{3}{16}$ of a ton of hay costs £$\frac{7}{8}$, what will $\frac{19}{20}$ of a ton cost?

87. How much will $\frac{5}{16}$ of a drum of figs come to, at the rate of $\frac{4}{5}$ of a dollar for $\frac{3}{4}$ of a drum?

88. Bought $48\frac{1}{3}$ lbs. of tea for \$$27\frac{3}{8}$: how much can be bought for $125?

89. Paid \$$35\frac{1}{2}$ for $\frac{4}{5}$ of an acre of land: how much can be bought for $7500?

90. If $29\frac{1}{2}$ yards of camlet make 3 cloaks, how many cloaks can be made of $737\frac{1}{2}$ yards? *Ans.* 75 cloaks.

91. If 57.35 acres of land produce 430.16 bushels of barley, how many bushels will 172.05 acres produce?

92. What will $730\frac{3}{4}$ yards of cloth cost, if you pay $112 for $14\frac{1}{2}$ yards?

93. If a cane 3 feet in length cast a shadow 5 feet long, how high is a steeple whose shadow is 175 feet?

94. Bought a hogshead of molasses for 4 firkins of butter, each containing 56 lbs., which was worth 10 cents a pound: what did the molasses cost per gallon?

95. Bought 15 yds. of silk at 7s. per yard, and 12 yds. of muslin at 3s. per yard, and paid the bill in cheese at 9d. per pound: how many pounds did it take to pay the bill?

96. If a cubic foot of pure water weighs 1000 oz., what will a pail of water weigh which contains $217\frac{1}{2}$ cubic inches?

97. If I pay $8400 for $\frac{5}{8}$ of a ship, what must I pay for the whole ship?

98. A farmer sold 174 sheep, which was $\frac{3}{5}$ of all he had; the remainder he divided equally between his two sons: how many did each receive?

99. A garrison having been besieged 108 days, found that $\frac{3}{5}$ of the provisions were consumed: how much longer would they last?

100. A garrison of 1520 men have 416955 lbs. of flour: how long will it last them, allowing each man $\frac{7}{8}$ lb. per day?

101. How long will 75240 gals. of water last a ship's company of 30 men, allowing each man $\frac{2}{5}$ gal. per day?

102. If 10 men can dig a cellar in 30 days, how long will it take 25 men to dig it?

103. If 6 men spend $48 in 7 weeks, how much will 24 men spend in 35 weeks? *Ans.* $960.

104. If 15 horses consume 70 bushels of oats in 27 days, how many bushels will 45 horses consume in 54 days?

105. If 6 men can build a wall 30 feet long, 6 feet high, and 3 feet thick, in 15 days, when the days are 12 hours long, how many days will it take 30 men to build a wall 300 feet long, 8 feet high, and 6 feet thick, working 8 hours a day?

106. A merchant in New York wished to pay £1500 in London: what will a bill of exchange cost him at 9 per ct. premium?

107. A broker in Boston sold a bill of exchange on Liverpool for £2500, 15s., at $9\frac{1}{2}$ per ct. premium: what did he get for it?

108. What will a bill on England for £3125, 12s. 6d. cost, when exchange is 10 per ct. above par?

109. A man wishing to remit $2550 to Ireland, bought a draft on London, at $12\frac{1}{2}$ per ct. advance: what was the amount of his bill in sterling money?

110. A farmer wishes to form a square field, which shall contain 1296 square rods: what is the length of its side?

111. A man owns a farm which contains 160 acres, and is in the form of a square: what is the length of its side?

112. What is the length of the side of a square field containing 40 acres?

113. What is the area of a triangle whose hypothenuse is 50 yards, and its perpendicular 30 yards?

114. What is the area of a triangle whose hypothenuse is 100 rods, and its base 60 rods?

115. Required the mean proportional between 49 and 81.

116. Required the mean proportional between 121 and 5.76.

117. What is the mean proportional between $\frac{4}{9}$ and $\frac{16}{49}$?

118. Required the mean proportional between $\frac{25}{36}$ and $\frac{81}{144}$.

119. A regiment containing 6912 soldiers, was so arranged that the number in rank was triple that in file: how many were there in each?

120. If a board is 8 in. wide, how long must it be to make a sq. ft?

121. How much silk $\frac{3}{4}$ yd. wide will it take to make a sq. yd.?

122. How much cambric $\frac{3}{4}$ yd. wide will it take to line 9 yds. of balzorine 1 yd. wide?

123. How many yds. of unbleached muslin $\frac{3}{4}$ yd. wide will it take to line 36 yds. of carpeting $1\frac{1}{4}$ yds. wide?

124. If it takes 10 yds. of broadcloth $1\frac{1}{2}$ yds. wide to make a cloak, how many yards of camlet $\frac{5}{8}$ yd. wide will make one?

125. How much will it cost to carpet a parlor 18 ft. square with carpeting $\frac{3}{4}$ yd. wide, which is worth $1.50 per yard?

126. A, B, and C, joined in a speculation; A put in $500, B $700, and C put in the balance; they gained $1200, of which C received $480 for his share: how much did A and B receive, and how much did C put in?

127. A, B, and C, gain $3600, of which A receives $6, as often as B receives $10, and C $14: what was the share of each?

128. The hour and minute hand of a clock are exactly together at noon: when will they next be together?

129. A farmer having lost $\frac{1}{3}$ of his sheep, and sold $\frac{1}{4}$ of them, had 500 left: how many had he at first?

130. If $\frac{1}{5}$ of a post stands in the mud, $\frac{1}{4}$ in the water, and 10 feet above the water, what is the length of the post?

131. Two persons start from the same place, one goes south 4 miles per hour, the other west 5 miles per hour: how far apart are they in 9 hours?

132. A messenger traveling 8 miles an hour, was sent to Mexico with dispatches for the army; after he had gone 51 miles, another was sent with countermanding orders, who could go 19 miles as quick as the former could go 16: how long will it take the latter to overtake the former; and how far must he travel?

ANSWERS TO EXAMPLES.

NOTE.—At the urgent request of several distinguished Teachers, who have received Thomson's Practical Arithmetic with favor, the publishers have issued an edition of it, containing the answers in the end of the book. It is hoped that pupils, who may use this edition, will have sufficient regard to their own improvement, never to consult the answer till they have made a *strenuous* and *persevering* effort to solve the problem themselves.

N. B.—The work without the answers is published as heretofore.

ADDITION.

EXERCISES FOR THE SLATE.—ART. **21.**

Ex. Ans.	Ex. Ans.	Ex. Ans.	Ex. Ans.
1, 2. Given.	21. 16840.	15. 582 a.	34. $512.
3. $786.	22. 220083.	16. $45.	35. 611 bu.
4. 8689.	23. 100003.	17. 98 cts.	$513.
5. 57757.	24. 134735.	18. $101.	36. $627.
6. 651465.	25. 104022.	19. $2788.	37. 630 lbs.
7. 8651761.		20. $102.	38. $3789.
8. 998943483	ART. **29.**	21. $846.	39. $1125.
9. 988.	1. $64.	22. 754 sh.	40. $2385 r.
10. 7673.	2. 46 lbs.	365 l.	$554 g.
11. 88765.	3. 48 yrs.	1119 b.	41. $1582.
12. 85879944.	4. $313.	23. $6821.	42. $1323.
13–15. Given.	5. $31.	24. $2324.	43. 525 m.
	6. 40 s.	25. $4900.	44. $4930.
ART. **27.**	7. $550.	26. $244.	45. 2234822.
16. 23770.	8. $2480.	27. 113 ts.	46. 4604345.
17. 161524.	9. $190.	28. 476 m.	47. 5067843.
18. 131570.	10. $278.	29. 73 yrs.	48. 4984097.
19. 1999990.	11. $58.	30. $1648.	49. 178346.
	12. 33 sch.	31. $34950.	50. 17069453.
ART. **28.**	13. $136.	32. $33700.	
20. 1913.	14. 64 m.	33. $3147.	

SUBTRACTION.

Ex. Ans.

ART. **34.**

1, 2. Given.
3. $232.
4. 413.
5. 353.
6. 418.
7. 3332.
8. 3231.
9. 32352.
10. 613134.
11. 531141.
12. 3151721.
13–15. Given.

ART. **38.**

16. 54182.
17. 124907.
18. 66104149.

ART. **39.**

19. Given.
20. 6121.
21. 2754087.
22. 932417.
23. 6834501.
24. 8960895.
25. 31090814

ART. **40.**

1. 13 yds.
2. $221.
3. 189 g.
4. 1003 bu.
5. $3791.
6. $1420.
7. $382.
8. $1079.
9. $374 bu.
10. $1989.
11. $479.
12. ———
13. $1291.
14. 53 m.
15. 93 m.
16. ———
17. 1706.
18. 67 yrs.
19. ———
20. $72320.
21. 427721.
22. 214412.
23. 1056109.
24. 194099.
25. 11763528.
26. 100 a.
27. $986.
28. $22.
29. $19.
30. 146 ts.
31. $1090.
32. $3838.
33. $5250.
34. $323.
35. 1933 a.
36. 565 men.
37. $773.
38. $18053.
39. $154.
40. $5491.
41. $6749.
42. $1695.
43. $2752.
44. $1913.
45. $332.
46. 12520 bu.
47. $1491.
48. $9699.
49. $21422.
50. $8000.

MULTIPLICATION.

ART. **47.**

1–4. Given.
5. 960 r.
6. 880 m.
7. 9096.
8. 88480.
9. 505505.
10. 9036906.
11. Given.

ART. **51.**

12. $664.
13. 1917 s.
14. $624.
15. $6153.
16. 5200 s.
17. 40030.
18. 608240.
19. 76342.
20. 41479110.
21, 22. Given.

ART. **53.**

23. 2268 s.
24. 3915 bu.
25. 19200 lbs.
26. $6394.
27–29. Given.
30. 507166416.

ART. **54.**

1. $2790.
2. $2552.
3. $9520.
4. 676 s.
5. 2511 s.
6. $13932.
7. $10955.
8. $3790.
9. $153900.
10. $180.
11. $414.
12. $945.
13. $1792.
14. $1664.
15. $2522.
16. $2090.
17. 4935 s.
18. 3071 bu.
19. 2944 qts.
20. $22224.
21. $1482.
22. $8991.
23. $10584.
24. $4096.
25. 35720 d.
26. 16425 d.
27. 90625 lbs.
28. 176175 lbs.
29. 78475 m.
30. $77970.

which have fallen under my observation, that I have adopted it in our Institute. Respectfully yours,

E. HOSMER.

Moravia, March 28, 1846.

From *J. C. Howard*, A. M., Principal of Norwich Academy, N. Y.

Mr. J. B. Thomson—Dear Sir,—I have examined, with much care and interest, your "Practical Arithmetic," and, without attempting to specify its various excellencies, I assure you, it approaches nearer to my idea of a complete work of its kind than any with which I am acquainted. I shall embrace the earliest opportunity to introduce it into the academy under my charge.

J. C. HOWARD.

Norwich, April 11, 1846.

From *S. W. Clark*, A. M., Principal of East Bloomfield Academy, N. Y.

Dear Sir, I have carefully examined your PRACTICAL ARITHMETIC. It is just such a text-book as we want—*clear*, *concise*, *lucid*, *logical*—a *Practical Arithmetic*, evidently written by a *practical teacher*. We shall introduce it as a text-book in this Institution at the commencement of our next term.

Respectfully yours, S. W. CLARK.

East Bloomfield, May 26, 1846.

From *R. M. Wanzer*, *Esq.*, Principal of Genoa Academy, N. Y.

From the actual use of Thomson's "Practical Arithmetic" in my School, I am persuaded that it is better adapted to give a pupil a *comprehensive and practical knowledge of figures* than any work of the kind with which I am acquainted.

R. M. WANZER.

Genoa, May 17, 1846.

From *Orson Barnes*, *Esq.*, late Superintendent of Onondaga Co., N. Y.

Dear Sir,—I have examined your First Lessons for Children; also, your Practical Arithmetic, and earnestly commend them to the attention of parents and teachers. In my estimation their introduction into our schools would prove a powerful auxiliary in developing and disciplining the intellectual faculties of the pupils, and essentially aid the teacher, in his arduous toils.

Respectfully yours, ORSON BARNES.

Baldwinsville, May 28, 1846.

From *Wm. A. Cropsey*, *Esq.*, Town Superintendent of Locke, N. Y.

I have examined Thomson's "Practical Arithmetic," and am fully convinced that it is a work of great merit. Its superiority will be readily seen in the excellent arrangement of the work, the clearness and simplicity of the rules and explanations, and the well selected examples, which are all admirably adapted to the capacities of the young. I think the arithmetics now in general use in this section, will soon be laid aside, and the "Practical Arithmetic" supply their places. I would be much pleased to see it introduced into every school in this town. Yours truly, WM. A. CROPSEY.

Locke, March 27, 1846.

From *Thos. J. Haswell*, A. M., Prin. of Chester Academy, N. Y.

J. B. Thomson, Esq.—Dear Sir,—I have examined your "Practical Arithmetic" and hesitate not to say it is the *best* I have seen. Its arrangement is natural—its rules are concise and lucid—its notes and observations appropriate and valuable—its examples abundant and happily selected; well calculated to interest the learner, and lead him to a thorough knowledge of the science. I shall introduce it immediately into my school. Your Algebra I have already in use, and shall introduce your Legendre, esteeming it as I do all the books of your series which I have seen, of superior merit.

Yours sincerely, THOMAS J. HASWELL.

Chester, July 14, 1846.

From *B. F. Everson*, *Esq.*, Principal of Public School, Union Springs, N. Y

J. B. Thomson Esq.—Dear Sir,—I have carefully examined your Practical

Arithmetic, and from the clearness and precision of its rules and explanations —its careful arrangement—the copiousness of its examples, both mental and for the "board"—the appropriateness of its suggestions and observations, and its easy plan of induction—I hesitate not to pronounce yours decidedly the most appropriate work for our schools I have ever noticed, and shall use my influence for its general adoption.

Yours truly,

BENJAMIN F. EVERSON.

Union Springs, May 28, 1846.

www.ingramcontent.com/pod-product-compliance
Lightning Source LLC
LaVergne TN
LVHW010159110826
845151LV00002B/558
* 9 7 8 1 4 2 5 5 3 5 7 7 3 *